Autogenous Shrinkage
of Concrete

Autogenous Shrinkage
of Concrete

Proceedings of the International Workshop
organized by JCI (Japan Concrete Institute)

Hiroshima
June 13–14, 1998

Edited by
Ei-ichi Tazawa
Hiroshima University
Japan

JCI

CRC Press
Taylor & Francis Group
Boca Raton London New York

CRC Press is an imprint of the
Taylor & Francis Group, an **informa** business
A TAYLOR & FRANCIS BOOK

CRC Press
Taylor & Francis Group
6000 Broken Sound Parkway NW, Suite 300
Boca Raton, FL 33487-2742

First issued in paperback 2019

ISBN-13: 978-0-419-23890-4 (hbk)
ISBN-13: 978-0-367-86390-6 (pbk)

British Library Cataloguing in Publication Data
A catalogue record for this book is available from the British Library

Library of Congress Cataloging in Publication Data
Autogenous shrinkage of concrete : proceedings of the international workshop, organised by the JCI
(Japan Concrete Institute),
Hiroshima, June 13–14, 1998 / edited by Ei-ichi Tazawa.
        p.   cm.
    Includes bibliographical references and index.
    1. Concrete—Expansion and contraction—Congresses.   I. Tazawa, Eiichi, 1936–   . II. Nihon
Konkurito Kogaku Kyokai.
TA440.A83   1999
620.1'361—dc21   98–45269
                                                         CIP

Publisher's Note
The publisher has gone to great lengths to ensure the quality of this reprint
but points out that some imperfections in the original may be apparent.

Visit the Taylor & Francis Web site at
http://www.taylorandfrancis.com

and the CRC Press Web site at
http://www.crcpress.com

CONTENTS

ORGANIZING COMMITTEE

Chairman Ei-ichi Tazawa, *Hiroshima University, Japan*
Members Paul Acker, *LAFARGE, France*
Akthem Al-Manaseer, *San Jose State University, USA*
Pierre-Claude Aïtcin, *University of Sherbrooke, Canada*
Klaas van Breugel, *Delft University of Technology, The Netherlands*
Jeffery J. Brooks, *University of Leeds, UK*
Masaki Daimon, *Tokyo Institute of Technology, Japan*
Walter H. Dilger, *The University of Calgary, Canada*
Ellis Gartner, *LAFARGE, France*
Seishi Goto, *Yamaguchi University, Japan*
Hamlin Jennings, *Northwestern University, USA*
Harald Justnes, *SINTEF, Norway*
Markku Leivo, *VTT Building Technology, Finland*
Swee Liang Mak, *CSIRO, Australia*
Harald S. Müller, *University of Karlsruhe, Germany*
Bertil Persson, *University of Lund, Sweden*
Toyokazu Shiire, *Kanagawa University, Japan*
R. Springenschmid, *Technical University of Munich, Germany*
Fuminori Tomosawa, *The University of Tokyo, Japan*
Taketo Uomoto, *The University of Tokyo, Japan*

TASK COMMITTEE

Chairman Ryoichi Sato, *Utsunomiya University*
Vice-chairman Etsuo Sakai, *Tokyo Institute of Technology*
Secretary Shingo Miyazawa, *Ashikaga Institute of Technology*
Members Takao Chikada, *Nippon Steel Chemical Co.*
Hiroshi Hashida, *Shimizu Co.*
Shunsuke Hanehara, *Chichibu Onoda Cement Co.*
Keiichi Imamoto, *Tokyu Construction Co.*
Tetsuro Kasai, *Tokai University*
Tatsuya Numao, *Ibaraki University*
Yoshiteru Ohno, *Osaka University*
Shyuichi Okamoto, *Taisei Co.*
Kazunori Takada, *Kajima Co.*
Nobuhumi Takeda, *Obayashi Co.*
Ei-ichi Tazawa, *Hiroshima University*
Hiroyuki Tanano, *Ministry of Construction*
Rokuro Tomita, *Nihon Cement Co.*
Kazuyuki Torii, *Kanazawa University*
Nobuteru Yamamoto, *Kyokuto Co.*
Asuo Yonekura, *Hiroshima University*

TECHNICAL COMMITTEE ON AUTOGENOUS SHRINKAGE OF CONCRETE (JAPAN CONCRETE INSTITUTE)

Chairman	Ei-ichi Tazawa, *Hiroshima University*
Secretary	Ryoichi Sato, *Utsunomiya University*
	Etsuo Sakai, *Tokyo Institute of Technology*
Members	Takao Chikada, *Nippon Steel Chemical Co.*
	Hiroshi Harada, *Chichibu Onoda Cement Co.*
	Hiroshi Hashida, *Shimizu Co.*
	Shunsuke Hanehara, *Chichibu Onoda Cement Co.*
	Masanori Higuchi, *Mitsui Construction Co.*
	Yousaku Ikeo, *Takenaka Co.*
	Kazumasa Inoue, *Takenaka Co.*
	Keiichi Imamoto, *Tokyu Construction Co.*
	Tetsuro Kasai, *Tokai University*
	Hiroyoshi Kato, *Tokuyama Co.*
	Toshiharu Kishi, *The University of Tokyo*
	Shingo Miyazawa, *Ashikaga Institute of Technology*
	Noriyuki Nishida, *Nishimatsu Construction Co.*
	Junichiro Niwa, *Tokyo Institute of Technology*
	Tatsuya Numao, *Ibaraki University*
	Yoshiteru Ohno, *Osaka University*
	Hideki Ohshita, *Chuou University*
	Shyuichi Okamoto, *Taisei Co.*
	Fujio Omata, *Sho-Bond Co.*
	Hiroshi Ohtani, *Tokyu Construction Co.*
	Kazunori Takada, *Kajima Co.*
	Nobuhumi Takeda, *Obayashi Co.*
	Satoshi Tanaka, *Nihon Cement Co.*
	Hiroyuki Tanano, *Ministry of Construction*
	Satoru Teramura, *Denkikagaku Kogyo Co.*
	Rokuro Tomita, *Nihon Cement Co.*
	Kazuyuki Torii, *Kanazawa University*
	Hitoshi Yamada, *Hazama Co.*
	Kouzaburo Yoshida, *Ube Industries*
	Tomonori Yoshida, *Tokuyama Co.*
Organizer	Japan Concrete Institute (JCI)
Co-sponsors	American Concrete Institute (ACI)
	Architectural Institute of Japan (AIJ)
	The Canadian Society for Civil Engineering (CSCE)
	Japan Cement Association (JCA)
	Japan Society of Civil Engineers (JSCE)
	National Ready Mixed Concrete Industry Association
	The Society of Materials Science, Japan (JSMS)

PREFACE

The phrase "autogenous shrinkage of concrete" was used by C. G. Lyman more than sixty years ago, when the phenomenon that concrete shrinks by itself without any change in mass or temperature was recognized for the first time. Since then, however, extensive research on the autogenous shrinkage had not been conducted for several decades. It was only from the beginning of this decade that the importance of the autogenous shrinkage was strikingly pronounced, when high strength concrete became feasible with comparative ease owing to the development of superplasticizer or useful mineral admixtures such as silica fume. At present this type of concrete is being used more frequently and in larger scale. Under these situations, researches on high strength concrete or structural members using high strength concrete have been extensively conducted in various countries.

In the course of investigation of high strength concrete a very strange phenomenon, that the concrete was cracked during under-water curing at constant temperature, was observed. Further, several facts were revealed in Japan with reference to the cracking of concrete. It was pointed out that cause of thermal cracking of mass concrete could not be explained only by temperature rise and mechanical properties, or that cracking of highly flowable concrete could not be explained only by drying shrinkage.

The present editor happened to experience with these strange cases of cracking. At one time the bending strength of sealed specimen of high strength concrete was found to decrease with increasing curing age. It was 1990 after six years of investigations that we realized these phenomena could only be explained by the autogenous shrinkage.

Investigations were initiated in various organizations since importance of the autogenous shrinkage was more widely recognized. Under these situations, a technical committee was organized in Japan Concrete Institute (JCI) in 1994. Collaboration on definition of technical terms, shrinkage mechanism, testing methods of shrinkage and stress, prediction of shrinkage strain and prediction of shrinkage stress was completed after three years. During this activity Standard Specification for Concrete was revised by a committee in Japan Society of Civil Engineers, prescriptions for the autogenous shrinkage were added to those of the drying shrinkage, and the importance of the autogenous shrinkage was socially approved.

On the other hand, in parallel with these trends high strength concrete has been widely investigated in European and North American countries, and many papers have been published on various aspects of autogenous shrinkage.

In view of these research activities, it was thought timely to have an International Workshop in order to exchange the latest informations with worldwide specialists on autogenous shrinkage and to share them for further development or breakthrough of crack control of high strength concrete or highly flowable concrete. The significance of this Workshop was understood by worldwide researchers and engineers. Twenty-six papers were submitted to this Workshop and more than one hundred specialists participated from seventeen countries. In the Workshop active discussions were achieved not only for each presentation but also for proposals of JCI committee on terminology, testing methods and predictions of chemical shrinkage, autogenous shrinkage and autogenous shrinkage stress. Many opinions were set forth particularly for the above proposals, partly because opinionaires were distributed in advance to specialists in this field. This final wrought-out includes these opinions, agreed items, unsolved problems and further targets, which might be useful for further investigation into the technical aspects of autogenous shrinkage. The Workshop, sponsored by JCI and co-sponsored by seven academic organizations, was successfully held. Sincere appreciation is conveyed to all of these organizations. The chairman would like to express his deep acknowledgement to all people who was involved in the Workshop. It is his great pleasure if this Workshop could be of any help to people working for the subject. Last but not least, the chairman thanks CSIRO for the generosity to tolerate Dr. Mak's paper in the book.

Ei-ichi Tazawa
Chairman

PART ONE
COMMITTEE REPORT

Technical Committee on
Autogenous Shrinkage of Concrete
Japan Concrete Institute

In this Committee Report, both "Terminology" and "Testing methods" are translations from the Japanese original, "Technical Committee Report on Autogenous Shrinkage of Concrete", which was published by JCI on November 1996. "State-of-the-art report" is a summary of the "Technical Committee Report on Autogenous Shrinkage of Concrete".

Autogenous Shrinkage of Concrete, edited by Ei-ichi Tazawa. Published in 1999 by E & FN Spon, 11 New Fetter Lane, London EC4P 4EE, UK. ISBN: 0 419 23890 5

1 Terminology

1.1 Autogenous shrinkage

Autogenous shrinkage is the macroscopic volume reduction of cementitious materials when cement hydrates after initial setting. Autogenous shrinkage does not include volume change due to loss or ingress of substances, temperature variation, application of an external force and restraint.

Autogenous shrinkage can be expressed as the percentage of volume reduction "autogenous shrinkage ratio" or one dimensional length change "autogenous shrinkage strain".

Note: The phenomenon of macroscopic volume reduction of cement paste, mortar and concrete caused by chemical shrinkage is referred as autogenous shrinkage. When autogenous shrinkage is measured, the initial length (volume) of specimens should be measured at the time of initial setting. Although autogenous shrinkage is essentially a three dimensional phenomenon, autogenous shrinkage strain(ε_{as}) is described as linear strain.

When ordinary concrete is subjected to drying, water evaporation results in drying shrinkage. In large structural elements made of concrete, temperature variation due to heat of hydration results in thermal strain. Under these conditions when the mass and the temperature of concrete are changed, the strain which has been obtained as drying shrinkage strain or thermal strain includes autogenous shrinkage strain under the corresponding temperature conditions.

Volume change that is generated when concrete is still fresh should be excluded in order to define autogenous shrinkage. Since autogenous shrinkage is generally used for prediction of cracking, the strain generated in a period when cementitous material is fresh is excluded. Therefore, the time of initial setting of cement is specified as the start point of autogenous shrinkage measurement.

1.2 Autogenous expansion

Autogenous expansion is the macroscopic volume increase of cementitious materials when cement hydrates after initial setting. Autogenous expansion does not include volume change due to loss or ingress of substances, temperature variation, application of an external force and restraint.

Autogenous expansion can be expressed as the percentage of the volume increase "autogenous expansion ratio" or one dimensional length change "autogenous expansion strain".

1.3 Autogenous volume change

Autogenous volume change includes autogenous shrinkage and autogenous expansion.

1.4 Chemical shrinkage

Chemical shrinkage is the phenomenon in which the absolute volume of hydration products is less than the total volume of unhydrated cement and water before hydration.   Chemical shrinkage is described by the following equation.(Notation in the equation is shown in Fig.1.1)

$$S_{hy} = \frac{(V_c + V_w) - V_{hy}}{V_{ci} + V_{wi}} \times 100 \tag{1}$$

S_{hy}: chemical shrinkage ratio(%)
V_{ci}: volume of cement before mixing
V_c : volume of hydrated cement
V_{wi}: volume of water before mixing
V_w : volume of reacted water
V_{hy}: volume of hydration products

Note: When hardened cement paste is considered as a composite of solid phase (unhydrated cement and hydration products), liquid phase (unhydrated water) and gas phase (air bubbles existing after mixing and the voids created by hydration), chemical shrinkage is considered to be the reduction of absolute volume of reactants, i.e. solids

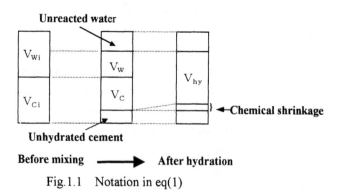

Fig.1.1   Notation in eq(1)

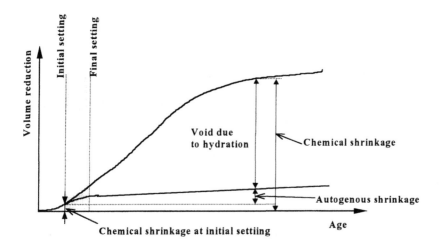

Fig. 1.2 Relation between chemical shrinkage and autogenous shrinkage

phase plus liquid phase. On the other hand, autogenous shrinkage is considered to be the reduction of the external volume since solid skeleton is formed.

After the skeleton of hydration products develops in the absence of an external source of water, the progress of hydration results in the formation of additional voids in the hardened microstructure of cement. Therefore, the macroscopic volume reduction, i.e. the autogenous shrinkage, is much less than the chemical shrinkage.

The relation between autogenous shrinkage and chemical shrinkage of cement paste without evaporation and any external source of water is schematically shown in Fig.1.2 and also described as eq.(2).

$$S_{hy} \doteqdot S_p + S_{as} + \Delta S_{hy} \tag{2}$$

Where,

S_{as}: autogenous shrinkage ratio(%)

ΔS_{hy}: chemical shrinkage ratio at the time of initial setting

S_p : the ratio of the volume of voids created by hydration to the volume of hardened cement paste(%)

In eq.(2), right and left sides are not explicitly equal each other. This is because the initial volume for calculating chemical shrinkage(S_{hy}) is specified as the volume of the mixture at the end of mixing, on the other hand, that for autogenous shrinkage(S_{as}) is specified as the volume at the initial setting.   Since the term ΔS_{hy} in the eq.(2) is very small, S_{hy} is approximated by the following equation for practical purposes.

$$S_{hy} \fallingdotseq S_p + S_{as} \tag{3}$$

The relation between autogenous shrinkage ratio and autogenous shrinkage strain is described by the following equation.

$$S_{as} = 300 \, \varepsilon_{as} \tag{4}$$

where,
ε_{as}: autogenous shrinkage strain

When the effect of gravity is not negligible, length change in the horizontal direction of cementitious material is different from that in the vertical direction.   The relation between chemical shrinkage and length change in the horizontal direction is schematically shown in Fig.1.3.   The relation between chemical shrinkage and length change in the vertical direction is shown in Fig.1.4(a) for mixtures without bleeding and Fig.1.4(b) for mixtures with bleeding.

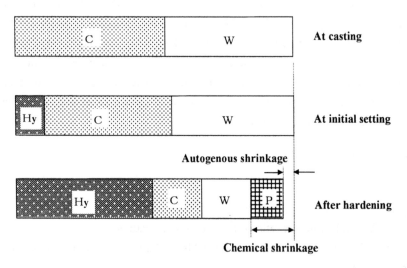

Fig.1.3   Relation between chemical shrinkage and autogenous shrinkage in the horizontal direction

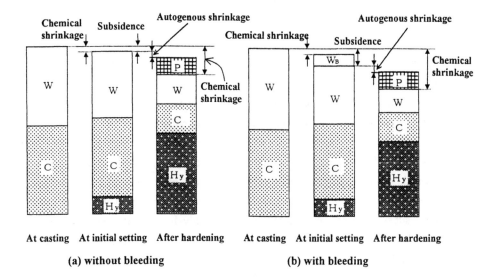

Fig.1.4   Relation between chemical shrinkage and autogenous shrinkage in the vertical direction

Where,
W :   unhydrated water
W_B :   bleeding water
C :   unhydrated cement
H_y :   hydration products
P :   voids generated by hydration

1.5. Self desiccation
    Hardened cement paste can be subjected to drying due to consumption of capillary water in the progress of cement hydration.   When this phenomenon happens it is called self desiccation.

Note: When the voids (P) created by hydration is not supplied with water from the surrounding environment, the hardened cementitious material can be substantially dried without evaporation.   This phenomenon is called self desiccation.   If a specimen with dense microstructures or a large cross section is used, water can penetrate into only the surface layer of the specimen even under water curing.   Then the central part of the specimen can be subjected to self desiccation.

1.6 Subsidence

Vertical length change in cementitious materials before initial setting, which is caused by bleeding, chemical shrinkage and so on, is called subsidence.

Note: In case of fresh concrete, in which the skeleton of hydration products has not been formed, cement particles are re-arranged by the effect of gravity as chemical shrinkage proceeds, then macroscopic shrinkage occurs in the vertical direction. Therefore, subsidence of concrete is caused not only by the difference in specific gravities of water and solid particles, which is manifested as bleeding, but also by chemical shrinkage.

2   Autogenous shrinkage and its mechanism

2.1   Introduction

Though autogenous shrinkage of concrete has been generally known since early times, no regard has been paid to it in the crack prevention for concrete structures compared with drying shrinkage of concrete. However, it is reported that concretes which have a low water to binder ratio and a large amount of binder such as high strength concrete and self-compacting concrete show greater autogenous shrinkage resulting in the occurrence of crack under certain circumstances[1,2]. With the development of the study of concrete, the results of investigations for autogenous shrinkage are now being accumulated. In the Standard Specification for Design and Construction of Concrete Structures prescribed by the Japan Society of Civil Engineers 1996[3], it was pointed out that the aoutogenous shrinkage should be taken into consideration as a cause of crack generation.

In this chapter, the cause of the autogenous shrinkage and its mechanism, various influencing factors, interactions with other deformations, and further different methods used for reducing autogenous shrinkage are explained.

At first, the properties and the autogenous shrinkage of self-compacting concrete, high strength concrete and massive concrete are summarized. Secondly, the relation between the chemical shrinkage of cement hydration and the autogenous shrinkage is clarified. The influencing factors on the autogenous shrinkage such as materials, mix proportions and curing conditions are summarized. As for the mechanisms of autogenous shrinkage, explanation is made on the basis of capillary tension due to self-desiccation, and further consideration is paid to the autogenous shrinkage in regard to microstructure and hydration products. Since autogenous shrinkage does not develop independently in actual concrete structure, relation between autogenous shrinkage of concrete and subsidence, plastic shrinkage, thermal strain during hardening, drying shrinkage, swelling, expansion due to expansive additives or carbonation shrinkage are investigated. Lastly, preventing or reducing methods for autogenous shrinkage of concrete are proposed including the problems to be solved in their practical use.

2.2   Concretes to which careful consideration should be given concerning autogenous shrinkage

As for the shrinkage of concretes that have low water to binder ratio or high binder content or high replacement ratio by fine granulated blast furnace slag, the development of autogenous shrinkage should be considered. High strength concrete, self compacting concrete, ordinary massive concrete should be included in this group. In the case of zero-slump concrete used for RCD, RCCP or for immediate stripping of concrete products, it is unknown whether autogenous shrinkage becomes larger or not because the water to binder ratio is low, but the unit volume of cement paste is small, and this should be the subject for future studies. Characteristics of mix-proportion for high strength concrete, self-compacting concrete   and massive concrete are shown in Table -2.1.

Table 2.1.   Materials and mix proportions of self-compacting concrete, high strength concrete and massive concrete

High strength concrete	Binder content : 450~600 kg/m³, Water to binder ratio : 0.25~0.40, Mineral admixture : Silica fume, Blast-furnace slag and Anhydrite, Chemical admixture : AE high range water reducing agent
Self-compacting concrete	Binder content : 350~500 kg/m³, Water to binder ratio : 0.30~0.60, Mineral admixture : Blast-furnace slag, Lime stone powder, Fly ash, Chemical admixture : AE high range water reducing agent, Viscosity controlling agent (cellulose type, acrylic type and glucose type)
Massive concrete	Binder content : 250~350 kg/m³, Water to binder ratio : 0.45~0.60, Cement : Low-heat cement, Blended cement, Mineral admixture : Blast-furnace slag, Fly ash

(1) High strength concrete

The specifications of the Japan Society of Civil Engineer provide that high strength concrete should have a strength not less than 60N/mm²[4]. And the use of concrete with design strength of 80N/mm² at most is recognized in the conventional guideline for design and construction of high strength concrete structures. However, now the guideline for design and construction of concrete structures using concrete with silica fume approves the use of concrete with design strength not more than 100N/mm²[5]. In general, though high strength concrete has been used for such concrete products as piles and poles, recently it is used as RC (reinforced concrete) in high-rise building[6]. For the production of high-strength concrete, admixtures such as superplasticizer, AE high range water reducing admixture, silica fume, granulated blast-furnace slag and anhydrite are used. Autoclave curing and steam curing are   sometime used as special curing method. In all these cases, a method that lowers the water to binder ratio to about 0.30 or less is adopted.

With the reduction of the water binder ratio autogenous shrinkage of concrete increases, and with the addition of silica fume it becomes larger. High strength concrete with water to binder ratio lower than 0.30 and replacement ratio of silica fume not less than 10% generates an autogenous shrinkage of 200~400x10⁻⁶.   The relative humidity at the center of a 400mm block made of high strength concrete with silica fume decreases to about 80% a week after placement as shown in Fig.-2.1[7].   This is called "self-desiccation".

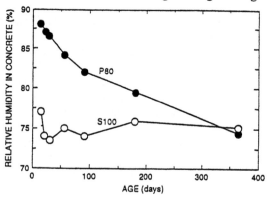

Fig. 2.1.   Relative humidity in high strength concrete with (S) or without (P) silica fume (P-80:design strength-80N/mm², S-100:design strength-100N/mm²)

Self-desiccation   is pronounced in high strength concrete, because the small amount of water used for mixing is rapidly consumed by early age hydration of cement. This is the main reason why large amount of autogenous shrinkage is observed for high strength concrete.

(2) Self -Compacting Concrete

Self-compacting concrete with the addition of AE high range water reducing agent that can be cast without vibration is now attracting attention. This is a concrete that has a remarkably high flowability and no segregation. According to the differences in the method used to prevent segregation, the self compacting concrete can be divided into three types, namely powder type, viscosity controlling agent type and combined type[2,8]. As for the powder type, granulated blast furnace slag, fly ash or limestone powder is used with cement by unit powder content of about 170~190 L/m^3. Water to powder ratio of this type of concrete is generally 0.30~0.40 by mass. For the viscosity controlling agent type of self-compacting concrete, water soluble compound of cellulose type, acrylic type or glucose type can be used as controlling agent, a unit powder content of about 350~500kg/m^3 and water to powder ratio of 0.3~0.6 by mass are used. The combined type of powder and viscosity controlling agent uses powder together with polysacharide polymer. Any type of self-compacting concrete shows greater autogenous shrinkage when the unit powder content increases. Though the types of mineral and chemical admixtures and cement makes some difference, autogenous shrinkage of self-compacting concrete with a unit powder content of 500kg/m^3 may be about 100~400x10^{-6}.

(3) Massive Concrete

As for conventional mass concrete except for high strength concrete or self-compacting concrete, materials used and mix proportion should satisfy strength requirement. Temperature rise due to heat of hydration should be kept as low as possible. Therefore, cement content per unit volume of mass concrete is kept low and moderate heat portland cement , low heat portland cement , blended cement or ternary system cement containing fly ash and or a granulated blast-furnace slag are used. Considering the unit binder content and water cement ratio, it quite unprobable that autogenous shrinkage of ordinary massive concrete becomes large. However, analysis of cracking of structures such as piers or anchorage shows that autogenous shrinkage of concrete in which a large amount of fine granulated blast-furnace slag has been replaced is about 100x10^{-6} which is too high to be neglected. It must be emphasized that autogenous shrinkage becomes large when a large amount of fine granulated blast furnace slag is used[9]. Therefore, when selecting the cementitious system for a massive concrete it is necessary to make a material design in which consideration is given to autogenous shrinkage besides temperature rise. Cast-in-place high strength concrete shows a remarkable temperature rise when it is used for piers or column, therefore it should be considered as a massive concrete.

2.3 Chemical shrinkage and autogenous shrinkage

Hydration of cementitious materials shows shrinkage, because the volume of hydrated products is smaller than the sum of the volumes of the unhydrated(original) cement and water. Table-2.2 shows an example of a volume change resulting from hydration of C$_3$S[10]. In the case of hydration of calcium silicate, many compositions of C-S-H with different Ca/Si were reported. Chemical shrinkage of these cases shows different values, because stoichiometric equations would be different for different C-S-H[11]. More detailed investigations are necessary for this point. Though some difference in

Table 2.2. Calculation of chemical shrinkage of C_3S

Hydration				Chemical shrinkage(%)
C_3S + 5.25H_2O → $C_{1.75}SH_{4.0}$ + 1.25CH				
Weight 228.3	94.6	230.3	92.6	
Density 3.12	1.00	1.90	2.24	3.16
Volume 73.2	94.6	121.1	41.3	
2C_3S + 6H_2O → $C_3S_2H_3$ + 3CH				
Weight 456.6	108.1	342.5	222.3	
Density 3.15	1.00	2.71	2.24	10.87
Volume 145.0	108.1	126.4	99.2	

shrinkage ratio may exist, the hydration of any clinker mineral results in the reduction of the volume as the result of hydration. As shown in Fig.2.2, the ratio of chemical shrinkage of portland cement is obtained as the sum of the ratio of chemical shrinkage of each clinker mineral[12].

Water to cement ratio and type of cement and admixture are the main factors which influence chemical shrinkage [12,13]. Cement having a large amount of C_2S such as a low heat portland cement generates small chemical shrinkage. Cements blended with additives such as granulated blast-furnace slag, fly ash or limestone powder also shows small chemical shrinkage. As for the cement mixed with granulated blast-furnace slag, chemical shrinkage decreases with the fineness of the granulated blast-furnace slag and its replacement ratio. Chemical shrinkage of this type is a phenomenon observed in the case

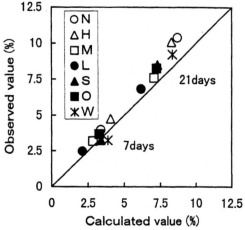

Type of cement	Mineral compound (%)			
	C_3S	C_2S	C_3A	C_4AF
N	49.7	23.9	8.8	9.4
H	64.6	9.7	8.8	8.5
M	48.4	31.2	2.0	12.2
L	27.8	56.3	2.4	7.6
S	58.0	18.5	1.9	13.7
O	56.2	20.7	1.8	14.0
W	67.2	11.2	11.6	0.6

Fig. 2.2. Relationship between measured values and calculated values for chemical shrinkage

where enough volume of water is supplied from outside.

To the contrary, autogenous shrinkage is a macroscopic volume change occurring after initial setting in a case where the supply of water from outside is not enough. As the hydration of cement progresses, pores are produced within the hardened cement paste due to the development of chemical shrinkage. As the hydration of cement progresses, capillary pore water and then gel water is consumed and menisci are produced in the capillary pores or fine pore of hardened structures due to the lack of water supply from outside, with the result that the hardened paste shows shrinkage due to the negative pressure. This mechanism may be explained with the aid of the capillary tension theory as in the case of drying shrinkage.

It is clear from definitions of chemical shrinkage and autogenous shrinkage that there is no direct correlation between chemical shrinkage and autogenous shrinkage though the latter is generated by the former as a main cause. Fig.2.3 shows the relation between chemical shrinkage and autogenous shrinkage for various types of cement[13]. It shows that the cement with smaller amount of aluminate phases gives smaller value of autogenous shrinkage, even if the chemical shrinkage is similar.

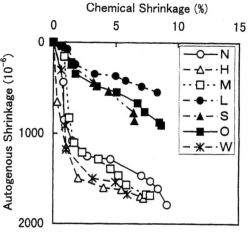

Type of	Mineral compound (%)			
cement	C_3S	C_2S	C_3A	C_4AF
N	51.5	22.9	8.9	8.8
H	62.8	11.8	8.0	8.5
M	42.9	35.9	3.2	11.9
L	22.9	58.8	2.5	9.4
S	61.0	16.9	1.7	14.6
O	57.6	20.5	2.2	13.7
W	61.1	16.5	12.1	0.6

Fig. 2.3. Relationship between chemical shrinkage and autogenous shrinkage

2.4 Mechanisms of Autogenous Shrinkage

Autogeneous shrinkage is caused by self-desiccation which is the result of the formation of fine pores in the microstructure of hardened cement body and the water consumption by the hydration reaction of cement. In order to understand the mechanism of autogenous shrinkage, it is necessary to understand (1)chemical shrinkage (2)microstructure (3)self desiccation.

(1) Chemical shrinkage

Cement minerals produce various types of hydrates when they hydrate. The total volume of solid and liquid phases decreases due to all of these reactions. The volume change of each phases can be calculated from crystal densities on the basis of stoichiometric calculation using the above mentioned equations.

(2) Microstructure

Once a skelton is formed after setting of cement, the hardened cement body cannot shrink any more as much as the volume reduction caused by hydration, that is, chemical shrinkage.

Therefore, further hydration is compensated by the formation of voids in the microstructure. While, the macroscopic volume change, namely autogenous shrinkage is dependent on the rigidity of microstructure that has close relation with the morphology of hydrates[14,15].

At the early stage of cement hydration, ettringite is formed on the surface of cement particles and in the pore solution. This ettringite is needle like crystal and is developed in the direction of needle axis.

Therefore, the formation of ettringite creates large volume of fine pores in the hardened body. Hydration reaction of silicate phases which continues slowly for long term produces fine and irregular-shaped C-S-H. The C-S-H is filled with fine pores made during the process of mixing and the process of setting and hardening. The formations of ettringite and C-S-H are strongly affected by the chemical composition of cement and the curing conditions. Therefore, the amounts of ettringite and C-S-H formed and the micro-pore structure vary in each hardened body. For example, it is known that the amounts of C-S-H formed significantly increase when blast furnace slag or silica fume is added.

(3)   Self-Desiccation

The amount of free water in hardened cement body gradually decreases due to the hydration reaction of cement minerals and fine pores are formed in hardened cement body.

Generally speaking, in a material with fine pores, the pore size in which water is able to exist at equilibrium with the atmosphere of certain relative humidity is variable and depends on the humidity. In a highly humid condition, water can exist as liquid in the larger pores. In hardened cement body, the amount of free water decreases and the micro-pores are formed by the hydration reaction of cement minerals. Therefore, the water vapor pressure reduces and the relative humidity in fine pore decreases. This phenomenon is called self-desiccation because of the decrease of relative humidity in the hardened cement body under the isolated condition without any loss of mass. The decrease of relative humidity is experimentally demonstrated in many cases[16,17,18,19,20,21]as shown in Fig. 2.1[7]. In the process of drying shrinkage of a hardened body, water starts to evaporate from larger pores. On the other hand, in the self-desiccation process, water is thought to be consumed in the place of hydrate formation, that is reaction front, and the place is suspected as fine pores in many cases. But, also in this case voids are created from larger pores, so local redistribution of water is inevitable. This process is often referred as capillary condensation. Water is more stable when it is put under the effect of higher energy of solids.

Self-desiccation is considered to be significant in the case of more amount of finer pores and less water in hardened cement body. In other words, the degree of self-desiccation is strongly related with microstructure formation.

(4)   Autogenous shrinkage

Self-desiccation occurs in hardened cement body due to hydration reaction. Comparing conventional drying shrinkage with autogenous shrinkage, there is a difference in the cause, the former is a evaporation of water towards the outer environments and the later is a consumption of water by hydration reaction. However, it is same in the point of decrease of humidity in hardened cement body. Therefore, the mechanisms of drying shrinkage, such as the capillary tension theory, is applicable to the mechanisms of autogeneous shrinkage. This is supported indirectly by the fact that the countermeasures for drying shrinkage, such as the reduction of surface tension by drying shrinkage reducing admixture[22] or the mixing of hydrophobic powders [23,24], are also effective as the countermeasures for autogenous shrinkage.

(5)   The difference between autogenous shrinkage and drying shrinkage

Although there are no differences between autogenous shrinkage and drying shrinkage in the point of the decrease of humidity in hardened cement body, autogenous shrinkage is different from drying shrinkage with regard to the mechanism of decrease in humidity. Drying shrinkage is caused by the diffusion of water into the outer environment. Therefore, drying shrinkage can be reduced by making the densely hardened body and prevent the diffusion of water towards the outer environment. Autogenous shrinkage is caused by the consumption of water by the hydration reaction of cement minerals. Since autogenous shrinkage is caused by self-desiccation, autogenous shrinkage increases with increase in the amount of finer gel pores and with decrease in   water cement ratio when the amount of water is decreased to make a low water cement ratio, the water in pores is quickly consumed by hydration reaction. Therefore, it is not a effective way to reduce autogenous shrinkage to prevent water diffusion into the outer environments.   In order to reduce autogenous shrinkage, it is necessary to reduce the fine micro-pores or to reduce the surface tension of water in the pores of hardened samples.

2.5   Factors influencing autogenous shrinkage

Autogeneous shrinkage of cement is influenced by mineral compositions of cement and their hydration ratio [12]. Autogenous shrinkage can be estimated by use of Eq.(2.1), this equation is derived from a regression analysis for autogenous shrinkage as a function of the hydration ratio of each mineral and each mineral content for various types of cement at a constant water-cement ratio of 0.3. Fig.-2.4 shows the relation between measured values and calculated values. By using Eq(2.1), autogenous shrinkage can be estimated from the mineral composition of various types of portland cement. In addition, it is clear from Eq.(2.1) that the C_3A and C_4AF a great influence on autogenous shrinkage.

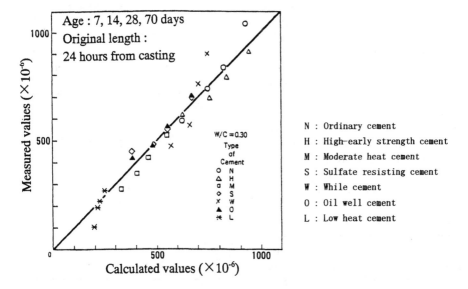

Fig. 2.4.   Relation between estimated values and measured values of autogenous shrinkage

$$\varepsilon_{as}(t)= -0.12 \cdot \alpha \, C_3S(t)(\%C_3S) - 0.070 \cdot \alpha \, C_2S(t)(\%C_2S) +$$

$$+ 2.256 \cdot \alpha \, C_3A(t)(\%C_3A) + 0.859 \cdot \alpha \, C_4AF(t)(\% \, C_4AF) \ \text{-------} \ (2.1)$$

Where, $\varepsilon_{as}(t)$: Autogenous shrinkage at age of 't'

$\alpha \cdot \cdot (t)$: Reaction ratio of each mineral at age of 't'

(%mass): Content ratio of each mineral

The influences of mineral admixtures and chemical admixtures on autogenous shrinkage of concrete are as follows. Autogenous shrinkage of concrete containing silica fume at low water-binder ratio [25,26] and the concrete with higher replacement ratio of fine blast furnace slag powder [23,27], is increased compared with that of ordinary concretes. On the other hand, autogenous shrinkage of concrete is reduced by using fly ash ,limestone powder , silica powder treated with water repellent or a drying shrinkage reducing agent [1,23,28,29]. As for an expansive additives, autogenous shrinkage of concrete with expansive additives occurs [23,29], but the final shrinkage of concrete is remarkably reduced compared with that of concrete without expansive additives. If autogenous shrinkage of expansive concrete is defined as shrinkage from the maximum length, the autogenous shrinkage is greatly influenced by the type of expansive agent and that of some agent is larger than the autogenous shrinkage of concrete with the same water to cement ratio and without expansive additives.

Autogenous shrinkage is influenced by mix proportion. As shown in Fig.2.5, autogenous shrinkage increases with a decrease in water to cement ratio or with an increase in the amount of cement paste[23,29,305]. This means that autogenous shrinkage is increased when unit content of binder increases or when the volume concentration of aggregate decreases. Autogenous

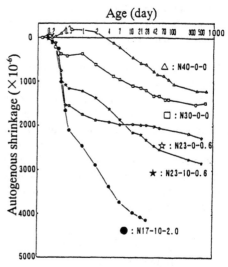

Fig. 2.5. Influence of water to binder ratio on autogenous shrinkage of cement

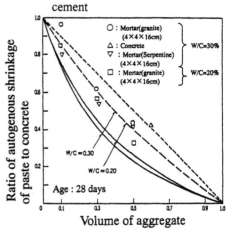

Fig. 2.6. Effect of the volume content of aggregates on autogenous shrinkage

shrinkage of concrete with different volumes of aggregate, can be estimated by Hobb's composite model, as shown in Eq.(2.4) [31], which was proposed to predict drying shrinkage of concrete by taking account the influence of aggregate restraint as shown in Fig.2.6[29,32]. Probably the influence of air content on autogenous shrinkage is the same as that of the volume of cement paste, but the detailed effects of air content on autogenous shrinkage has not been made clear. It is reported that autogenous shrinkage is nearly constant under an air content from 0.1 to 0.3[33]. The relationship between autogenous shrinkage and unit content of powder or replacement ratio of admixture is influenced largely by types and combinations of powder or binder[27].

$$\varepsilon_c / \varepsilon_p = 1 - Va \quad \text{----- (2.2)}$$

$$\varepsilon_c / \varepsilon_p = (1 - Va) / \{(Ea/Ep - 1) \, Va + 1\} \quad \text{---------- (2.3)}$$

$$\varepsilon_c / \varepsilon_p = \frac{(1 - Va) \, (Ka/Kp + 1)}{1 + Ka/Kp + Va \, (Ka/Kp - 1)} \quad \text{--------- (2.4)}$$

where, ε_c : Autogenous shrinkage of concrete

ε_p : Autogenous shrinkage of cement paste

Va : Volume concentration of aggregate
Ea : Modulus of elasticity of aggregate
Ep : Modulus of elasticity of cement paste
Ka : Bulk modulus of elasticity of aggregate
Kp : Bulk modulus of elasticity of cement paste
Ka : $Ea/3 \, (1 - 2\mu)$
Kp : $Ep/3 \, (1 - 2\mu)$

μ : Poisson's ratio

Curing conditions influence autogenous shrinkage such that initial shrinkage becomes higher, latter shrinkage becomes lower at high temperature [25]. The influence of curing temperature on autogenous shrinkage of concrete can be estimated by maturity of curing condition. And, the effect of manufacturing process, or compaction etc. on autogenous shrinkage is hardly studied.

2.6 Autogenous shrinkage and other volume changes
Autogenous shrinkage does not generally occur by itself but with other kind of volume changes in concrete structures. In this section, the relation between autogenous shrinkage and other kind of volume change, such as volume change due to heat of hydration, drying shrinkage and swelling and so on, is discussed.
(1) Volume change before setting
Subsidence of fresh concrete has been considered in connection with segregation. As shown in Fig.-2.7[34], however, chemical shrinkage(corresponding to the bleeding

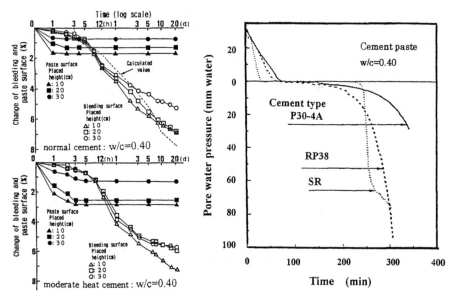

Fig. 2.7. Change of bleeding and paste Surface.

Fig.2.8. Negative pressure for sealed samples.

surface in the figure) and subsidence (corresponding to the paste surface in the figure) occurs simultaneously during the first few hours. This suggests that chemical shrinkage should be considered as one of the causes of subsidence. Main cause of subsidence is a gravitational force, and thus it always occurs to vertical direction and no subsidence occur to horizontal direction. At early stage subsidence is always larger than the chemical shrinkage when bleeding is observed on top surface of concrete.

It has been said that plastic shrinkage is caused by rapid evaporation of water from concrete surface before setting. As shown in Fig.2.8[35], however, negative pressure is observed in pore water about 2 hours after casting although cement paste is sealed to prevent evaporation. It can be said that the negative pressure is caused by chemical shrinkage and that a driving force of shrinkage in concrete without evaporation may exist. Therefore chemical shrinkage should be considered as one of the factors influencing plastic shrinkage.

(2) Volume change due to heat of hydration

In massive concrete structures, both autogenous shrinkage strain and thermal strains due to heat generation during cement hydration are simultaneously generated. It is convenient to estimate thermal strain and autogenous shrinkage strain separately. It has been reported that autogenous shrinkage of concrete under varying temperature could be estimated by maturity concept[9]as shown in Fig.-2.9. The authors have also reported that the principle of superposition (Eq.(2.5)) is valid for autogenous and thermal strains, and that thermal strain could be obtained by subtracting the estimated autogenous shrinkage strain from the observed strain. The calculated thermal strain obtained by this method has linear relation with the temperature variation during heating and cooling periods, therefore thermal strain can be expressed with a constant coefficient of thermal expansion, as shown in Eq.(2.6)(see Fig.-2.9).

$$\Delta \varepsilon = \Delta \varepsilon ' + \Delta \varepsilon a \qquad \text{------- (2.5)}$$

where, $\Delta \varepsilon$:total strain, $\Delta \varepsilon '$:thermal strain, $\Delta \varepsilon a$:autogenous shrinkage strain

$$\Delta \varepsilon ' = \alpha \Delta T \quad (2) \qquad \text{------ (2.6)}$$

where, α :coefficient of thermal expansion($1/°C$) , ΔT :temperature change ($°C$)

It has been suggested, however, that maturity concept is not valid for some concretes with mineral admixture[36]. Investigations in this field are necessary in more detail.

(3) Drying shrinkage

When concrete is subjected to drying condition after curing, drying shrinkage occurs simultaneously with autogenous shrinkage. The relation between the strain of sealed and dried specimens is shown in Fig.-2.10, where the original length was measured one day after casting[37]. It can be seen that shrinkage of dried specimens is much larger than that of sealed specimen for concrete with 0.40 water to cement ratio. On the other hand, for 0.17 water/binder mixture, the values for both type of specimens is almost the same. This suggests that the most part of drying shrinkage is not attributed to evaporation but to autogenous shrinkage for concrete with very low water-cement ratio. A schematic illustration for the relation between autogenous shrinkage and drying shrinkage is shown in Fig.-2.11[3].

A prediction equation for estimating shrinkage strain of concrete has been proposed by Japan Society of Civil Engineers(JSCE)[3]. The JSCE equation has been

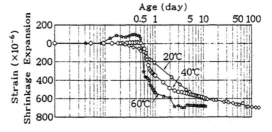

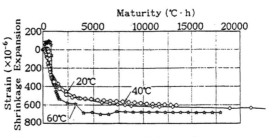

Fig. 2.9.　Autogenous shrinkage of concrete (water to cement ratio:0.20)

derived under conditions in which drying shrinkage is dominant and it is a function of water content of the concrete and the ratio of the surface area to the volume of the member. It should be noted that this equation can not be applied for concrete with water-cement ratio smaller than 0.40, concrete under water curing and concrete subjected to drying from an age earlier than 3 days. For these concretes a significant part of the total shrinkage is caused by autogenous shrinkage.

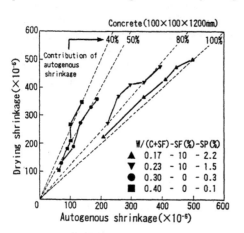

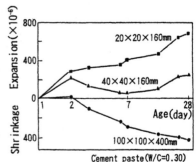

Fig.2.10. Relation between drying shrinkage and autogenous shrinakge for concrete

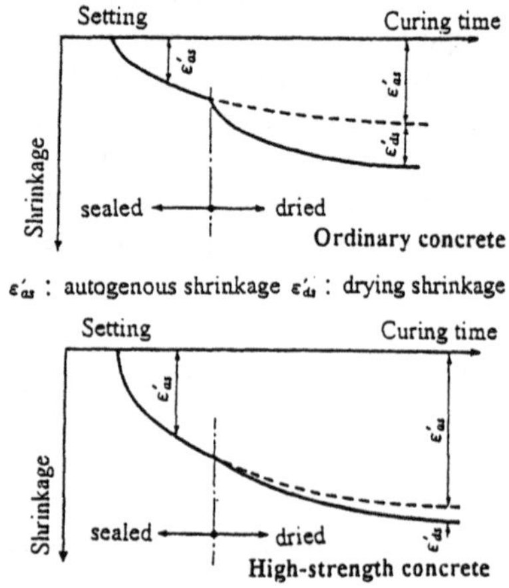

Fig. 2.11 Schematic illustration of the shrinkage of concrete

(4) Other kind of volume changes
a. Swelling

Ordinary concrete generally increases its volume in moist conditions. A cement paste with low water-cement ratio or large dimensions, however, decreases its length although it is cured under water as shown in Fig.-2.12 and Fig.-2.13[38]. This is because curing water permeates only into the external parts of the specimen while

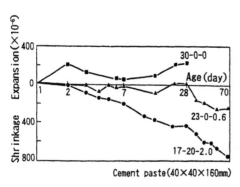

Fig. 2.12. Length change of cement paste under water(effect of W/C).

Fig. 2.13. Length change of cement paste under water(effect of size).

internal parts are subjected to self-desiccation. In this case, the mean length change of the specimen is influenced by both swelling and autogenous shrinkage which occur simultaneously in the cross section.

b. Early expansion due to hydration
It has been reported that a little expansion can be observed at early ages for belite rich cement with or without blast furnace slag and this expansion may be attributed to the formation of ettringite[15].

c. Expansive additive
Autogenous shrinkage occurs even in concrete with expansive additives. Length change of cement pastes with different type of expansive additives are shown in Fig.-2.14[38]. After the expansion of concrete at early ages, shrinkage is observed under sealed condition for cement paste with expansive additives. Final shrinkage of cement paste with expansive additives is generally lower than plain cement paste, although the effect differs with the type of additive. Mechanism of autogenous shrinkage due to the hydration of expansive additives and influence of type of additives on autogenous shrinkage have not been made clear yet.

d. Carbonation shrinkage
In concrete with low water-cement ratio carbonation may occur only in the external part of a concrete member, therefore, the combination of carbonation and autogenous shrinkage does not have to be considered in engineering practice.

2.7 Reduction of autogenous shrinkage
Based on the mechanism and the influencing factors of autogenous shrinkage, reduction methods of autogenous shrinkage are investigated. There are summarized in Fig.-2.15.

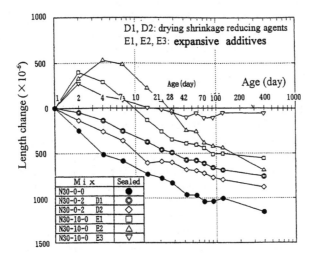

Fig. 2.14. Influence of drying shrinkage reducing agent and expansive additives on autogenous shrinkage of concrete.

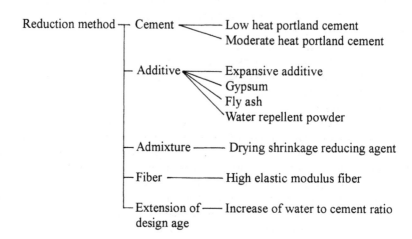

Fig. 2.15. Reducing methods of autogenous shrinkage

(1) Materials

Autogenous shrinkage of concrete can be reduced by using a portland cement having a higher a C_2S content and lower contents of C_3A or C_4AF[38]. Therefore, autogenous shrinkage of concrete made with a moderate heat portland cement and a low heat portland cement are greatly reduced compared to that with normal portland cement. However, it is not easy to evaluate quantitatively the effect of these cements on the reduction of autogenous shrinkage because autogenous shrinkage depends also on compressive strength of concrete. Fig.-2.16 shows a relationship between curing temperature and the water to cement ratio which is necessary to obtain a compressive strength of 40, 60 and $80N/mm^2$ at the age of 28day and 91day[39]. The calculated value of water to cement ratio to obtain the design strength of concrete is shown in Table-2.3. When compressive strength of concrete are considered at the same age, autogenous shrinkage could be reduced by using a low heat portland cement since the water to cement ratio of low heat portland cement is slightly less than that of normal portland cement. If the design compressive strength is based on the strength at 91-day, it is not necessary that autogenous shrinkage with low heat portland cement is considered because the water to cement ratio of low heat portland cement is slightly larger than that of normal portland cement. The increase of gypsum content in cement causes autogenous expansion and reduces autogenous shrinkage as a result, though the investigation for stability of this cement is necessary .

By using expansive additives, the total shrinkage of concrete is reduced compared to that of normal type as shown in Fig.-2.14[29]. In self-compacting concrete, it is indicated that 30 percent replacement of fly ash to portland cement greatly reduces autogenous shrinkage[30]. However, when fly ash is replaced, an extension of design age is required due to decrease in the rate of strength development. Therefore when selecting admixture, it is necessary to consider the balance between strength development and the effect of reducing autogenous shrinkage. The reducing effect of drying shrinkage reducing agent on autogeneous shrinkage is shown in Fig.-2.14 [29].

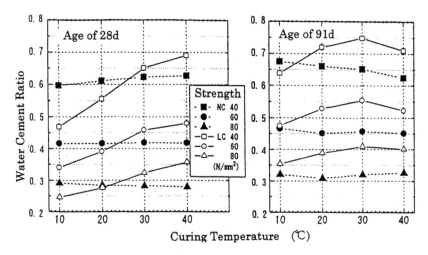

Fig. 2.16. Relation between the curing temperature and water to cement ratio to obtain the design strength

Table 2.3   Water to cement ratio to obtain the design strength

Design strength (N/mm²)	Normal portland cement (curing time:28d)	Low heat portland cement (curing time:28d)	(curing time :91d)
60	0.42	0.39	0.53
80	0.28	0.27	0.39

The addition of drying shrinkage reducing agent can   reduce the surface tension of capillary pores water and hence can reduce autogenous shrinkage without changing the other properties of hardened concrete.   The combination of expansive additive and drying shrinkage reducing agent could be highly effective to reduce autogenous shrinkage.   It is confirmed that powder treated by water repellent[29] and fiber of high elastic modulus[41] are also effective to reduce autogenous shrinkage.

(2)   Mix proportion
It is thought that the factor in mix proportion which effects greatly on autogenous shrinkage is water to cement or binder ratio.   However, increase of water to cement or binder ratio, which is effective to reduce autogenous shrinkage, is not practical because water to cement ratio is determined by requirements on the strength and durability .

2.8   Reference

1.   Tazawa,E. and Miyazawa,S.(1992) Autogenous shrinkage caused by self desiccation in cementitious material, 9th International Congress on the Chemistry of Cement, pp.712-718.
2.   Japan Concrete Institute(1994) Report of the Technical Committee on Superworkable Concrete,Part 2 (in Japanese).
3.   Japan Society of Civil Engineers(1996) Standard Specification for Design and Construction of Concrete Structure (in Japanese).
4.   Japan Society of Civil Engineers(1980) Guideline for Design and Construction of High Strength Concrete (in Japanese).

5. Japan Society of Civil Engineers(1995) Guideline for Design and Construction of Concrete with Silica Fume (in Japanese).

6. Nagataki,S. and Sakai,E.(1994)Applications in Japan and South East Asia, High Performance Concretes and Applications (edited by Shah,S.P. and Ahmad,S.H.),Edward Arnold,London,pp.375-397.

7. Mak,S.L. and Torii,K.(1995) Strength development of high strength of ultra high-strength concrete subjected to high hydration temperature, Cement & Concrete Research,Vol.25, No.8,pp.1791-1802.

8. Japan Concrete Institute(1993) Report of the Technical Committee on Superworkable Concrete, Part 1 (in Japanese).

9. Tazawa,E., Matsuoka,Y., Miyazawa,S. and Okamoto,S.(1994) Effect of autogenous shrinkage on self stress in hardening concrete, Proc. Int'l Symp. On Thermal Cracking in Concrete at Early Ages,Munchen,pp.221-228.

10. Tazawa,E.(1969) Influence of curing time on shrinkage and weight loss of hydrating portland cement, Proc. of Japan Society of Civil Engineers, No.156,pp39-52.

11. Japan Concrete Institute(1996) Report of the Technical Committee on Modeling for Cement and Concrete (in Japanese).

12. Tazawa,E., Miyazawa,S.and Sato,T.(1993) Influence of Cement Composition on Autogenous Shrinkage, JCA Proceedings of CEMENT & CONCRETE, No.47, pp.528-533 (in Japanese).

13. Tazawa,E.,Miyazawa,S. and Kasai,T.(1995) Chemical shrinkage and autogeous shrinkage of cement paste, Cement & Concrete Research,Vol.25,No.2,pp.288-292

14. Tazawa, E. Miyazawa,S., Miura, T. and Tani,S.(1994) The autogeneous shrinkage of cement paste prepared with various cement, 21st Seminar of Cement and Concrete, Vol.1, pp69-74 (in Japanese).

15. Takahashi, T., Nakata, H.,Yoshida, K.and Goto, S(1996) Influence of hydration on autogenous shrinkage of cement paste, Concrete Research and Technology (Japan Concrete Institute), Vol.7, No.2, pp.137-142 (in Japanese).

16. Hooton, R.D., Sato, J.A. and Mukherjee, P.K(1992) A new method for assessing frost damage in non-air entrained hydraulic structure, ACI-131, G.M. Idorn International Symposium - Durability of Concrete -,Vol.1, pp.338-357.

17. McGratti, P. and Hooton, R.D.(1991) Self-desiccation of portland cement and silica fume modified mortar, American Ceramic Society, Ceramic Transactions, Vol.16, pp.489-500.

18. Hirao, H. Uchikawa, H. and Hanehara, S(1996) Influence of humidity and hardened structure on the autogenous shrinkage of hardened cement paste, Proceeding of the Japan Concrete Institute, Vol.18, No.1, pp705-710 (in Japanese).

19. Wittmann, F.(1968) Surface tension shrinkage and strength of hardened cement paste, Materiaux et Constructions, Vol.1, No.6, pp.547-552.

20. Paillere, A.M., Buil, M. and Serrano, J.J.(1990) Effect of fiber addition on the autogenous shrinkage of silica fume concrete, discussion, Authors' closure, ACI Material, p.82, Jan-Feb..

21. Sellevold, E.J., Justnes, H., Smeplass, S. and Hansen E.A (1995)Selected properties of high performance concrete, Advances in Cement and Concrete,

pp.562-609.
22. Okamoto, S and Matsuoka, Y(1994)A study on autogenous shrinkage of cement paste from the stand point of pore structure, Proceeding of the 49th annual conference of the Japan Society of Civil Engineers, Vol.V, pp.1089-1019 (in Japanese).
23. Tazawa, E. and Miyazawa,S(1994) Effect of cementitious materials and mix proportion on the autogenous shrinkage of cement based material, Journal of materials, concrete structures and pavement (Japan Society of Civil Engineers), No.502, V-25, pp43-52 (in Japanese).
24. Tazawa, E. and Miyazawa, S(1994) Autogenous shrinkage by hydration of cement, Concrete Journal (JCI), Vol.32(9), pp.25-30 (in Japanese).
25. Tazawa,E., Miyazawa,S., Shigekawa,K.(1991)Macroscopic Shrinkage of Hardening Cement Paste due to Hydration, JCA Proceedings of CEMENT & CONCRETE, No.45, pp.122-127, (in Japanese)
26. Imamoto,K., Ohtani,H.(1995) A Study on Shrinkage of High-Ultra High strength Concrete, Proceedings of the Japan Concrete Institute, Vol.17, No.1, pp.1061-1066 (in Japanese).
27. Tazawa,E., Miyazawa,S., Sato,T., Hashimoto,S.(1992) Autogenous shrinkage of cement paste by use of blast furnace slag fine powder, Proceedings of 19th Cement and Concrete Research Convention, pp.23-28 (in Japanese).
28. Chikamatu,R., Takeda,N., Kamata,F., Sogo,S.(1993)Effect of Admixtures on Autogenous Shrinkage of Concrete, Proceedings of the Japan Concrete Institute, Vol.15, No.1, pp.543-548 (in Japanese).
29. Tazawa,E.(1994) Autogenous Shrinkage of Cement Paste Caused by Hydration, CEMENT & CONCRETE, No.565, pp.35-44.
30. Momotani,T., Abe,M., Sasahara,A., Yasuda,M., Hasuo,K.(1992) A Study on Mechanical Properties and Durability of High Fluidity Concrete, Summaries of Technical Papers of Annual Meeting Architectural Institute of Japan, A-1, pp.297-298 (in Japanese).
31. Hobbs,D.W.(1974) Influence of Aggregate Restraint on the Shrinkage of Concrete, Journal of ACI, Vol.71, No.9, pp.445-450.
32. Miura,T., Tazawa,E., Miyazawa,S., Hori,A.(1995) Influence of Blast Furnace Slag on Autogenous Shrinkage of Concrete, Proceedings of the Japan Concrete Institute, Vol.15, No.1, pp.359-364 (in Japanese).
33. Miyazawa,S.(1992) Self-stress of Concrete due to Autogenous Shrinkage and Drying Shrinkage, Doctoral Thesis, Hiroshima University, pp.57-61 (in Japanese).
34. Kasai, T. and Tazawa, E. (1988) Degree of hydration of cement estimated from measurement of shrinkage due to chemical reaction, Bulletin of the Faculty of Engineering, Hiroshima University, Vol.37,No.1, pp.23-29 (in Japanese).
35. Sellevold,E., Bjontegaard,O. Justnes,H. and Dahl,P.A. (1994) High Performance Concrete, Early Volume Change and Cracking Tendency, RILEM International Symposium on Thermal Cracking in Concrete at Early Ages, pp.229-236.
36. Matsunaga, A., Yoneda, S.,Takeda, N. and Sogo, S.(1996) Thermal cracking resistance of concrete used admixtures with different autogenous shrinkage properties,Vol.18,No.1,pp1287-1292 (in Japanese).
37. E.Tazawa and S.Miyazawa: Experimental study on mechanism of autogenous

shrinkage of concrete, Cement and Concrete Research, Vol.28, No.8, pp.1633-1638, 1995

38. Tazawa, E. and Miyazawa, S. (1997) Influence of constituent and composition on autogenous shrinkage of cementitious materials, Magazine of Concrete Research, Vol.49, No.178, pp.15-22.

39. Tanaka,S.,Maruoka,M.,Takeuchi,M. and Tomita,R.(1996) Strength development of high strength concrete with belite rich cement subjected to thermal history, Proceedings of the Japan Concrete Institute, Vol.18, No.1, pp.237-242 (in Japanese).

40. Ogawa,A., Sakata,K. and Tanaka,S.(1995) A study on reducing shrinkage of highly-flowable concrete, 2nd CANMET/ACI International Symposium on Advances in Concrete Technology Proceedings, pp55-72.

41. Mmiyazawa,S.,Kuroi,T. and Shimomura,H.(1997) Autogenous shrinkage stress of fiber reinforced mortar, JCA Proc. of CEMENT & CONCRETE, Japan Cement Association, pp.560-565 (in Japanese).

3   Autogenous Shrinkage Stress and Its Estimation

3.1   Introduction

High performance concrete is achieved by reducing water-binder ratio and/or enhancing flowability.   Cracking in high performance concrete ,which is often used to increase durability of structures, should be relatively important compared with that in ordinary concrete.   It has been realized that autogenous shrinkage affects significantly crack development in concrete structures with improvement of concrete qualities, such as high strength and high fluidity.

Cracking due to autogenous shrinkage is difficult to control, because densifying the microstructure of concrete by reducing the water-binder ratio utilizing mineral admixtures, such as fly ash and granulated blast-furnace slag, is an intrinsic part of the performance enhancement.   The relatively weak shrinkage-reducing effect of curing also makes such crack control difficult.   The purpose of performance enhancement includes not only increasing the strength but also eliminating the factors adversely affecting the durability of concrete, such as temperature cracking and defective consolidation during placing.

Autogenous shrinkage strain is significantly observed in high strength concretes with low water-binder ratio, in which thermal stress due to hydration heat of cementitious materials concurrently develops.   It is therefore necessary to evaluate both stresses in such concretes in order to control cracking.   Stress due to drying shrinkage combined with autogenous shrinkage should be estimated in case of concrete subjected to drying.

Temperature due to hydration heat rises rapidly and strength also develops rapidly in high strength concrete with a low water-binder ratio.   Because such a high-strength concrete is generally used for concrete structures smaller than so called massive ones, the temperature after the peak value drops rapidly.   Therefore, during temperature drop, tensile stress is remarkably produced by the simultaneous effects of autogenous shrinkage and temperature drop induced shrinkage.   In order to analyze the stress accurately, more exact modeling of mechanical properties of concrete after placing, such as elastic modulus and creep coefficient must be achieved.

Accordingly, the committee investigated ① testing methods for autogenous shrinkage stress from the standpoint of minimizing thermal stress, ②methods of estimating autogenous strain required for estimating the autogenous stress, ③ modeling of mechanical properties of concrete, ④analysis method of autogenous shrinkage stress, ⑤evaluation of hydration heat and drying shrinkage induced stresses combined with autogenous shrinkage stress, ⑥the effect of autogenous shrinkage on cracking observed in full-sized reinforced concrete columns and frames used high-strength concrete.

Autogenous shrinkage stress develops depending on shrinkage strain, mechanical properties such as elastic modulus and creep and restraining conditions.   Main factors for generation of autogenous shrinkage stress are tabulated in Table 3.1, compared with those of thermal stress due to hydration effect and drying shrinkage stress.

Table 3.1.   Factors for autogenous shrinkage, hydration heat and drying shrinkage induced stresses

	Autogenous shrinkage stress	Thermal stress	Drying shrinkage stress
Duration of stress generation	during hydration	within 2 weeks after casting	during drying
Restraint by r.b.	◎ or ○	×	◎ or ○
Restraint of existing concrete or foundation, etc.	◎ or ○	◎ or ○	○ or ×
Nonlinear distribution of shrinkage or temperature	○ or ×	◎ or ○	◎ or ○
Water-binder ratio	◎	○	◎
Curing and/or ambient temperatures	○	◎	○
Size	○	◎	◎
Cause of shrinkage	self desiccation	temperature change	evaporation

r.b : reinforcing bar, ◎ : significant, ○ : slightly significant, × : negligible

3.2   Test method for autogenous shrinkage stress

Restrained stress in concrete specimens sealed and protected from evaporation was measured by using the restraining system of "Testing method on cracking of concrete due to restrained drying shrinkage", which is a draft of Japanese Industrial Standards [1].   However, autogenous shrinkage stress could not be purely generated because thermal stress was simultaneouly generated due to temperature difference between the restraining frame and the concrete specimen.   Therefore, the committee proposed a testing method for measuring autogenous shrinkage stress, in which a reinforcing bar was embedded into the specimen concrete as a restraining body.   The details of the proposed "Test method for autogenous shrinkage stress of concrete" are described in Chapter 4.

3.3   Method of prediction for autogenous shrinkage stress
(1)   Concept

Stress generation due to autogenous shrinkage can be explained by the fact that shrinkage strain is restrained by restraining bodies. This mechanism is basically not different from the development of hydration heat-induced stress as well as drying shrinkage stress.

A prediction of the stress due to these sorts of volume changes is made in the following three steps:

1. Prediction of the strain due to temperature change, autogenous shrinkage, and drying shrinkage
2. Modeling of elastic modulus and creep coefficient varying with age
3. Stress analysis taking restraining conditions into account

Thermal strain is generally determined by multiplying the temperature change by the thermal expansion coefficient.   Estimation equations for drying shrinkage have been already provided in some standard specifications, though accurate estimations are still difficult.   On the other hand, autogenous shrinkage is affected by a number of factors, such as water-cement ratio, type and content of admixtures, type of cement, and temperature, and few estimation equations have been established yet taking all of these factors into account.   As predicting autogenous shrinkage strain is indispensable for stress analysis, an empirical equation proposed by Tazawa and Miyazawa was recommended by the committee. Though it has been pointed out that autogenous shrinkage distributes nonlinearly within a section similarly to drying shrinkage [2], the nonlinearity is not dealt with in this report.

In order to estimate autogenous shrinkage stress as accurate as possible, the committee emphasized the following investigations.

1. Modeling of elastic modulus, creep coefficient, and tensile strength of concrete for several months after placing, because autogenous shrinkage stress develops rapidly at very early ages, and develops continuously and slowly for a few months or more.
2. Development of creep analysis method and its verification for stresses produced by simultaneous action of autogenous shrinkage and temperature change, because temperature develops significantly even in relatively small structures.
3. Evaluation of restraint stress by reinforcing bar in heavily reinforced high-strength concrete members.

(2) Equation of autogenous shrinkage strain

The committee recommends an equation for predicting autogenous shrinkage strain given below. The equation consists of the rate and ultimate value of autogenous shrinkage strain as a function of water-binder ratio, and coefficient to express effects of types of cement and mineral admixture.

$$\varepsilon_c(t) = \gamma \cdot \varepsilon_\infty \cdot \beta(t) \qquad \text{---------(3.1)}$$

where,

$$\varepsilon_{c0} = 3070 \exp\{-7.2(W/B)\} \qquad \text{---------(3.2)}$$
$$\beta(t) = 1 - \exp\{-a(t-t_o)^b\} \qquad \text{---------(3.3)}$$

where,

γ : coefficient to express effects of types of cement and mineral
     admixture (in the case of ordinary portland cement, $\gamma = 1$)

ε_∞ : ultimate value of autogenous shrinkage strain (x 10^{-6})

W/B : water-binder ratio

a, b : constants

t_o : initial setting time (days)

t : age (days)

t and t_o are adjusted by Eq.(3.4) according to the elevated or reduced temperature:

$$t \text{ and } t_o = \Sigma \Delta t_i \cdot \exp[13.65 - 4000/(273 + T(\Delta t_i)/T_o)] \qquad \text{---------(3.4)}$$

where,

Δt_i : number of days during which concrete temperature is T°C (days)

T_o : 1°C

Figure 3.1 shows the relationship between the autogenous shrinkage strain calculated by the equations and that obtained by the experiments. Water-cement ratio and aggregate content of the concrete are 0.2~0.56 and 0.55~0.65, respectively. The calculated values range from 0.6 to 1.4 of the measurements, which shows that the equations could predict autogenous shrinkage strain with satisfactory accuracy.

However, the effects of the temperature and type and content of mineral admixtures on the rate and the ultimate value of autogenous shrinkage is to be investigated more.

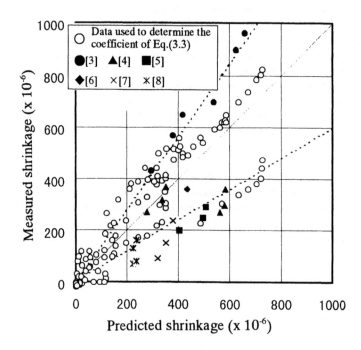

Fig. 3.1.   Comparison of measured and predicted shrinkage strains

(3)   Mechanical properties of concrete

In order to analyze autogenous shrinkage stress more accurately, it is necessary to model the mechanical properties of high-strength concrete as well as self-compacting concrete containing mineral admixtures especially at early ages.   In contrast to conventional concrete with a normal strength, however, few studies have been conducted regarding the mechanical properties of high-performance concrete at early ages.   In this section, previous studies are reviewed and models of the mechanical properties for the analysis of shrinkage stress are investigated.

a.   Compressive strength and Young's modulus

Though the equation of the CEB-FIP MODEL CODE 1990 (MC90) [9] is widely applicable, it cannot be applied to concrete at early ages before one day as indicated in the Fig.3.2 (a).   The equations for compressive strength and elastic modulus from final setting time to 28 days are proposed, which is modified by incorporating the effect of setting time into the equations of CEB Model 1990, as shown in Eqs. (3.5) and (3.6).   The calculations by Eq. (3.6) are indicated in Fig. 3.2(b).

$f_c(t)=f_{c28} \cdot \exp[s_f\{1-((28-a_f)/(t-a_f))^{0.5}\}]$ ----------(3.5)

where,

$f_c(t)$: compressive strength at temperature adjusted age t (N/mm²)

f_{c28} : compressive strength at 28 days of concrete cured in water at 20°C (N/mm²)

s_f : coefficient depending upon type of cement

a_f : coefficient depending upon setting time(days)

  t : temperature adjusted concrete age(days)

$E_c(t)=E_{c28} \cdot \exp[s_E\{1-((28-a_E)/(t-a_E))^{0.5}\}]$ ----------(3.6)

where,

$E_c(t)$: modulus of elasticity at temperature adjusted age t (N/mm²)

E_{c28} : modulus of elasticity at 28 days of concrete cured in water at 20°C (N/mm²)

s_E : coefficient depending upon type of cement

a_E : coefficient depending upon setting time (days)

  t : temperature adjusted concrete age(days)

    Young's modulus can be also roughly estimated by Young's modulus-compressive strength relationship proposed by the ACI-363[10] or New RC[11], but they overestimate Young's modulus when compressive strength is below 30 N/mm².

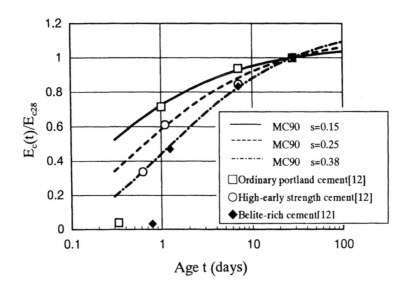

Fig.3.2 (a)   Comparison of Young's modulus development of high-strength concrete measured and those predicted by MC 90

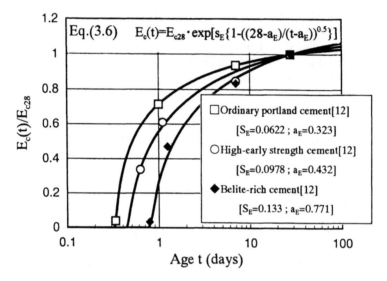

Fig.3.2 (b)   Comparison of Young's modulus development of high-strength concrete measured and those predicted by modified MC 90 equation(3.6)

b.   Relationship between splitting tensile strength and compressive strength

Relationships between splitting tensile strength and compressive strength calculated in accordance with Eq.(3.7) proposed by ACI-363, Eq.(3.8) by Noguchi and Tomosawa [13] and Eq.(3.9) by JSCE [14] are compared with experimental results obtained by previous studies, as shown in Fig. 3.3.   The ACI-363 equation and Noguchi-Tomosawa equation for high-strength concrete are considered to represent the relationship between the splitting tensile strength and the compressive strength satisfactorily.   When a more conservative evaluation is intended for the crack development, the Noguchi-Tomosawa equation and JSCE equation is considered appropriate, though the latter was proposed for ordinary concrete.

Figure 3.4 shows the relationship between the splitting tensile strength and the tensile stress at cracking due to restraint of the shrinkage strain.   The tensile stress at cracking is around 70% of the splitting tensile strength on average as shown in the figure.

ACI-363 equation ;
$$f_t = 0.59 \times f_c^{0.5}$$
----------(3.7)

Noguchi-Tomosawa equation ;
$$f_t = 0.291 \times f_c^{0.637}$$
----------(3.8)

JSCE equation ;
$$f_{tk} = 0.23 \times f_{ck}^{2/3} \quad (f_{tk}: \text{characteristic value})$$
----------(3.9)

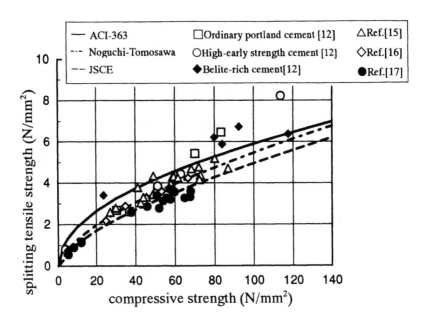

Fig. 3.3.   Relationship between splitting tensile strength and compressive strength

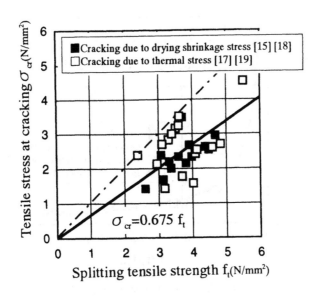

Fig. 3.4.   Relationship between tensile strength and the stress at cracking

c.   Creep

High-strength concrete using ordinary portland cement with water-cement ratio of 0.3, those using high-early strength portland cement or belite-rich cement, containing 10% silica fume with a water-binder ratio of 0.23 were made to investigate creep properties [12].   Loading was carried out at three different ages.   The equations for the notional creep coefficient Φ_0 and the rate β_H of MC90 shown in Eq.(3.10) were modified to Eqs.(3.11) and (3.12) by regression analysis.   Figures 3.5 and 3.6 show the relationship between modified Φ_0 and $f_c(t)/f_{c28}$ and the relationship between modified β_H and $f_c(t)/f_{c28}$, respectively.   The creep coefficient predicted by using modified Φ_0 and β_H is drawn in Fig. 3.7 with the measurements.   The applicability of this equation should be expanded by accumulating experimental data.

Creep equation of CEB-FIP MODEL CODE 1990

$$\Phi(t,t_0)= \Phi_0(((t-t_0)/t_1)/(\beta_H+(t-t_0)/t_1))^{0.3} \qquad\qquad \text{---------(3.10)}$$

where,

$\Phi(t,t_0)$: creep coefficient at temperature adjusted age t (days) after loading at temperature adjusted age t_0(days)

t_1 : 1 day

$\Phi_0 = 0.725(f_c(t)/f_{c28})^{-0.933}$ ---------(3.11)

$\beta_H = 0.490\exp(6.39[f_c(t)/f_{c28}])$ ---------(3.12)

$f_c(t)$: compressive strength at t days (N/mm²)

f_{c28} : compressive strength at 28 days (N/mm²)

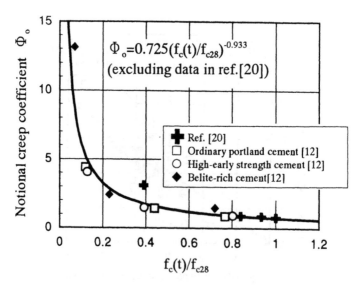

Fig. 3.5.   Relationship between modified Φ_0 and $f_c(t)/f_{c28}$

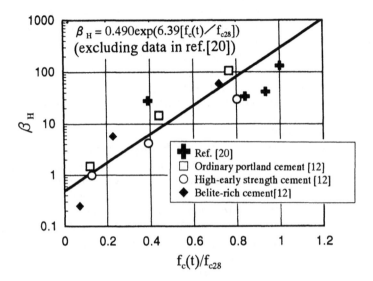

Fig. 3.6.   Relationship between modified β_H and $f_c(t)/f_{c28}$

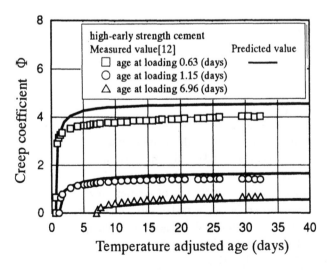

Fig. 3.7.   Comparison of measured and predicted creep coefficients

(4) Method of predicting autogenous shrinkage stress

The factors affecting autogenous shrinkage stress are the amount of autogenous shrinakge, elastic modulus, and creep, which are basically the same as those affecting the thermal stress and drying shrinkage stress, because the amount of shrinkage and mechanical properties of concrete influence dominantly the restraining stress. As autogenous shrinkage stress develops with the time like temperature stress due to hydration heat of cement and drying shrinkage, appropriate analysis method is needed.

In this section, the applicability of step-by-step method [21] is investigated based on the principle of superposition.

a. Step-by-step method

The concept of stress analysis by the step-by-step method is shown in Fig.3.8 and the analysis procedure for reinforced concrete axial members is given by Eqs.(3.13)~ (3.15) [5]. This method enables to estimate rapid development of stress and stress relaxation due to creep. Autogenous shrinkage stress computed, based on the assumption of linear strain distribution through the cross section, using creep coefficient obtained by Eqs.(3.10)~(3.12) and Young's modulus by Eq.(3.6), is compared with stress measured in Fig.3.9. The specimen for analysis is 1000mm in length, 100mm in width and in depth which is restrained by steel bars. The percentage of cross section of steel bar to that of specimen concrete is about 8%. Good agreement between both shows the validity of step-by-step method for the analysis of autogenous shrinkage stress of high-strength concrete.

Stress-strain relation of concrete ;

$$\sigma_c(t_{i+1/2}) = E_e(t_{i+1/2},\ t_i)[\ \varepsilon_c(t_{i+1/2}) - \varepsilon_{c,cr}(t_{i-1/2}) - \varepsilon_f(t_{i+1/2})] \qquad \text{---------(3.13)}$$

Equilibrium condition ;

$$A_c \cdot \sigma_c(t_{i+1/2}) + A_s \cdot \sigma_s(t_{i+1/2}) = 0 \qquad \text{---------(3.14)}$$

Strain compatibility ;

$$\varepsilon_c(t_{i+1/2}) = \varepsilon_s(t_{i+1/2}) \qquad \text{---------(3.15)}$$

where,

$E_e(t_{i+1/2},\ t_j) = 1/J(t_{i+1/2},\ t_j)$

$J(t_{i+1/2},\ t_j) = 1/E_c(t_i) + \Phi(t_{i+1/2},\ t_j)/E_{c28}$

$E_c(t_i)$: Young's modulus of concrete at the middle of i th time interval

$\Phi(t_{i+1/2},\ t_j)$: creep coefficient at time $t_{i+1/2}$ for loading at age t_j

$\sigma_c(t_{i+1/2})$: stress of concrete at the end of i th time interval

$\sigma_s(t_{i+1/2})$: stress of steel bar at the end of i th time interval

$\varepsilon_c(t_{i+1/2})$: actual strain of concrete at the end of i th time interval

$\varepsilon_s(t_{i+1/2})$: actual strain of steel bar at the end of i th time interval

$\varepsilon_f(t_{i+1/2})$: sum of temperature strain and autogenous shrinkage strain of concrete at end of i th time interval

$\varepsilon_{c,cr}(t_{i-1/2}) = \Sigma\ \Delta\ \sigma(t_j)\ J(t_{i+1/2},\ t_j) - J(t_{i+1/2},\ t_j)\ \sigma_c(t_{i-1/2})$

$\Delta\ \sigma_c(t_i) = \sigma_c(t_{i+1/2}) - \sigma_c(t_{i-1/2})$

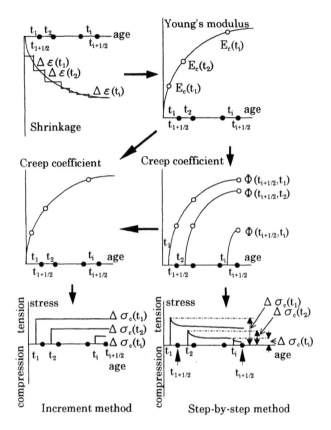

Fig. 3.8.   Concept of stress analysis by step-by-step method

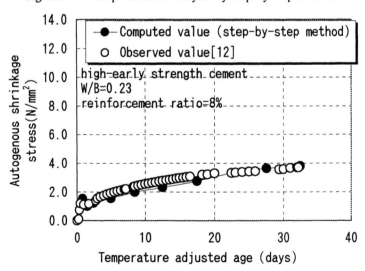

Fig. 3.9.   Comparison of autogenous shrinkage stress computed by step-by-step method with that measured.

b. Increment method

Increment method is usually applied to predict thermal stress of massive concrete structures, which sums up stress developed at each time interval. This method can not estimate stress relaxation with time. For the sake of practical performance, Young's modulus of concrete considering the effect of creep is recommended for massive concrete in JSCE Code [14]. Applicability of effective Young's modulus method to predict autogenous shrinkage stress in high-strength concrete is investigated in Fig.3.10. Modified Young's modulus with time obtained by Eq.(3.6) and a modified creep coefficient by Eqs.(3.10) ~ (3.12) are used. As indicated in the figure, analysis method, in which effective Young's modulus is used, overestimates the development rate of stress and coincidence between analysis and experiment is dependent on the loading age. The appropriate age at loading for creep coefficient to be used in the analysis should vary depending on the type of concrete. Equations for increment method are as follows ;

$$\sigma_c(t_i) = \Sigma \Delta \sigma_c(t_j) \qquad \text{----------(3.16)}$$
$$\Sigma \Delta \sigma_c(t_j) = \Delta \varepsilon_c(t_j) \cdot E_e(t_j) \qquad \text{----------(3.17)}$$
$$E_e(t_j) = E_c(t_j)/[1+\Phi(t_j, t_o) \cdot E_c(t_o)/E_{c28}] \qquad \text{----------(3.18)}$$

where,

$\sigma_c(t_{i+1/2})$: concrete stress at the end of i th time interval

$\Delta \varepsilon_c(t_j)$: increment of restraint strain between $t_{j+1/2}$ and $t_{j-1/2}$

$\Phi(t_j, t_o)$: creep coefficient at time t_j for loading at age t_o

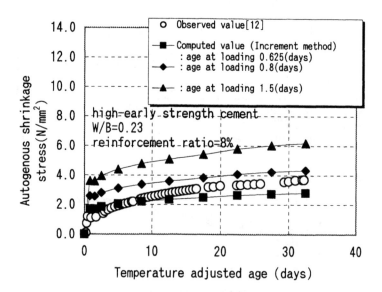

Fig. 3.10. Comparison of autogenous shrinkage stress computed by increment method with effective Young's modulus with that measured

(5) Autogenous shrinkage stress in hardening concrete subjected to temperature change

When estimating cracking in massive concrete structures, the stress produced by steel bar restraint has been neglected, because of the small difference of temperatures between concrete and steel bar and their similar coefficients of thermal expansion. However, concrete is restrained and stressed by reinforcing bar in the same way as drying shrinkage when concrete shrinks autogenously, because autogenous shrinkage is caused by self-desiccation. This section deals with a method for measuring stress in concrete produced by simultaneous action of autogenous shrinkage and temperature change due to cement heat hydration and the effects of the temperature change on the autogenous shrinkage stress are discussed.

a. Method for measuring autogenous shrinkage stress in hardening concrete subjected to temperature change

Autogenous shrinkage stress was measured in accordance with the draft of testing method "Cracking test of hydrated concrete by restraining thermal deformation caused by heat of hydration" proposed by Japanese Industrial Standards(JIS) [23] as illustrated in Fig.3.11. Restraining frame is made of steel pipe in which temperature controlled water is circulated. The temperature controlling system of the frame enables to give any degree of restraint to specimen concrete. Temperature simulated from hydration heat of cement is given to the specimens by controlling ambient temperature in the room.

Autogenous shrinkage stress is obtained by subtracting thermal stress from total stress. Total stress, thermal stress and autogenous shrinkage stress are obtained by the following equations (3.19),(3.20) and (3.21);

$$\sigma_c(t_{i+1/2}) = \Sigma \, \Delta \sigma_c(t_j) = \Sigma \, (A_s \cdot E_s \cdot \Delta \varepsilon_s(t_j)/A_c) \qquad \text{----------(3.19)}$$

where,

$\sigma_c(t_{i+1/2})$: total stress of specimen concrete at the end of i th time interval

$\Sigma \, \Delta \sigma_c(t_j)$: increment of stress of specimen concrete between $t_{j+1/2}$ and $t_{j-1/2}$

$\Delta \varepsilon_s(t_j)$: increment of stress related strain of the frame measured between $t_{j+1/2}$ and $t_{j-1/2}$

E_s : Young's modulus of the frame

A_s : cross sectional area of steel pipe frame

A_c : cross sectional area of specimen concrete

$$\sigma_{cT}(t_{i+1/2}) = \Sigma \, \Delta \sigma_{cT}(t_j) = \Sigma \, (K \cdot E_e \cdot \alpha_c \cdot \Delta T(t_j)) \qquad \text{----------(3.20)}$$

where,

$\sigma_{cT}(t_{i+1/2})$: total temperature stress of specimen concrete at the end of i th time interval

$\Sigma \, \Delta \sigma_{cT}(t_j)$: increment of temperature stress of specimen concrete between $t_{j+1/2}$ and $t_{j-1/2}$

α_c : thermal expansion coefficient of specimen concrete

$\Delta T(t_j)$: temperature change between $t_{j+1/2}$ and $t_{j-1/2}$

E_e : apparent Young's modulus of specimen concrete considering the effect of creep which is obtained from inclination in measured stress-temperature relationship

K : degree of restraint and = $[\Delta \varepsilon_r(t_j) - \Delta \varepsilon_{c,r}(t_j)]/\Delta \varepsilon_r(t_j)$

$\Delta \varepsilon_r(t_j)$: increment of strain of specimen concrete without restraint between $t_{j+1/2}$ and $t_{j-1/2}$

$\Delta \varepsilon_{c,r}(t_j)$: increment of strain of specimen concrete restrained by the frame between $t_{j+1/2}$ and $t_{j-1/2}$

$$\sigma_{as}(t_{i+1/2}) = \sigma_c(t_{i+1/2}) - \sigma_{c,T}(t_{i+1/2}) \qquad \text{----------(3.21)}$$

where,

$\sigma_{as}(t_{i+1/2})$: autogenous shrinkage stress of specimen concrete at the end of i th time interval

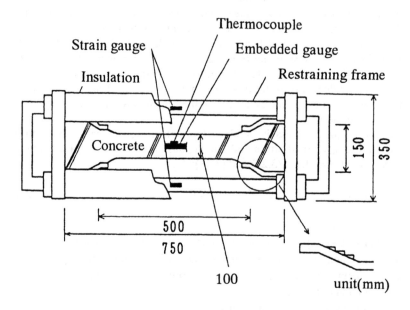

Fig.3.11.   Cracking test of hydrated concrete by restraining thermal deformation caused by heat of hydration (draft of JIS)

Table 3.2 lists the ratio of the autogenous shrinkage stress to the total stress obtained by Eq.(3.21), in which the effects of types of cement and mineral admixture, water-binder ratio and concrete temperature are investigated.   When high-strength concrete with water-binder ratio of 0.3 was subjected to a temperature history due to hydration heat, the ratio of the autogenous shrinkage to the total stress at cracking depended on the type of binders, and tended to be lower when belite-rich cement was used.

Table 3.2. Effect of autogenous shrinkage stress in hardening concrete subjected to temperature change on cracking in accordance with the draft of JIS

No.	Type of binder	W/B	$\Delta T max$ (°C)	γ
1	BC+BS	0.30	30.0	0.71
2			1.0	0.95
3	BC	0.30	38.0	0.41
4			1.0	0.95
5	OP+BS	0.30	29.8	1.71
6			10.0	1.31
7	MC+BS	0.54	20.1	0.20
8	HC+SF	0.23	9.7	0.72
9	BC+SF	0.23	6.8	0.58
10	OP+BS		6.0	0.57
11			5.0	0.50
12	OP	0.25	10.5	0.62
13			11.0	0.16
14	OP+SF		9.0	0.65
15			9.5	0.42

$\Delta T max$: maximum temperature rise in concrete placed at 20 (°C)
γ = ratio of autogenouse shrinkage stress to total stress at cracking
OP: ordinary portland cement
MC: moderate-heat portland cement
HC: high-early srengtht portland cement
BC: belite-rich cement
BS: blast-furnace slag
SF: silica fume

b. Effects of ambient temperature on autogenous shrinkage stress

Autogenous shrinkage stress of specimen concretes, which were restrained by embedded steel bars, depends on ambient temperature and, however, its dependence is varied by the type of binder, as indicated in Fig. 3.12 [5].

In the case of concrete made of ordinary portland cement without mineral admixture, autogenous shrinkage stress was lower under a higher ambient temperature. When blast-furnace slag with fineness of 8000cm²/g was used, however, the stress developed more rapidly at early age and thereafter tended to reach a lower value at the ambient temperature of 40°C than that at 20°C.

When silica fume was used, a higher ambient temperature led to a higher autogenous shrinkage stress. The development rate and the magnitude of autogenous shrinkage stress are concluded to depend on the ambient temperature as well as the type of binders.

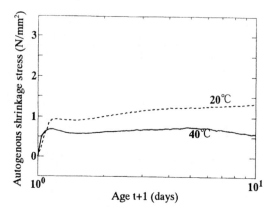

(a)   concrete made of ordinary portland cement without mineral admixture

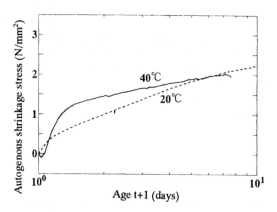

(b)   concrete made of ordinary portland cement and blast-furnace slag

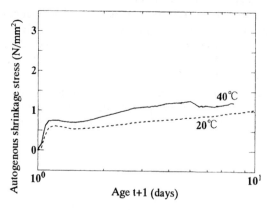

(c)   concrete made of ordinary portland cement and silica fume

Fig. 3.12.   Effect of ambient temperature on autogenous shrinkage stress

(6) Shrinkage stress of concrete under drying after curing

a. Shrinkage strain and shrinkage stress

When concrete is exposed to the air after curing, shrinkage stress increases with higher rate corresponding to rapid development of shrinkage strain. The step-by-step method overestimates shrinkage stress after drying, as shown in Fig. 3.13, which suggests that the mechanical properties including, especially, creep coefficient should be improved considering the effect of drying.

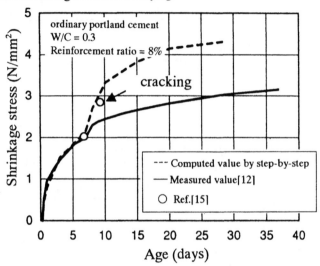

Fig.3.13. Effect of drying after curing on shrinkage stress

b. Shrinkage stress of concrete to consider autogenous shrinkage effect

The lower water-binder ratio of self-compacting concrete becomes, the more autogenous shrinkage stress develops as well as at the shorter days after drying cracking occurs, as shown in Fig.3.14. Table 3.4 gives the results of a restraining test in accordance with the JIS draft for drying shrinkage cracking [1], in which the effects of type of binder and duration of curing are examined [24] [19]. Mix proportion of concrete is tabulated in Table 3.3. Curing is performed by sealing the surface of the specimen in the mould. The temperature was maintained at 20°C and the relative humidity was 60% in the room during curing and drying. This table reveals that the percentage of the autogenous shrinkage stress to the total restraint stress of concrete exposed to the air after curing at cracking tends to increase as the curing period is longer, depending on the age at cracking and the materials used.

Table 3.3. Mix proportion of concrete

TC	G_{max} mm	W/B	s/a	Air %	Unit weight(kg/m³)								
					W	C	MC	BS	FA	V	L18	S	G
W1		0.32	0.52	2.0	160	179	-	165	166	-	-	857	824
W2	20	0.45	0.52	2.0	190	150	-	139	139	0.5	-	857	824
MS		0.36	0.51	3.5	172	-	513	-	-	-	28	828	827
S6		0.56	0.51	3.5	172	-	308	200	-	-	17	828	827
OP		0.55	0.51	4.5	165	200	-	-	-	-	-	927	924

C : ordinary portland cement, MC : moderate-heat portland cement, BS : blast-furnace slag, FA : fly-ash,
V : viscous agent, L18 : lime-stone powder

Table 3.4. The effect of drying after curing on cracking

TC		t_c	f_c	f_t	Tensile stress (N/mm²)			σ_d/f_t	σ_r/f_t	σ_{as}
		day	N/mm²	N/mm²	σ_{as}	σ_d	$\sigma_r=\sigma_{as}+\sigma_d$			$/\sigma_r$
W1	A	11.7	41.6	3.13	0.84	0.84	1.69	0.27	0.54	0.5
W2		11.8	31.2	2.60	0.48	0.94	1.42	0.36	0.55	0.34
MS	A	53.3	54.1	3.70	0.46	3.02	3.48	0.82	0.94	0.13
	B	51.1	47.7	3.41	0.40	2.98	3.38	0.87	0.99	0.12
	C	54.7	36.6	2.88	0.25	3.18	3.43	1.10	1.19	0.07
S6	A	11.4	51.4	3.58	0.95	1.39	2.34	0.39	0.65	0.41
	B	13.1	43.0	3.19	0.10	2.98	3.08	0.93	0.97	0.03
OP	A	31.6	40.3	3.07	0.43	1.96	2.39	0.64	0.78	0.18

TC: type of concrete

t_c: age at cracking (day)

A, B and C : duration of curing (A=7days, B=2days, C=16 hrs).

f_c: compressive strength at cracking predicted from compressive strength-age relation

f_t: splitting tensile strength predicted by $f_t = 0.29f_c^{0.637}$(N/mm²)

σ_{as} : autogenous shrinkage stress

σ_d : Stress developed after drying,

σ_r : total restrained stress

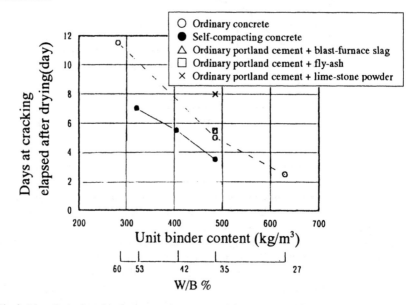

Fig.3.14. Relationship between days at cracking elapsed after drying and water-binder ratio

(7) Autogenous stress in full-sized concrete members

This section reports experimental and numerical investigations on the autogenous shrinkage stress generated in full-sized members made of ultra high-strength concrete.

a. Reinforced concrete column made of ultra high-strength concrete

The autogenous shrinkage stress in a reinforced concrete column with reinforcement ratio of 2.97% was measured, which was 850mm x 850mm in cross section and 2200mm in height. A plain concrete block having the same cross section was also prepared to measure the autogenous shrinkage strain. The dimension and configuration of the column, arrangement of reinforcing bars and location of measurements of temperature as well as strain are shown in Fig. 3.15.

Concrete was produced using belite-rich cement and silica fume which replaced 10% in mass of cement, with the water-binder ratio being 0.23. The compressive strength of concrete cured in water with the temperature of 20 °C was approximately 110 N/mm^2 at 28 days. The material and mix proportion of concrete are presented in Tables 3.5 and 3.6.

The temperature histories observed in the column are shown in Fig. 3.16. The maximum concrete temperatures at the center and the edge in the cross section of the column were 60.7°C and 54.0°C, respectively.

Fig. 3.17 indicates development of autogenous shrinkage strain measured by embedded gauge in the concrete block, which reached approximately 600 x 10^{-6} at 3 days. Autogenous shrinkage stress measured in the column was approximately 2.2 N/mm^2 at 3 days as shown in Fig. 3.18. The figure also indicates that the step-by-step method based on the superposition principle shown in Fig. 3.8 accurately evaluates the autogenous shrinkage stress produced by reinforcement restraint, in which the stress due to the temperature difference between center and edge was neglected.

Table 3.5. Materials

Material	type/characteristics
Cement	belite-rich cement specific gravity=3.20 fineness = 4160cm^2/g
Silica fume	specific gravity=2.20 fineness = about 200000cm^2/g
Fine aggregate	pit sand specific gravity=2.64 absorption=1.16%
Coarse aggregate	quartzite crushed stone specific gravity=2.63 absorption=0.58%
Chemical admixture	air entrained and high-range water reducing agent

Table 3.6. Mix proportion

Material	Proportion
Water-binder ratio W/B	0.23
Sand-coarse aggregate ratio s/a	0.45
Water kg/m^3	165
Cement kg/m^3	645
Silica fume kg/m^3	72
Fine aggregate kg/m^3	689
Coarse aggregate kg/m^3	839

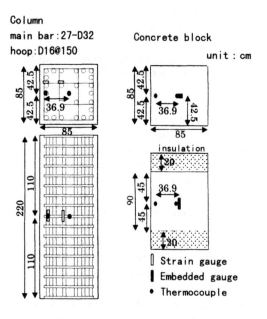

Fig.3.15.   Details of test column

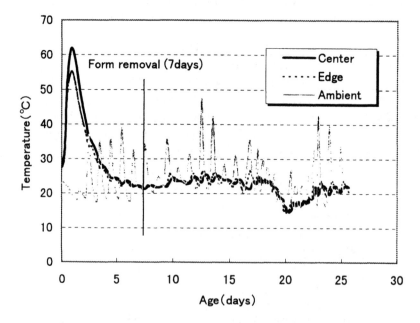

Fig.3.16.   Temperature histories in reinforced concrete column

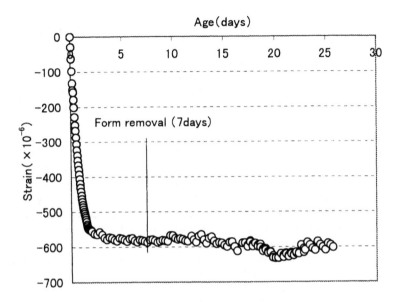

Fig.3.17. Development of autogenous shrinkage strain measured in concrete block

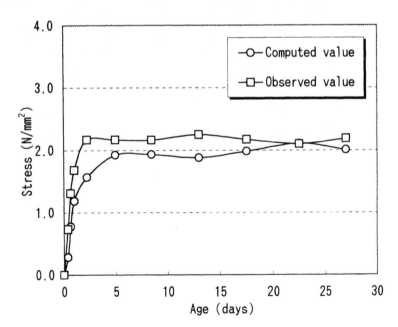

Fig.3.18. Autogenous shrinkage stress measured in full-sized reinforced concrete
column compared with that computed based on step-by-step method

b.   Reinforced concrete indeterminate structure made of ultra high-strength concrete

The autogenous shrinkage stress of a reinforced concrete indeterminate structure was numerically investigated, using two dimensional finite element method to which the step-by-step method was applied.   The reinforced concrete frame illustrated in Fig.3.19 was made of concrete with water binder ratio of 0.23 in which high-early strength cement and silica fume replaced by 10% in mass of the cement were used. The footing of the column was constructed directly on the foundation.   The details of the cross sections of the beam and the column are also shown in Fig.3.19.   The element mesh also shown in Fig.3.20.

The temperature histories at the middle height of the column are shown in Fig. 3.21. The maximum concrete temperatures were 99.0°C at the center and 85.0 °C at the edge in the cross section, respectively.   The autogenous shrinkage strain with time measured by embedded gauge in the concrete block shown in Fig.3.19 is expressed in Fig.3.22.   The creep coefficient was determined by modifying the CEB-FIP Model Code 1990(MC90) based on the experiment shown in Fig.3.7.   Young's modulus with time was also determined in accordance with MC90.

The structure should be cracked not only by autogenous shrinkage but also by hydration heat of cement, depending on kind of restraint, because the structure is indeterminate and temperature rise is very high.

Temperature change induced stress is produced due to restraint by foundation and nonlinear distribution of temperature in the cross section.   Autogenous shrinkage stress is caused by reinforcing bar restraint in addition to the restraint of foundation.

The analysis revealed that the percentages of the autogenous shrinkage stress due to reinforcement, autogenous shrinkage stress due to the restraint of foundation, thermal stress due to the restraint of foundation and thermal stress due to the nonlinear distribution of temperature were 24%, 69%, -77% and 84% at 0.75days at the location ① shown in Fig.3.20, respectively, as well as 34%, 98%, -114% and 82% at 0.63 days at the location ② in the same figure, respectively.   At 5.8 days when temperature was stabilized and autogenous shrinkage reached ultimate value, they were 28%, 82%, 49% and -59% and 24%, 70%, 58% and -52% at the same locations, respectively.

This suggests that the autogenous shrinkage has a significant effect on the cracking of reinforced concrete indeterminate structure made of ultra high-strength concrete.

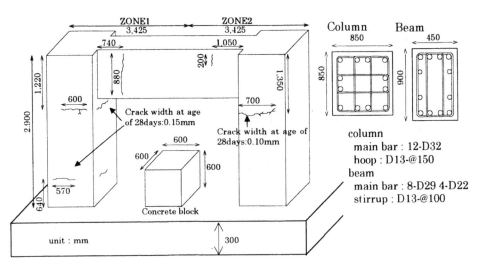

Fig. 3.19 Reinforced concrete frame

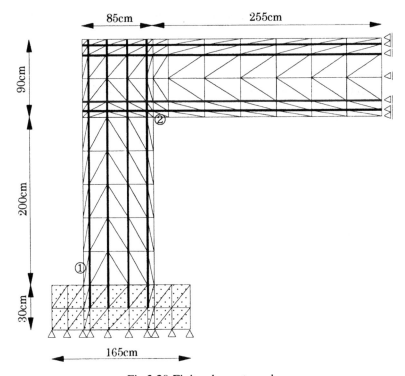

Fig.3.20 Finite element mesh

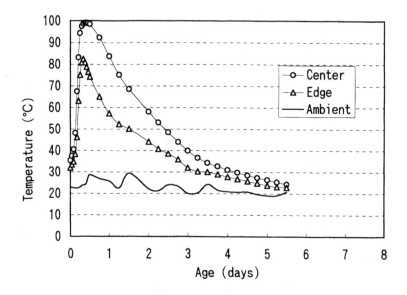

Fig.3.21.   Temperature histories at middle height of the column

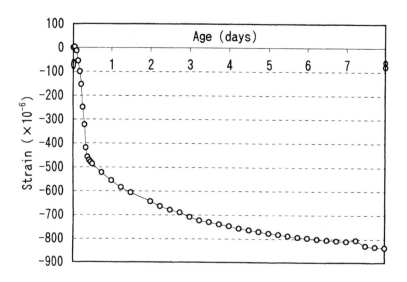

Fig.3.22.   Autogenous shrinkage strain in concrete block

3.4 References

1. Draft of Japanese Industrial Standards "Testing Method on Cracking of Concrete due to Restrained Drying Shrinkage".
2. Tazawa, E. and Miyazawa, S. (1992) Tensile and Flexural Strength of Cement Mortal Subjected to Non-Uniform Self-Stress, Magazine of Concrete Research, 44, No.161, pp.241-248.
3. Japan Concrete Institute(1996) Report of Technical Committee on Autogenous Shrinkage, pp.82-90.
4. Imamoto, K. and Ohtani, H. (1996) A Study on Autogenous Shrinkage of Ultra High Strength Concrete, Proceedings of the Japan Concrete Institute, Vol.18, No.1, pp.255-230.
5. Tsutsui, H., Sato, R. and Xu, M. (1996) A Study on Stress due to Autogenous Shrinkage in High-strength Concrete, Proceedings of Cement and Concrete, JCA, No.50, 1996, pp.478-483.
6. Yasuda, M., Abe, M., Sasahara, A. and Momotani, T. (1996) Experimental Study on Shrinkage and Crack of High Fluidity Concrete, Proceedings of the Japan Concrete Institute, Vol.18, No.1, pp. 147-152.
7. Koyanagi, M., Nakane, J. and Huchita, Y. (1996) Thermal Cracks Caused by Hydration-Heat of High Strength Concrete, Proceedings of the Japan Concrete Institute, Vol.18, No.1, pp. 1299-1304.
8. Japan Concrete Institute(1996) Report of Technical Committee on Super Workable Concrete (1) .
9. CEB-FIP MODEL CODE 1990, Thomas Telford.
10. ACI Committee 363 (1984) State-of-the-Art Report on High-Strength Concrete, Journal of American Concrete Institute, July-August, pp.364-411.
11. Japanese Architectural Standard Specification (1997), pp.165-169.
12. Japan Concrete Institute (1996) Report of Technical Committee on Autogenous Shrinkage, pp.203-210.
13. Noguchi, T., Tomosawa, F.(1995) Relationship between Compressive Strength and Various Mechanical Properties of High-Strength Concrete, Journal of Structural and Construction Engineering (Transactions of AIJ), Vol.472, pp.11-16.
14. Japan Society of Civil Engineers(1996) Standard Specification for Design and Construction of Concrete Structures.
15. Momotani, T., Abe, M., Sasahara, A., Yasuda, M. (1995) A Study on Mechanical Properties and Durability of High Fluidity Concrete (Part 9. Tests of Autogenouse Shrinkage and Cracking of Concrete due to Restrained Drying Shrinkage), Summaries of Technical Papers of Annual Meeting Architectural Institute of JAPAN, Materials and Construction, A, pp. 297-298.
16. Hisaka, M., Mano, S., Masuda, Y., Kanda, A. (1990) Influence of Water-Cement Ratio on Drying Shrinkage and Cracking of Concrete, Extended Abstracts; Annual Meeting of CAJ, No.44,pp.782-787.
17. Matsunaga, A., Yoneda, S., Takeda, N. and Sogo, S. (1996) Thermal Cracking Resistance of Concrete Used Admixtures with Different Autogenous Shrinkage Properties, Proceedings of the Japan Concrete Institute, Vol.18, No.1, pp. 1287-1292

18. Sujino, K. (1996) Generation Mechanism of Autogenous Shrinkage Stress in Massive Concrete, A Master's Thesis, Hiroshima University.
19. Nishida, N., Fukutome, K. and Shimomura, T. (1992) The Effect of Water Content on Drying shrinkage Cracking Resistance of Concrete, Proceedings of the 47th Annual Conference of the Japan Society of Civil Engineers, Part 5,pp.946-947.(the unpublished data included)
20. Larrard, F. (1990) Creep and Shrinkage of High-Strength Field Concrete, High Strength Concrete, ACI SP-121, pp.577-598.
21. NEVILLE, A. M., DILGER, W. H. and BROOKS, J. J. (1983) Creep of Plain and Structural Concrete, Construction Press, Longman, pp246-255.
22. Sato, R., Xu, M., Yang, Y. (1997) Stress of High-Strength Concrete due to Autogenous Shrinkage Combined with Hydration Heat of Cement, ACI's International SP.172-44, pp.837-852.
23. Draft of Japanese Industrial Standards "Cracking Test of Hydrated Concrete by Restraining Thermal Deformation Caused by Heat of Hydration".
24. Shimomura, T. and Uno, Y. (1995) Study on Properties of Hardened High Performance Concrete Stripped at Early Ages, Journal of Materials, Concrete Structures and Pavement, Japan Society of Civil Engineering, No.508, pp15-22.

4 Testing method

I   Test method for chemical shrinkage of cement paste

1 Scope
This test method covers the determination of chemical shrinkage of cement paste.

2 Apparatus
(1) Glass vessel
The capacity and height shall be respectively 50 ml or more and 75 mm or less.
The sample vessel (capacity: 50 ml and height: 70mm) is recommended.
(2) Measuring pipette
The capacity and accuracy shall be respectively 5 ml and 0.05 ml.
(3) Silicon plug or rubber plug
(4) Paraffin wrap
(5) Mortar mixer
(6) Washing bottle
(7) Lime saturated water
(8) Waterproof adhesive

3 Preparation of cement paste

3.1 Materials
Bring cement and water to a temperature of $20 \pm 3°C$ before mixing.

3.2 Water cement ratio of cement paste
Water cement ratio shall be 0.50.

3.3 Mixing of cement paste
Cement paste shall be mixed in room of $20 \pm 3°C$. For mixing cement paste, Hobart mortar mixer shall be used. Mixing time shall be three minutes.

4 Procedure (see Fig. 1)
(1) Put measuring pipette in central part of the silicon plug, and then the nose of measuring pipette shall be stuck out of the silicon plug by about 2 mm.   Furthermore, fix the measuring pipette in the plug with waterproof adhesive.
(2) Measure mass of vacant sample vessel with an accuracy of 0.1g.   This value shall be taken as $M_1(g)$.
(3) Place cement paste in the sample vessel and tap it in order to let out air bubble. Measure the total mass from cement paste and the vessel with an accuracy of 0.1g. This value shall be taken as $M_2(g)$.
(4) Fill the sample vessel with lime saturated water, paying attention not disturb the paste surface.
(5) Insert perpendicularly the silicon plug putting measuring pipette into the sample vessel.   At this stage, it shall be confirmed that the top surface of lime saturated water

rises in the inside of measuring pipette.

(6) Add lime saturated water from an upper inlet of the measuring pipette by the squirt or washing bottle to the measuring range.

(7) Seal the upper part of the measuring pipette with paraffin wrap. Furthermore, fix and seal the joint between the sample vessel and the silicon plug with waterproof adhesive. Store this specimen at a temperature of 20 ± 3°C. A minimum of three specimens shall be prepared for each test.

(8) Record the initial reading of water level in the pipette by accuracy of 0.05ml. This value shall be taken as H_0(ml) when starting the measurement.

(9) Measure the reading of water level in the pipette at due age (T_n). These values shall be taken as H_n(ml). In case that the decrease of the water in the pipette exceeds measuring range, supply lime saturated water from an upper inlet of the pipette.

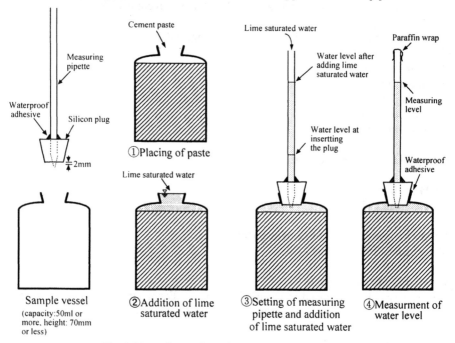

Fig. 1 Measuring method of chemical shrinkage

5 Calculation

5.1 Calculation for the initial volume of cement paste
Calculate the initial volume of cement paste by the following equation.

$$V_P = \frac{(M_2 - M_1) \times (W/C + 1/G_c)}{W/C + 1} \quad \text{(ml)}$$

where:

        W/C: Water cement ratio of cement paste (=0.50)

        G_c: Specific gravity of the cement

V_P: Volume of cement paste

5.2 Calculation for chemical shrinkage ratio
Calculate the chemical shrinkage ratio at any age by the following equation.

$$S_{hyn} = \frac{H_n - H_0}{V_P} \times 100 \qquad (\%)$$

where:

S_{hyn}: Chemical shrinkage ratio at age n

The chemical shrinkage ratio shall be the average value of tested specimens.

6 Report
The report shall include the followings:
(1) Source and identification of the cement that is used
(2) Mixing method of the cement paste
(3) Ambient temperature throughout the test period
(4) Mass and volume of the cement paste
(5) Chemical shrinkage ratio at each measuring age
    An example is presented in Fig. 2.
(6) Other reports
When water cement ratio is not 0.50:
(a) Water cement ratio
(b) Size of sample vessel
(c) Thickness of cement paste in sample vessel

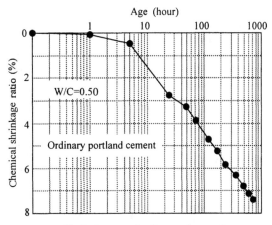

Fig.2 An example of measuring result

II   Test method for autogenous shrinkage and autogenous expansion of cement paste, mortar and concrete

1 Scope
This test method covers the determination of autogenous shrinkage and autogenous expansion of cement paste, mortar and concrete specimens.

2 Test specimens

2.1 Size of test specimens
For cement paste and mortar, specimens shall be $40 \times 40 \times 160$ mm.   For concrete, the width and height of test specimens shall be the same, and shall be more than 3 times the maximum size of coarse aggregate.   The length of test specimens shall be more than 3.5 times the width or the height.   For concrete containing coarse aggregate having maximum size less than 30 mm, the dimensions of the test specimen is $100 \times 100 \times 400$ mm (or 500 mm) .

2.2 The number of test specimens
A minimum of three specimens shall be prepared for each test.

3 Apparatus

3.1 Mold
The mold for the specimens shall be made of steel and shall be rigid.   The mold shall have a hole $3 \sim 5$ mm in diameter at the center of each end plate, so that gauge plugs can be set through the holes.

3.2 Gauge plug
The gauge plugs are used as marking points for length change measurement, which are to be in contact with the spindles of dial gauges.   Examples of gauge plugs are shown in Fig. 1 and Fig. 2.

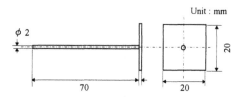

Fig. 1   A gauge plug for concrete

3.3 Dial gauge[*1]
The dial gauges shall conform to JIS B 7503 and shall be accurate to 0.001 mm.

> *1 Other sensors with the same accuracy as the dial gauge may be used.

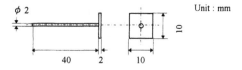

Fig. 2   A gauge plug for cement paste and mortar

3.4 Length comparator
For length change measurement after mold removal, comparator method or contact

gauge method conforming to JIS A 1129 shall be used.

4 Procedure

4.1 Change in length of specimen until mold removal (see Fig. 3)

(1) A polytetrafluoroethylene(PTFE) sheet(thickness 1 mm) shall be placed on the bottom of the mold and a polystyrene sheet (thickness 3 mm) shall be placed inside each end plate of the mold so that free movement of the specimen is not restrained by the mold. Polyester film (thickness 0.1 mm) shall be placed on the PTFE sheet, on the polystyrene sheet and on the both sides of the mold so that the specimen is not in contact with the mold.

(2) Insert the gauge plugs into the mold through the holes of end plates, where the principal axes of the gauge plugs shall coincide with longitudinal axis of the test specimen. Gauge plugs shall be installed so that they extend into cement paste and mortar specimens 15 ± 5 mm, and into concrete specimens 30 ± 5 mm. Measure the original distance between the innermost ends of the gauge plugs (L) by accuracy of 1 mm.

(3) Place the mixture in the mold and compact it[*2], taking care not to move the gauge plugs by vibration.

  *2 When compaction by vibration may cause segregation in the mixture, such as high fluidity concrete, place it by the other proper method.

(4) Finish the surface of the mixture, then cover the surface with polyester film (thickness 0.1 mm) and also with wet cloth in order to prevent water evaporation from the mixture. Store the specimens at a temperature of 20 ± 2 ℃.

(5) At the time of initial setting[*3], put the spindles of the dial gauges on the gauge plugs so that their principal axes coincide with the principal axis of the specimens, and record the initial reading of dial gauges (X_{oa} and X_{ob}).

  *3 The time of initial setting shall be determined by Test Method JIS A 6204 for mortar and concrete, and JIS R 5201 for cement paste.

(6) Record the readings of the dial gauges at the age of 24 hours[*4] (X_{ia} and X_{ib}). If desired, record readings of the dial gauges at the specified ages between the initial setting time and the age of 24 hours. Then remove promptly the test specimens from the molds and the test as shown in 4.2 shall be continuously conducted.

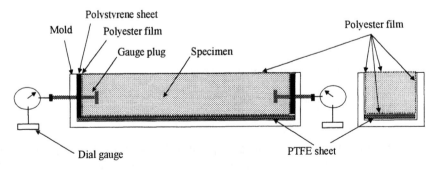

Fig.3   Measurement of length change before mold removal

*4   If the specimens cannot be removed from the molds at the age of 24 hours due to delayed setting, they will have to be removed at the proper age.   In this case, readings of the dial gauges just before mold removal shall be taken as X_{ia} and X_{ib}.

(7)   Measure the temperature of the mixtures at the center of specimens at every length measurement.   Thermal strain should be subtracted from the observed strain assuming that thermal coefficient of expansion is $20\times10^{-6}/℃$ for cement paste and $10\times10^{-6}/℃$*5 for concrete.

*5   Coefficient of thermal expansion of mortar should be determined taking the volume fraction of aggregate into account.

4.2 Change in length of specimen after removal from the mold

(1)   After the test described in chapter 4.1, remove the specimens promptly.   Measure the mass of the specimens and seal all the surfaces of the specimens with aluminum adhesive tape (thickness 0.05 mm) *6.

*6   If contact chips are to be put on the surface of the specimens, strip the aluminum tape of minimum area, then attach them directly on the specimens.

(2)   Measure the length of the specimen.

(3)   Put the specimens in vinyl bags and seal them, then store them in a room at a temperature of 20 ± 2 ℃. Change in mass of the specimens shall be 0.05 % or less*7 during the test period after form removal.

*7   Relative humidity   from 60% to 80% is preferable to minimize change in mass of specimens.

(4)   Measure changes in length and mass of the specimens at the ages of 3, 7, 14 and 28 days*8.   If desired, the test may be continued until the later ages.

*8   The length change measurements shall be conducted by Test Method JIS A 1129, "Test method for length change of mortar and concrete".

5 Calculation

Express autogenous shrinkage strain and autogenous expansion strain as the linear strain (ΔL) by the following equations and show the average value of tested specimens.
Before mold removal,

$$\Delta L = \Delta L_1$$

After mold removal,

$$\Delta L = \Delta L_1 + \Delta L_2$$

where:
ΔL: Length change :
ΔL_1: Length change before mold removal

$$\Delta L_1 = \frac{(X_{ia}-X_{0a})+(X_{ib}-X_{0b})}{L}$$

ΔL_2: Length change after mold removal, which shall be determined by Test Method JIS A 1129 "Test method for length change of mortar and concrete" and shall not expressed in percentage.

L: Distance between the innermost ends of gauge plugs measured in 4.1 (2)

X_{0a} , X_{0b} : Initial reading of dial gauge [*9]

X_{ia} , X_{ib} : Reading of dial gauge at time i [*9]

*9 Unit of L, X_{0a}, X_{0b}, X_{ia} and X_{ib} shall be the same.

6 Report,

The report shall include the followings:

(1) Source and identification of each material used

(2) Mix proportion

(3) Preparation of specimens

(4) Initial setting time

(5) Age of mixture at form removal[*10]

(6) Size of the specimens and effective gauge length

(7) Ambient temperature throughout the test period

(8) Ambient temperature at time of measurements

(9) Length change at each measuring time

(10) Mass of the specimens at each measurement

*10 In case that the specimens can not be removed at the age of 24 hours.

Reference Documents

Japan Industrial Standards:

JIS A 1129 Test method for length change of mortar and concrete

JIS A 6204 Chemical admixture for concrete

JIS B 7503 Dial gauge

JIS R 5201 Test method for physical properties of cement

III. Test Method for Autogenous Shrinkage Stress of Concrete

1 Scope
This test method covers the determination of the stress caused by restraining autogenous shrinkage of concrete.

2 Apparatus

2.1 Mold
(1) The mold shall consist of a bottom plate, side plates and end plates. The inside dimension of the mold shall be $100 \times 100 \times 1500$mm as shown in Fig.1.
(2) The molds for test specimens shall be made of steel and shall be rigid. The bottom plate, side plates and end plates shall be joined with screws. Grease shall be applied to the joints of the mold in order to prevent water from leaking. A polytetrafluoroethylene (PTFE) sheet (thickness 1mm) shall be placed on the bottom of the mold and a foamed polystyrene sheet (thickness 3mm) shall be placed inside each end plate of the mold so that free movement of the specimen is not restrained by the mold. Polyester film (thickness 0.1mm) shall be placed on the PTFE sheet, on the polystyrene sheet and on the both sides of the mold so that the specimen is not contact with the mold.
(3) The mold shall have a hole 34mm in diameter at the center of each end plate, so that a reinforcing bar can be set through the holes. The axis of the reinforcing bar shall be coincide with a longitudinal axis of the test specimen[1].

   *1 The gap between the reinforcing bar and the hole shall be filled with fat clay.

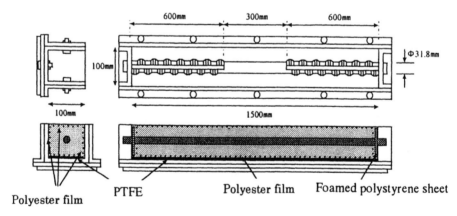

Fig1. Mold and specimen

2.2 Apparatus for restraining
(1) A deformed bar 32mm in nominal diameter and 1600mm in length shall be used.
(2) The ribbed edge of the bar shall be lathed within 150mm from the center (totally 300mm), and the lathed cross section of the bar shall be 28.5 ± 0.5mm in diameter.

2.3 Measurement apparatus

The apparatus to measure the strain of the reinforcing bar shall have an accuracy of more than 1×10^{-6}.

2.4 Strain gauge

(1) A self temperature compensated strain gauge[2] shall be used. Before the test, it shall be confirmed that residual strain does not occur through temperature history.

> [2] A three wire type strain gauge is preferable in order to prevent the influence of ambient and concrete temperature changes.

(2) The strain gauge adhered to the lathed bar shall be protected by waterproofing materials. The lathed portion of the bar (300mm length) shall be bound up trebly with a teflon seat[3] (thickness 0.1mm) in order to prevent bonding between concrete and reinforcing bar. The boundary between the ends of teflon seat and those of bar shall be sealed with a vinyl tape.

> [3] Bind up the bar carefully in order to minimize the decrease in the area of concrete cross section.

2.5 Adhesive agent

An adhesive agent shall be used, in order that the strain gauge will completely adhere to the reinforcing bar during the test.

3 Test specimen

3.1 Dimension of test specimen

The dimension of the test specimen is shown in Fig.1. The maximum size of coarse aggregate shall be smaller than 25mm.

3.2 The number of test specimens

A minimum of three specimens shall be prepared for each test.

3.3 Placing and curing

(1) The concrete shall be placed according to JIS A 1132 (Method of making and curing concrete specimens). Place the concrete in the mold and compact it[4], taking care not to damage the gauge.

> [4] When compaction by vibration may cause segregation in the concrete, compact it by the other proper method.

(2) Finish the surface of the concrete, then cover the surface with polyester film (thickness 0.1mm) and also with wet cloth in order to prevent water evaporation from the concrete.

(3) The concrete shall be stored at a temperature of $20 \pm 3\,^{\circ}\mathrm{C}$.

(4) Remove the molds[5] after 24 hours and promptly seal all the surfaces of the specimens with aluminum adhesive tape (thickness 0.05mm).

> [5] If the molds are not able to be removed from the specimens at the age of 24 hours due to delayed setting, they may be removed at the proper age.

3.4 Storage of test specimen

(1) Each specimen shall be stored under the same conditions during the test. There shall be sufficient space allowed between the test specimens. The test specimens

shall be laid on the floor. Two rollers shall be placed between the specimens and the floor in order to prevent friction. They shall be placed at 40cm distance from both ends of the specimen.

(2) The specimens shall be stored in a room at $20\pm2°C$, 80%R.H[*6].

  *6 A room below 80%R.H. is allowed in the case that the water evaporation from the concrete specimen is prevented.

4 Measurement method

4.1 Measurement point
The longitudinal strain and temperature shall be measured at the midspan of the reinforcing bar. In order to measure the longitudinal strain, two gauges shall be placed symmetry to the axis of the reinforcing bar.

4.2 Measurement time
The strain of the reinforcing bar shall be measured at every one hour up to 1 day, then every four hours up to 3 days, and then every 24 hours up to the end of the test.

5 Calculation
Express autogenous shrinkage stress by the following equation using the average value of the strains of the reinforcing bar.

$$\sigma c = (Es \times \varepsilon s \times As) / Ac$$

where
σc: autogenous shrinkage stress of concrete (N/mm^2)
Es: elastic modulus of reinforcing bar (N/mm^2)
εs: strain of reinforcing bar
As: cross section area of the reinforcing bar (mm^2)
Ac: cross section area of the concrete (mm^2)

6 Report
The report shall include the followings.
(1) Source and identification of all materials used in the mixture
(2) Mix proportion
(3) Room temperature and relative humidity during the test
(4) Strain of reinforcing bar at the time of measurement
(5) Temperature of concrete at the time of measurement
(6) Calculated autogenous shrinkage stress at the time of measurement
(7) The time of occurrence of concrete cracks
(8) The location of the cracks
(9) The autogenous shrinkage stress at the time of occurrence of the concrete cracks
(10) The autogenous shrinkage strain at the time of measurement
(11) Others (compressive strength, tensile strength etc.)

Discussions

T.A.Hammer

First comment: We propose to call all external volume changes, taking place without any mass change as autogenous deformation. The autogenous deformation is of course mainly caused by the chemical shrinkage, which is fundamental. However, it is also influenced by possible re-absorption of bleed water and water absorbed in the aggregate which may reduce the autogenous shrinkage and even turn it into swelling. Autogenous deformation may be measured both volumetric and linear.

The second comment is that one should distinguish between the fundamental point of view and design point of view. From the design point of view, we suggest that we should distinguish between the plastic phase, or what we call the initial phase, in which we have typical cracks due to plastic shrinkage, and the thermal phase or hardening phase. In the thermal phase autogenous deformation will generate stresses that can be taken into account during design. These two phases should be considered separately and when we should use different kinds of test methods. In your suggestion this deviation point was around the time of initial setting. And then we may run into some problems; how do we define the initial setting and how do we test and confirm initial setting which is very consistency dependent, etc.

E.Tazawa

I agree with your comment. We have to divide plastic stage deformation and hardened state deformation. We don't have objection for that. It completely coincides with our idea. You mentioned the difficulty in defining the initial setting time. It depends on the procedures we take. There are many different ways to determine the initial setting point. The deformation after the initial setting is related to stress. The deformation until the initial setting time is related to the deformability of fresh concrete. The cracking during fresh stage is caused by the lack of deformability not by stress. After the initial setting time the cracking is caused by the stress, which is quite different. Anyhow we have to determine the initial setting time. From that time we have to take care of stress behavior of solid. As you proposed, we have to have some agreement how to measure the initial setting time.

You noticed that you have some experience of expansion caused by bleed water. Normally, we have settlement to the vertical direction during fresh stage, but somehow belite cement or belite rich cement, we have expansion during very early stage. The reason for that is not understood yet. So I recommend some chemist to analyze this aspect in very short time. We often observe this phenomenon. Sometime we have expansion due to early age generation of ettringite at fresh stage. That may be the second phase of this expansion. We don't have very large expansion at very early stage. You mentioned that absorption of aggregate may provide another reason of expansion. As for this point, we have one paper presented by Dr. Justnes. In the concluding session, we will introduce one more contribution for that phenomenon. Absorption of water by aggregate reduces autogenous shrinkage. We thank you very much for your good comments. We can reflect your comments on the test methods and the definition.

T.A.Hammer   Your paste measurements showed that the case for w/c=0.40 gave an expansion during the first hours, while lower w/c did not. Do you have an explanation?

S. Miyazawa   We used $20 \times 10^{-6}/^{\circ}C$ as the coefficient of thermal expansion for eliminating thermal strain due to hydration heat, but the coefficient is larger for cement paste at early ages. This may be one reason for expansion. Another reason is thought to be ettringite formation at early ages. Cement paste with low water-cement ratio made with medium heat cement or belite rich cement usually shows expansion at early ages.

S.L.Mak   We have found in trying to superpose autogenous shrinkage and temperature components, the constant thermal expansion coefficient did not work well. It may be right to say that it is probably a higher value at early age. If you use a relation of thermal expansion coefficient with equivalent age, for instance, exponential type of equation, from initial higher value and dropping down to a constant value later on, superpositioning law it works quite well. Early age thermal expansion is very hard to measure. In fact, we often find out relationship by doing backward calculation. More than anything else. Just a comment.

E.Tazawa   These data are taken for different temperatures, 20°C 40°C and 60°C constant (see Fig.2.9), we have different autogenous shrinkage for different time. If we plot these data using maturity concept, we've got these lines here. These three curves are not exactly the same but look more similar, so we assumed a single line in average of these three lines. Then if you use that curve, the calculated value of autogenous shrinkage turned out to be this one. This green line shows the autogenous shrinkage which is measured at constant temperature 20°C. If we subtract this actual deformation which was taken in increasing temperature. We got the final result like this line, exactly the same line for ascending and descending temperature at different time. We have some histerisis property. It was believed that this kind of histerisis always occurs for thermal deformation of cement paste. If you use the autogenous shrinkage which occurs concurrently with thermal deformation, we have very straight relation for thermal deformation and the two lines for ascending and descending temperature coincide each other.

K.van Breugel The maturity concept can be used to present autogenous shrinkage of concrete cured at different temperatures as a function of maturity or equivalent age. When the temperature range considered is not too large, we can get nice results, suggesting that we have to do with a "natural law" or "physical law". This, however, is not the case. Particularly at high temperature, where the temperature will affect the microstructure quite dramatically and hence the thermodynamic equilibrium in the gradually developing matrix, a strong correlation between autogenous shrinkage and maturity becomes questionable. It is on the microstructural level that we have to consider the thermodynamic equilibrium in the pore system, this equilibrium being the major issue in view of autogenous deformations of hardening concrete. The mutually interrelated mechanisms and processes involved in microstructural development and the associated deformation behavior in hardening concrete do not necessarily obey our "crude" maturity rules. I think we have to re-consider the applicability of traditional maturity concept when dealing with very complex phenomena like autogenous shrinkage.

E.Tazawa We have some objection within the committee for the validity of the maturity concept. As you notice, when we have different hydration products, we could have different coefficient of thermal expansion. It is natural. So, your point is very correct, I believe. If we cure a specimen in autoclave, for example, we will have different cement hydrate which has different coefficient of thermal expansion. Curing at very high temperature would destroy basic concept by itself. In the concluding session, I will have a chance to talk about the special curing effect, autoclave curing and also curing under high temperature pressurized water. These two curing conditions yield quite different results using the same temperature history and the same materials also. This means that the effect of hydration products is very important for the deformation during curing and also the deformation after the curing. I don't have any objection to your comments. It is quite true that you mentioned.

F.Tomosawa

I also quite agree to your opinion. Maturity theory is, I think, too empirical to explain the autogenous shrinkage or the development of strength and rigidity of concrete. We have to make progress on the hydration process.

S.L.Mak

Steel fibers generally tend to work better restraining drying shrinkage stress. Some of the plastic synthetic fibers tend to work better at early age in restraining plastic cracking. You plot the relationship with the stiffness, but the stiffnesses for three different types of fibers. Fibers behave quite differently in terms of some hydrophilic and some hydrophobic, mechanism interlock and randomness orientation. I am not sure whether or not relationship with stiffness is as clear cut as that. Do you have any data for one single type of fiber with different stiffness. The relationship you showed we almost think by increasing stiffness of fiber regardless of type could reduce autogenous shrinkage?

S.Miyazawa

The fibers have length of 30 mm and the shape are almost round, not deformed, and distribution and anchoring properties may be not much different. Details of the fibers used in the experiment will be presented at the Sidney Diamond Symposium in this summer.

Y.Tsuji

Concerning the cracking frame test, if the frame is cracked at one day or so, after cracking you have estimated the stress of the frame. As Prof. Tomosawa said, during the hydration, temperature is different for different portions of the beam. How do you consider dependence of degree of autogenous shrinkage, it depends very much on the temperature of beam portion.

R.Sato

Our computation was focused on the first cracking. After cracking, stress is decreased and moment distribution should be changed. The dependence of temperature on autogenous shrinkage in the frame is considered by temperature adjusted age.

F.Tomosawa

Have you measured the stress of reinforcement? How much remaining stress of reinforcement do you have?

R.Sato

In this full size indeterminate structure, computed concrete stress was very high. The reinforcement ratio of the column may be 1 to 1.3%. In this case, reinforcement must be highly compressed.

F.Tomosawa
So we can calculate.   Stress in reinforcement is not so harmful for structural performance of the members, how do you think?

R.Sato
The effect of reinforcement stress on structural performance is cracking.   On other effects more investigation is needed.

M.S.Akman
The shrinkage stress can not be measured directly on reinforced structural element.   If the given values are obtained from the test achieved on separate little specimen, an error certainly occurs because the surface shrinkage measured on large elements is not the same.

R.Sato
We obtained stress in concrete by measuring the strain of re-bar. We did not measure directly the stress of concrete. But usually in mass concrete structures, we measure directly concrete stress by means of stress meters. In this column concrete stress is calculated based on equilibrium requirement. So concrete stress is averaged value in the cross section. The development length from both end to the middle hight measured the strain is long enough to keep the strain compatibility between steel and concrete.

E.Tazawa
We measured strain of steel.   As for steel, we can neglect the effect of creep so we multiply modulus of elasticity to the strain of steel which was measured at the central position of the actual structure.   The force which is generated in the steel is correct, so that force is distributed to concrete and we assume uniform stress in the concrete.   But concrete stress is not uniform all through the section, it depends on situation and on distribution of microcracking which exists in different way throughout concrete section.

PART TWO

CHEMICAL SHRINKAGE AND AUTOGENOUS SHRINKAGE

1 THE INFLUENCE OF CEMENT CHARACTERISTICS ON CHEMICAL SHRINKAGE

H. Justnes
SINTEF Civil and Environmental Engineering,
Cement and Concrete, Trondheim, Norway
E.J. Sellevold
Norwegian University of Science and Technology, Faculty of Civil Engineering,
Department of Structural Engineering, Trondheim, Norway
B. Reyniers, D. Van Loo, A. Van Gemert, F. Verboven and D. Van Gemert
Katholieke Universiteit Leuven, Faculteit Toegepaste Wetenschappen,
Departement Burgerlijke Bouwkunde, Heverlee, Belgium

Abstract
The total and external chemical shrinkage were determined for 10 different portland cements with a wide range of mineral composition and fineness. The total chemical shrinkage was determined by a dilatometry technique monitoring the change in water level in a graded pipette. Measurement of <u>true</u> external chemical shrinkage is very difficult to carry out because of the disturbance of bleeding. To avoid this, some of the samples were placed in thin walled, elastic rubber containers and rotated continuously under water, while most were measured at low w/c. The external chemical shrinkage was equal to the total chemical shrinkage until it "flattened" out – demonstrating that a semi-rigid skeleton was established which led to empty contraction pores during further hydration (i.e. shrinkage). The fineness of the cements dominated the initial rate of chemical shrinkage (i.e. hydration rate) and the "flattening out level" (i.e. fraction of external chemical shrinkage) seems to be quite independent of fineness. The induction period prior to setting is shown to not be "dormant", but rather quite active in terms of volume changes the first hour. The influence of the different cement minerals on the chemical shrinkage is discussed as well.
Keywords: Chemical shrinkage, cement characteristics

1 Introduction

There is currently wide interest in the early volume change of concrete since it is considered a main "driving force" to early age cracking. Early age cracking is a relevant practical problem, particularly in connection with high performance concrete. The present paper is a continuation of previous work [2, 3, 4] concerned with developing

Autogenous Shrinkage of Concrete, edited by Ei-ichi Tazawa. Published in 1999 by E & FN Spon,
11 New Fetter Lane, London EC4P 4EE, UK. ISBN: 0 419 23890 5

methods to measure both the chemical shrinkage associated with cement hydration and the external manifestation of the phenomenon.

Total chemical shrinkage during the hydration of cement is caused by the smaller volume of the reaction products (e.g. CSH gel and CH) compared with the reactants (e.g. alite and water). As a rule of thumb, the total chemical shrinkage at 100 % hydration is about 6.25 ml/100 g cement (i.e. 25 % of the chemically bound water corresponding to a w_n of 0.25 g/g cement [1]). The external chemical shrinkage equals the total chemical shrinkage until the skeleton of hydration products bridging the unreacted cement grains is strong enough to resist the contracting forces. At this point (5-9 h depending on cement composition, fineness, w/c etc.), the external chemical shrinkage rate slows down drastically and the shrinkage vs. time curve flattens out. Thereafter, the second manifestation of total chemical shrinkage; the formation of internal contraction pores, is dominating.

The terminology in the present paper is total chemical shrinkage, which is the sum of external chemical shrinkage and the volume of empty contraction pores at all stages. In literature, there is a terminology confusion: Total chemical shrinkage may be named *water absorption, volume contraction, autogenous volume change* or *Le Chatelier shrinkage* after the first scientist who examined the shrinkage of cement paste. External chemical shrinkage is also named *external volume change, bulk shrinkage* and *autogenous shrinkage*.

2 Experimental

2.1 Components
The chemical composition, potential minerals according to Bogue and the finenesses according to Blaine of the 10 different portland cements studied are given in Table 1.

All the components for the cement slurry were kept at room temperature (20±1°C). Cement (±1 g) and distilled water (±0.01 g) were weighed to make the slurry. The water was added at once and this moment was taken as the zero point on the time scale. The slurry was mixed for 2 minutes at gear 1 and 1 minute at gear 2 in a Hobart mixer. The bowl with the slurry was put on a vibration table in order to remove some of the entrained air. However, lengthy vibration was avoided to prevent "bleeding". Slurry used for the total and the external chemical shrinkage test for the same cement was taken from the same batch.

2.2 Total chemical shrinkage
Cement slurry was put into three 50 ml Erlenmeyer flasks, forming a 1 cm thick layer in each. The mass of sample was determined by differential weighing. The remaining empty volume was then filled with distilled water, avoiding turbulence. A silicon rubber stopper was put in each flask, and a pipette was filled with water and stuck through a hole in the stopper. A graded pipette of 0.2, 0.5 or 1 ml was chosen depending on the expected volume change. The flasks were put in a water bath at 20±1°C. The position of the meniscus in the pipette was read hourly for 48 h. The decrease of the water column in ml is directly the total chemical shrinkage (a drop of liquid paraffin on top prevented evaporation), which was expressed as ml/100g cement (mean value from three parallels). The method relies on the assumption that all contraction pores are filled with water,

which probably is correct at this early age and with such a thin sheet of cement slurry. The method is described and discussed in detail elsewhere [2, 3, 4].

Table 1 Cement characteristics

Characteristic	A	B	C	D	E	F	G	H	I	J
Chemical (%)										
C	63.29	63.32	63.16	63.61	63.44	64.08	64.32	64.56	64.62	64.82
S	20.80	20.25	20.28	20.84	20.92	22.64	22.13	22.83	21.98	22.00
A	4.54	4.76	4.89	5.04	4.60	4.23	4.05	3.34	3.48	3.53
F	2.24	3.67	3.61	3.30	3.54	3.34	3.39	4.95	4.88	4.78
Š	3.40	3.62	2.98	2.70	3.06	2.66	3.07	2.18	2.18	1.75
M	2.90	2.24	2.21	-	1.80	1.49	1.03	0.76	1.45	1.42
Alkalis	1.04	1.05	1.13	1.06	0.94	0.57	0.51	0.55	0.57	0.59
Mineral (%)										
C_3S	57.4	56.3	57.0	46.9	55	48.0	53	53.6	59.5	61.2
C_2S	16.4	15.6	16.0	24.4	19	28.7	24	25.0	18.2	16.9
C_3A	7.9	6.4	6.9	7.8	6.2	5.6	5.0	0.5	1.0	1.3
C_4AF	6.6	11.2	11.0	10.0	10.8	10.2	10.3	15.1	14.9	14.5
CŠ	5.8	6.2	5.1	4.6	5.2	4.5	5.2	3.7	3.7	3.0
Fineness (m^2/kg)	461	438	309	361	358	413	418	305	290	303

2.3 External chemical shrinkage

Three elastic rubber bags (i.e. condoms) were filled with cement slurry while the condoms were inserted in a 100 mm plastic tube of 50 mm inner diameter. Each condom was closed by twisting the upper part and tying it with a thin copper wire, and sealed by spraying silicon glue into the open end. The excess end part was cut off and the total mass determined.

The filled and sealed condoms in their tubes were kept in a water bath of 20±1°C after the tubes had been turned perpendicularly to their axes in order to let all air bubbles escape. The tubes were kept on an ordinary rotating table modified to function under water (i.e. a chain transfer between the motor and the rollers placed under water). During the first 10 h, the condom was weighed every hour under water. According to Archimedes principle, an external shrinkage will lead to a reduction in buoyancy, which will be registered as a weight increase. Each condom was weighed in a basket under water hanging on a scale in a separate water bath with no stirring to avoid turbulence. The transfer between the bath for rotation and the bath for weighing took place under water at all times by placing the tube in a water filled glass under water in one bath and taking it out of the glass under water in the second bath. This was done to avoid trapping of any air bubbles in the transfer process.

After the last weighing under water at 48 h, the condoms were wiped dry and weighed in air. Finally, the condom, including copper wire and silicon glue, were stripped off and weighed in order to calculate the net weight of the cement slurry after subtracting the weight of the tube. The external shrinkage is presented as the mean value of three parallel measurements and given in ml/100 g cement or vol%. In the present study, many of the samples were not rotated. In such a case the tube in the preceding procedure was excluded and the condoms carefully weighed as such (called "static" in the figures). The method is described and discussed in detail elsewhere [2, 3, 4].

3 Results and discussion

The total chemical shrinkage curves from 1 to 10 hours are shown in Fig. 1, while the external chemical shrinkage curves are plotted from 10 min to 60 min in Fig. 2 and from 1 to 15 hours in Fig. 3 for some of the portland cements in Table 1. The total chemical shrinkage curves for the cements in Fig. 1 are plotted from 1 to 48 h on a logarithmic time scale in Fig. 4, and the external chemical shrinkage curves for the cements in Fig. 3 are plotted from 1 to 48 h in Fig. 5. All the external chemical shrinkage measurements presented in Figs. 2, 3 and 5 were obtained <u>without</u> rotation.

Fig. 2 reveals that there is a considerable difference in chemical shrinkage the first hour for the different cements, and also that there must be a considerable part "missed" for the first 10 minutes. If one focuses on the typical high strength cement F, a chemical shrinkage of 0.14 ml/100 g cement is observed from 10-60 min. Extrapolation of this curve estimates 0.06 ml/100 g cement the first 10 min, resulting in a total of 0.20 ml/100 g cement for the first hour after water addition. The volume changes the first hour in Fig. 2 demonstrate that the induction period, or so called "dormant" period, prior to setting is much more active than the general impression gained by other methods (e.g. isothermal calorimetry).

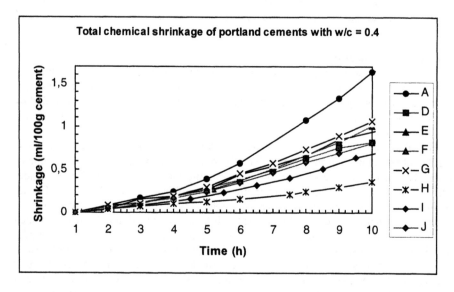

Fig. 1 Total chemical shrinkage development for different portland cement pastes with w/c = 0.40 from 1 to 10 h.

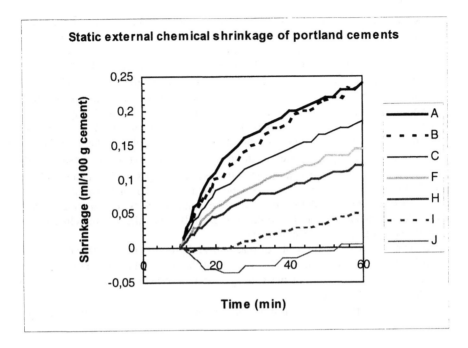

Fig. 2 External chemical shrinkage development for different portland cement pastes with w/c= 0.40 from 10 to 60 min. Samples not rotated.

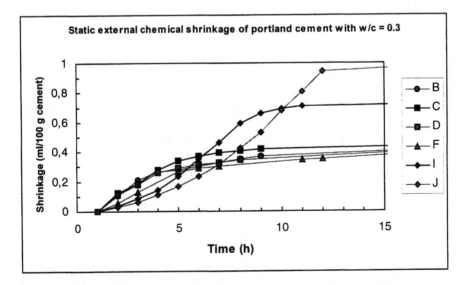

Fig. 3 The external chemical shrinkage development for different portland cement pastes with w/c = 0.30 from 1 to 15 hours. Samples not rotated.

The different behaviour of the different cements can partly be explained by their difference in fineness and partly by their difference in mineralogy. The fineness of the cements (i.e. surface area) decreases in the order A (461 m^2/kg) > B > (438) > G (418) > F (413) > D (361) > E (358) > C (309) > H (305) > J (303) > I (290). The C$_3$A/SO$_3$ content for the same order of cements where 7.9/3.40, 6.4/3.62, 5.0/3.07, 5.6/2.66, 7.8/2.70, 6.2/3.06, 6.9/2.98, 0.5/2.18, 1.3/1.75, 1.0/2.18 %, respectively. The chemical shrinkage of the cements the first hour in Fig. 2 follows the order A $\approx$ B > C > F > H > I > J, and there was no direct and obvious correlation with cement characteristics. However, increased fineness and increased content of C$_3$S and C$_3$A clearly contributes to increasing early shrinkage. So does increasing SO$_3$ content, but the compound in which it is contained determines the net effect; as the clinker mineral Aphthitalite (K$_3$N$\bar{S}_2$) or calcium sulphate added to the mill which may end up as gypsum (C$\bar{S}$H$_2$), hemihydrate (C$\bar{S}$H$_{0.5}$) or anhydrite (C$\bar{S}$). All these sparingly soluble calcium sulphates have different solubility (9.52 g/l for hemihydrate and 2.58 g/l for gypsum).

For instance, the dissolution of a salt like Aphthitalite will lead to a shrinkage more or less corresponding to the volume of crystals, since the volume change of the liquid is negligible at low concentrations.

Knowing the density, ρ (g/ml), of reactants and products of a chemical reaction, it is possible from the molar weight, M (g/mol), of the involved compounds to calculate the volume change, ΔV (ml), pr mass, m (g), reactant remembering the basic relations n = m/M (mol) and ρ = m/V.

For instance the recrystallisation of hemihydrate to gypsum will per gram hemihydrate;

CaSO$_4$·½H$_2$O   +   1½ H$_2$O =   CaSO$_4$·2H$_2$O

m = 1.00 g	0.19	1.19
M = 145.15 g/mol	18.02	172.17
n = 6.89 mmol	10.33	6.89
ρ = 2.74 g/ml	1.00	2.32
V = 0.365 ml	0.186	0.511

give a volume change of ΔV = 0.511-(0.365+0.186) = -0.040 ml (i.e. shrinkage) per gram hemihydrate.

In this way one can also estimate the shrinkage of the initial ettringite formation;

C$_3$A   +   3 C$\bar{S}$H$_2$ +   26 H   =   C$_6$A$\bar{S}_3$H$_{32}$

m = 1.00 g	1.91	1.73	4.64
M =270.20 g/mol	172.17	18.02	1255.26
n = 3.70 mmol	11.10	96.20	3.70
ρ = 3.03 g/ml	2.32	0.998	1.78
V = 0.330 ml	0.823	1.733	2.607

ΔV = 2.607 - (0.330+0.823+1.733) = -0.273 ml/g C$_3$A, while the chemically bound water is 1.73 g/g reacted C$_3$A, meaning that the chemical shrinkage is about 16 % of the chemically bound water.

The shrinkage of the alite reaction can be estimated in a similar manner, but the magnitude is strongly dependent of the composition and density of the CSH-gel formed. Jennings et al [5] estimated the density of a CSH gel of composition $C_{1.7}SH_4$ to 2.01 g/ml at an early age, which leads to the chemical shrinkage;

C_3S	+	5.3 H	=	$C_{1.7}SH_4$	+	1.3 CH

m = 1.00 g	0.418	0.995	0.422
M = 228.32 g/mol	18.02	227.2	74.09
n = 4.38 mmol	23.21	4.38	5.69
ρ = 3.15 g/ml	0.998	2.01	2.24
V = 0.317 ml	0.419	0.495	0.188

ΔV = (0.495+0.188) - (0.317+0.419) = -0.053 ml/g C_3S, while the chemically bound water is 0.42 g/g C_3S (comment: high value, may include physically bound water), meaning that the chemical shrinkage is about 13 % of the chemically bound water (comment: low value due to high value of chemically bound water).

The total chemical shrinkage for the different cements in Fig. 4 were obtained for w/c = 0.40, while the static external chemical shrinkage for the different cements in Fig. 5 was measured at w/c = 0.30 to avoid the effect of bleeding at 0.40. It has been shown earlier [3] that the w/c effect is negligible for total and external chemical shrinkage in the w/c range 0.3 – 0.5, and that external shrinkage matches the total chemical shrinkage until the "knee point" where the flattening out starts for the external chemical shrinkage. The effect of bleeding when performing external chemical shrinkage measurements at w/c = 0.40, can be seen from Fig. 6, where the cements D and G in Table 1 have been tested without (i.e. static) and with rotation.

The total chemical shrinkage in Fig. 4 until 48 h of course reflects the cement hydration, which can be seen to be faster for the cements with the combined highest surface and highest alite (C_3S) content according to the idealised longer term reaction;

2 C_3S	+	6.5 H	=	$C_3S_2H_{3.5}$	+	3 CH

m = 1.00 g	0.257	0.770	0.487
M = 228.32 g/mol	18.02	351.49	74.09
n = 4.38 mmol	14.24	2.19	6.57
ρ = 3.15 g/ml	0.998	2.50	2.24
V = 0.317 ml	0.257	0.308	0.217

ΔV = (0.308+0.217) - (0.317+0.257) = -0.049 ml/g C_3S, which is dependent of the uncertain density of the amorphous CSH-gel $C_3S_2H_{3.5}$ and its real composition.

It is important to note that the chemical shrinkage of the C_3A reaction to ettringite is much higher ($\approx$ 5 times) than the reaction of C_3S. The overall chemical shrinkage of portland cement is in general considered to be about 0.06 ml / g cement (or about 25 % of the chemically bound water of 0.25 g/ g cement reacted [1]) and is a weighted sum of the contribution from the hydration of each of its minerals at a given time.

Looking at the total chemical shrinkage after 48 h in Fig. 4, it is seen that the cements form 3 groups; {A}, {D, E, F, G} and {H, I, J}, with high, medium and low values of fineness multiplied with C_3S content, respectively. Cement B is expected to fall in the high group (Fig. 2) and C in the medium group, but they were not measured.

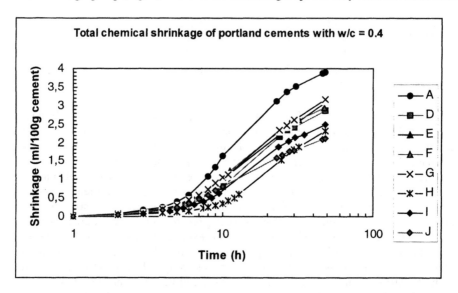

Fig. 4 Total chemical shrinkage development for different portland cements with w/c = 0.40 from 1 to 48 hours.

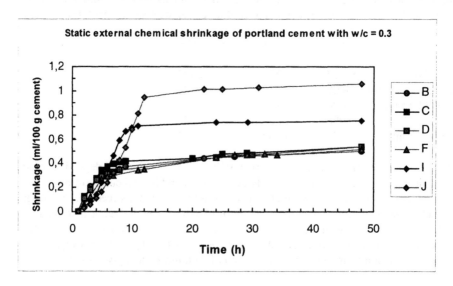

Fig.5 External chemical shrinkage development for different portland cements with w/c = 0.30 from 1 to 48 hours. The samples were not rotated.

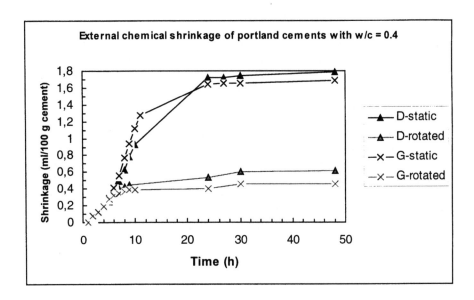

Fig.6 External chemical shrinkage development for two different portland cements (D and G) with w/c = 0.40 from 1 to 48 hours with (i.e. "rotated") and without (i.e. "static").

The external chemical shrinkage in Fig. 5 until 48 h shows that the flattening out level is about equal for the tested cements, except for cement I and J that have a much higher level. Cement I and J are also the coarsest cements with low C_3A content, and it could be explained by a higher degree of hydration required to obtain a hydrate network strong enough to resist the contraction forces. On the other hand, it could also be an effect of bleeding, which is known to be high in these cements. The sensitivity of the method to bleeding can be seen from Fig. 6, where finer cements (D and G) are tested at w/c = 0.40 without (i.e. static situation leading to bleeding) and with rotation.

4   Conclusions

Total and external chemical shrinkage initiates upon the contact between cement and water. Thus, the period prior to setting is not "dormant" in terms of volume changes.

The two quantities are identical until a solid skeleton is formed (6-10 hours). Further hydration results in empty pores (self-desiccation) and greatly reduced external chemical shrinkage.

The early reactivity depends largely on fineness, and the contents of the most reactive phases; C_3A, C_3S, alkalis and sulphates.

5   References

1. Copeland, L.E. and Hayes, J.C.(1953) The determination of non-evaporable water in hardened Portland cement paste. *ASTM Bul.*, No. 194, pp. 70-74.
2. Justnes, H., Reyniers, B., Van Loo, D. and Sellevold, E.J. (1994) An Evaluation of Methods for Measuring Chemical Shrinkage of Cementitious Paste. *Nordic Concrete Research*, No. 14, pp. 45-61.
3. Justnes, H., Van Gemert, A., Verboven, A. and Sellevold, E.J. (1996) Total and External Chemical Shrinkage of Low W/C-ratio Cement Pastes. *Advances in Cement Research*, Vol. 8, No. 31, pp. 121-126.
4. Justnes, H., Van Gemert, A., Verboven, F., Sellevold, E.J. and Van Gemert, D. (1997) Influence of Measuring Method on Bleeding and Chemical Shrinkage Values of Cement Pastes. *Proc. 10th International Congress on the Chemistry of Cement*, Gothenburg, Sweden, 2-6 June, paper 2ii069, 8 pp.
5. Jennings, H.M., Dalgleish, P.L. and Pratt, J. (1981) *Journal of American Ceramic Society*, Vol. 64, No. 10, pp. 567-572.

2 EXPERIMENTAL ASSESSMENT OF CHEMICAL SHRINKAGE OF HYDRATING CEMENT PASTES

S. Boivin, P. Acker, S. Rigaud and B. Clavaud
Concrete Department, LAFARGE Laboratoire Central de Recherche, St Quentin, Fallavier, France

Abstract
The present article deals with a continuous method for measuring the chemical shrinkage of cement pastes. This experimental approach reveals possible measurement artefacts which will be presented and discussed, such as size effects. Two types of measurement are tested: (1) dilatometry method, (2) weighing method. Results for each method are compared and conclusions on the choice of measurement are drawn. Our results finally highlight the influence of the large and rapid changes in permeability which are especially characteristic of the hydration of low water to cement ratio cement pastes.
Keywords: chemical shrinkage, cement paste, cement hydration.

1 Introduction

Chemical shrinkage, i.e. the reduction in total solid plus liquid volumes produced by the chemical hydration reaction, can be considered to be the driving mechanism of autogenous shrinkage. For this reason, its experimental assessment is an important input in the study of autogenous deformations. It especially allows the determination of their initial rate, which is a key parameter for early age cracking problems.

The measure of chemical shrinkage was initially developed in 1900 with very diluted cement pastes (Le Chatelier [1] used water to cement ratio of 1 to 5). Since then, the measure has been used by several authors, either as a means to follow hydration of cement paste (among them, Powers [2], Rey [3], Geiker [4], Paulini [5]) or in combination with autogenous shrinkage measurements (among them, Buil [6], Tazawa et al. [7], Justnes et al. [8]). In the latter case, measurements of chemical shrinkage were performed with low water to cement ratio (w/c) cement pastes (since autogenous

Autogenous Shrinkage of Concrete, edited by Ei-ichi Tazawa. Published in 1999 by E & FN Spon, 11 New Fetter Lane, London EC4P 4EE, UK. ISBN: 0 419 23890 5

shrinkage is of particular interest for w/c<0.4). This necessitates some caution in the way the measure is done, as will be shown in this article.

After some brief definitions of the terminology that will be used, the present article presents a continuous method for measuring the chemical shrinkage of cement pastes. This experimental approach reveals possible measurement artefacts among which size effects or effect of w/c, which will be presented and discussed.   Two types of measurement are tested: (1) dilatometry method, (2) weighing method. Results for each method are compared and conclusions on the choice of measurement method are drawn. Our results finally highlight the influence of the large and rapid changes in permeability which are especially characteristic of the hydration of low w/c cement pastes.

2   Terminology

In order to understand the distinction that is made in this article between chemical shrinkage and autogenous shrinkage, it is necessary to first recall the distinction that needs to be made between absolute volume and apparent volume.

The apparent volume of a cement paste sample is its external volume, i.e. it is the sum of the volumes of its different components either solid, liquid or gaseous. On the other hand, the absolute volume is defined as the sum of the volumes of only the solid and liquid phases (see Figure 1).

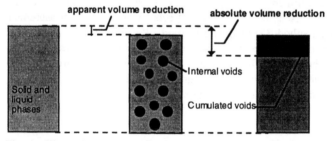

Fig. 1. Illustration of the definition of apparent and absolute volumes.

Chemical shrinkage is a reduction in absolute volume whose effect on apparent volume (autogenous shrinkage) will mainly depend on the porosity and on the rigidity of the material.

2.1 Chemical shrinkage
The chemical and physical changes which occur during hydration are accompanied by a reduction in absolute volume, i.e., the combined volume of the liquid and solid components after hydration is less than the initial volumes of water and anhydrous cement. Le Chatelier [1] first observed this phenomenon under the following conditions: a diluted paste of cement and water was introduced in a flask and covered with water extending into a narrow neck; the experiment then consisted in recording the gradual drop of the level of water; it was the first experimental assessment of the absolute volume reduction caused by hydration.

2.2 Autogenous shrinkage

Autogenous shrinkage (self-desiccation shrinkage) is the apparent volume reduction resulting from hydration. The driving mechanism of this shrinkage is chemical shrinkage ([6, 9]). Depending on the rigidity and on the porosity of the material, the chemical shrinkage will have different macroscopic effects (see Figure 2):

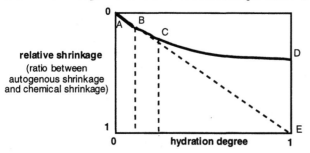

Fig. 2. Schematic evolution of autogenous shrinkage as a function of hydration degree (from Acker [10]).

As long as the material behaves as a suspension, autogenous shrinkage is proportional to the degree of hydration (part AB). When a continuous rigid skeleton is formed throughout the material, the deformations imposed by chemical shrinkage become more and more restrained (parts BC and CD). From that point on, autogenous shrinkage will only be a small part of chemical shrinkage. This was experimentally confirmed in [8].

3   Materials and experimental methods

3.1 Materials

The ordinary Portland Cement used in this study was produced by LAFARGE Ciments in France (Saint Pierre la Cour CPA CEMI 52.5 CP2). The cement characteristics are detailed in Table 1.

The cement pastes, containing only cement and water, were prepared with a Kenwood mixer, with a 5 minute long mixing procedure. The end of mixing was taken as the zero point on the time scale.

Table 1. Cement composition

Chemical composition (%)

CaO (C)	SiO_2 (S)	Fe_2O_3 (F)	Al_2O_3 (A)	SO_3 (S)	MgO (M)	Alkalis K_2O	Na_2O	Loss on ignition
64.85	20.31	2.91	4.93	2.98	0.9	0.64	0.17	1.5

Bogue composition (%)				Specific Surface Area: 3330 cm^2/g.
C_3S	C_2S	C_3A	C_4AF	Water demand: 28,3 %
55.7	14.5	8.1	8.9	

3.2 Experimental methods

Two types of measurement of chemical shrinkage were tested: (1) dilatometry method, (2) weighing method.

3.2.1 Dilatometry method

This method is the one originally developped by Le Chatelier [1]. It consists in following the decrease in level of a water column above a cement paste. The material used here to apply this method is very simple and was inspired by Justnes et al. [8].

After mixing, the cement paste was placed under vibration (0.27 mm magnitude) into a 125 ml erlenmeyer flask. Different thicknesses of paste were used (from around 0.7 cm up to 3.5 cm). Distilled water was then gently added. A silicon rubber stopper with a graded burette of 25 ml (tolerance of reading ± 0.1ml) was fixed under water (to avoid the entrapment of air bubbles) to the flask. The flask was then placed in a thermo-regulated bath at 20°C, itself placed in a thermo-regulated lab. A drop of oil was added on top to prevent evaporation. Stability of the level of water (potentially jeopardized by small variations in ambiant temperature or evaporation) has been checked for over two weeks.

The decrease in the level of water was frequently recorded. Chemical shrinkage is expressed in mm^3 per gram of anhydrous cement.

3.2.2 Weighing method

The measure with the dilatometry method was performed in a non-automatic way. Readings were frequent but information on the curve could not be totally complete (night periods are missing for example). Buil [6] did develop an automatic measure with a dilatometric type of apparatus. However, this type of measure appeared too complex to be developed in this program. In order to get continuous informations, a weighing method was preferred. This type of method had been used in the past by Rey [3]. More recently, Geiker [4] and Paulini [5] also used weighing methods.

The principle of the method used in this study was described by Rey [3]. The experimental set-up used in this study is represented in Figure 3.

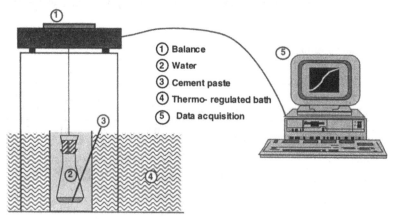

Fig. 3. Experimental set-up used for the weighing method.

The same material as for the dilatometry method is used here (erlenmeyer flask, silicon rubber stopper), as well as the same procedure to fill the erlenmeyer flask with cement paste. A very small hole was prepared in the rubber stopper. Once distilled water had been added on top of the cement paste, the stopper was fixed under water and the flask was placed in a container filled with distilled water, itself placed in a thermo-regulated bath. The flask was finally hanged to a balance (10 mg. precision).

The apparent weight of the flask is the weight of the cement paste minus the buoyancy force (water does not participate to the measure). This weight increases during hydration. The recorded increments of weight correspond, apart from water density changes, to the quantities of water penetrating into the paste as they are measured in the dilatometric method.

Weight increase is recorded continuously through a data acquisition system. Chemical shrinkage is expressed in mm^3 per gram of anhydrous cement.

4 Results

4.1 Dilatometry method
The punctual measures first performed with the dilatometry method allowed us to determine the global shape of the curves on different sizes of samples and on two w/c (0.3 and 0.5).

4.1.1 w/c = 0.3; size effect

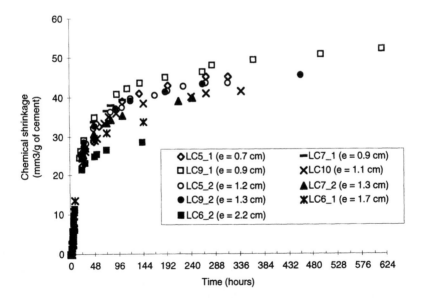

Fig. 4. Chemical shrinkage measured with dilatometry method: w/c=0.3.

The results obtained with the dilatometry method on low w/c cement pastes (w/c=0.3) are represented in Figure 4. The nomenclature used in the legend is the following: LCX-Y, where X indicates the mix number and Y the sample number; « e » represents the thickness of the layer of cement paste introduced in the erlenmeyer flask (e varied between 0.7 and 2.2 cm).

An important size effect is noticeable in Figure 4. For the type of cement paste that has been used here, there is a critical thickness (about 1 cm) above which, at a given time, the water floating on the surface of the paste can not infiltrate the whole sample thickness, or at least not sufficiently fast to avoid the creation of internal voids in the middle or at the bottom of the sample. The kinetics of water infiltration is slower than the kinetics of self-desiccation. This is most probably due to the reduced permeability of the paste during hydration. From that point on, the pores created by self-desiccation cannot all be filled with water; the measured chemical shrinkage is then less than the real one (see section 4.3 for further explanation on this effect). Figure 5 gives a simplified view of this size effect.

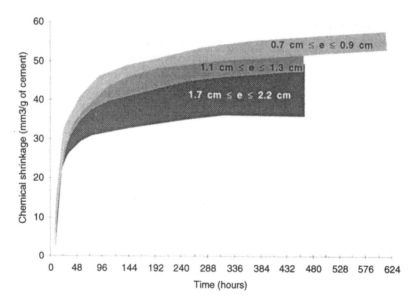

Fig. 5. Size effect on the measure of chemical shrinkage for w/c = 0.3.

4.1.2 w/c = 0.5

The size effect observed for w/c=0.3 was not observed for w/c=0.5 (on the time scale used). Figure 6 illustrates the results obtained for w/c=0.5 and with sample thickness varying from 1 to 3.5 cm. In that case, the permeability of the paste remains sufficiently high to allow the water to penetrate into the created voids during hydration.

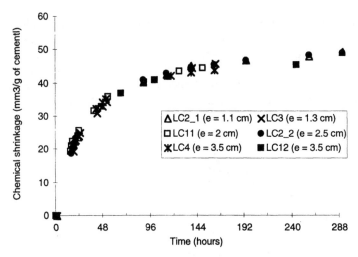

Fig. 6. Chemical shrinkage for w/c = 0.5.

4.2 Weighing method

Differents tests have been performed with the weighing method in order (1) to verify the good correlation between measures obtained with the two methods (dilatometry and weighing methods), (2) to refine the curve description, allowing a better comprehension of the phenomena.

4.2.1 w/c = 0.5

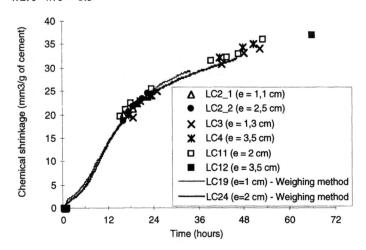

Fig. 7. Comparison between measures by dilatometry (points) and by weighing (curves).

The comparison of punctual measures by dilatometry and continuous measure by the weighing method indicates a very good correlation of the two types of results (see Figure 7).

4.2.2 w/c = 0.3

The size effect which was observed by dilatometry has been also observed with the weighing method. It was then possible with the continuous measure to determine the time at which the size effect appeared for the studied mixes. Figure 8 presents a zoom on the obtained curves (same type of mix - 3 different thicknesses: 1, 2 and 3 cm). The size effect appeared at about 1320 min. (22 hours) after the end of mixing. Before that time, the curves are superimposed. After that time, the curves artificially flatten for sample thicknesses above 1 cm.

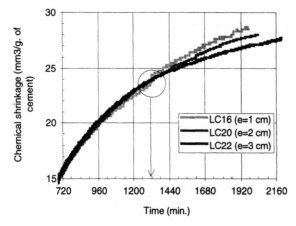

Fig. 8. Determination of the time after which the size effect appears.

4.3 Influence of w/c on the chemical shrinkage measure

From the present set of experiments, the w/c parameter was found to only play a role on the hydration kinetics.

Figure 9 presents curves obtained by isothermal micro-calorimetry for the two mixes, w/c=0.3 and w/c=0.5. An offset of about 5 hours in the peaks of heat flux is noticeable between the two types of mixes, the lower w/c having a more rapid and earlier hydration.

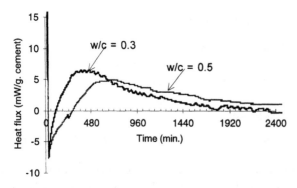

Fig. 9 Heat flux measured by isothermal micro-calorimetry for w/c=0.3 and w/c=0.5.

This difference in hydration kinetics between the two mixes can also be observed on the chemical shrinkage curves as a function of time: there is first an offset between the two mixes (the slope of the w/c=0.3 being steeper); then, the curves merge to reach a common plateau (see Figure 10: on this figure, the upper curve is obtained for w/c=0.3; the lower curve and the points were obtained for a w/c=0.5 either by dilatometry or by weighing). If the curves on Figure 10 were plotted as a function of the hydration degree rather than as a function of time, they should be superimposed.

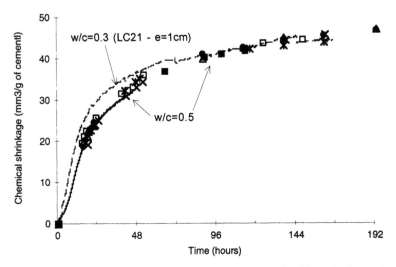

Fig. 10. Comparison of chemical shrinkage curves obtained for w/c=0.3 and w/c=0.5.

This comparison between w/c=0.3 and w/c=0.5 confirms the fact that for about 1 cm cement paste thickness, the measured chemical shrinkage for w/c=0.3 is no longer underestimated since it finally reaches the same value as for w/c=0.5 (see Section 4.1.1). This result on the role played by the w/c is in agreement with Ref. [8] but not with Ref. [4] in which it was observed that decreasing the w/c lead to a decrease in the ultimate value of chemical shrinkage. This observed effect may come from a size effect for the lower w/c (as described previously).

4.4 Chemical shrinkage / autogenous shrinkage

Coupling chemical shrinkage measure with a measure of autogenous shrinkage allows to confirm that apparent volume variations coincide with absolute volume variations until the mineral skeleton is sufficiently rigid to restrain those deformations ([8, 10]).

Autogenous shrinkage has here been measured by a weighing method similar to the one used in [8] (but without rotation of the sample). The cement paste is introduced in a thin latex membrane, which is then sealed and immersed in water. Recording the weight of the sample allows the determination of its volume variations (mass is constant; only the buoyancy force changes during the test). As was shown in [8], this measure is very sensitive to bleeding. It was used in our study only for low w/c cement pastes which did not present any bleeding.

Comparison of chemical shrinkage and autogenous shrinkage for a w/c=0.3 is shown in Figure 11.

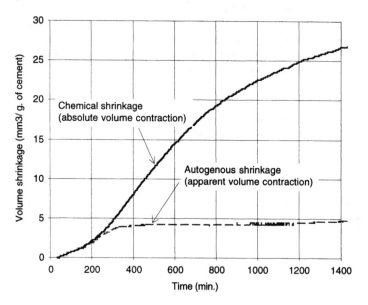

Fig. 11. Comparison of chemical shrinkage and autogenous shrinkage for w/c=0.3.

The two curves begin to diverge at about 200 min. (3.3 hours). At about 300 min. (5 hours), the autogenous shrinkage curve flattens and the two curves completely diverge. The evolution of the Young's modulus of the paste was experimentally determined by an acoustic measure with a setup developed in Ref. [11] (see Figure 12). At about 5 hours, the Young's modulus is equal to 2 GPa and rapidly increases. For this type of mix, this rigidity is high enough to restrain deformations imposed by the chemical hydration reactions.

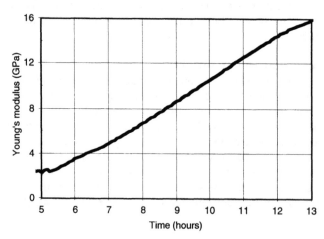

Fig. 12. Evolution of the Young's modulus as a function of time for w/c=0.3.

5   Discussion - Future work

Two methods for measuring chemical shrinkage of cement pastes were tested and found to give identical results. The dilatometry method is a very simple one that allows to run in parallel a great number of tests on the same mix. However, the measure is not automatic. On the other hand, the weighing method allows a continuous measure of chemical shrinkage as long as a balance is dedicated to the performed test. Depending on what type of result is expected from the test, one of these methods will be preferred. If only the plateau value of chemical shrinkage is to be found, the dilatometry method seems perfectly adequate. Otherwise, if information such as initial slope is looked for, a continuous measure such as the weighing method will be preferred.

The influence of dissolved ions from the paste into the water for the set-up presented here is under study and has not been fully assessed yet. This is still an important aspect of the measure that may imply certain modifications in the setup in order to gain accuracy in the measure.

6   Conclusions

Two types of chemical shrinkage measurements were performed in this study. The following conclusions can be drawn:

1. The dilatometry method and the weighing method give results in perfect agreement.
2. An important size effect is noticeable for low water to cement ratio cement pastes. For these types of cement pastes, there is a critical thickness above which pores can be created in the sample without being filled with water; the measured chemical shrinkage is then underestimated. The critical thickness will depend on the type of material used. In our case, for a w/c=0.3, it was estimated to be about 1 cm. No critical thickness was found for w/c=0.5 within the time measures, and it is not certain whether there is one beyond the sizes or at older ages than those used in this study.
3. The size effect does not appear instantaneously but after a certain time, which needs to be determined in function of the material used. In our case, for a w/c=0.3, this size effect appeared about 22 hours after the end of mixing. This lag corresponds to the time after which the permeability of the paste is too reduced to let the water penetrate beyond the critical thickness.
4. The w/c ratio only influences the measure in terms of hydration kinetics. Plateau values of chemical shrinkage are the same independently of the w/c.
5. It is confirmed that chemical shrinkage and autogenous shrinkage coincide as long as the paste is not rigid enough to restrain the deformations imposed by hydration reactions. In our case, for w/c=0.3, chemical shrinkage and autogenous shrinkage fully diverge at about 5 hours, which corresponds to a Young's modulus of the paste of about 2 GPa. In other words, when that value is reached for the Young's modulus of the paste, the cement paste cannot re-arrange itself any more (or at least not fast enough) and internal voids are created within the paste.

Finally, the size effect observed in the chemical shrinkage measurement shows that water brought on the surface of the material (by water curing for example) only penetrates on a small thickness and does not prevent self-desiccation of the material beyond that point. It is still difficult to extrapolate this remark to the case of concretes, since their porosity and permeability are not directly comparable to those of pure cement pastes of same w/c.

7 References

1.   Le Chatelier H. (1900) Sur les changements de volume qui accompagnent le durcissement des ciments. *Bull. Société de l'Encouragement pour l'Industrie Nationale*, 5eme série, tome 5.
2.   Powers T.C. (1935) Absorption of Water by Portland Cement Paste during the Hardening Process. *Industrial and Engineering Chemistry*, Vol. 27, N°7.
3.   Rey M. (1950) Nouvelle méthode de mesure de l'hydratation des liants hydrauliques. Publication technique du CERILH, N°31.
4.   Geiker M. (1983) *Studies of Portland Cement Hydration - Measurements of Chemical Shrinkage and a Systematic Evaluation of Hydration Curves by means of he Dispersion Model*. Institute of Mineral Industry, Technical University of Denmark.
5.   Paulini P. (1992) A weighing method for cement hydration, *Proceedings of the 9th International Congress on the Chemistry of Cement*, Vol 4.
6.   Buil M. (1979) Contribution à l'étude du retrait de la pâte de ciment durcissante. *Rapport de recherche LPC*, N°92.
7.   Tazawa E., Miyazawa S. and Kasai T. (1995) Chemical Shrinkage and Autogenous Shrinkage of Hydrating Cement Paste. *Cement and Concrete Research*, Vol 25, N°2.
8.   Justnes H., Van Gemert A., Verboven F., Sellevold E.(1996) Total and external chemical shrinkage of low w/c ratio cement pastes. *Advances in Cement Research*, Vol.8, N°31.
9.   Hua C. (1995) Analyses et modélisations du retrait d'autodessiccation de la pâte de ciment durcisssante. *Etudes et Recherches des Laboratoires des Ponts et Chaussées*, Série Ouvrages d'Art OA15.
10.   Acker P. (1988) Comportement mécanique du béton: apports de l'approche physico-chimique. *Rapport de Recherche LPC*, N°152.
11.   Boumiz A. (1995) *Etude comparée des évolutions mécaniques et chimiques des pâtes de ciment et mortiers à très jeune âge - Développement des techniques acoustiques*. Thèse de doctorat, Université Paris 7.

3 RELATIONSHIPS BETWEEN AUTOGENOUS SHRINKAGE, AND THE MICROSTRUCTURE AND HUMIDITY CHANGES AT INNER PART OF HARDENED CEMENT PASTE AT EARLY AGE

S. Hanehara, H. Hirao and H. Uchikawa
Chichibu Onoda Cement Corporation, Chiba, Japan

Abstract
This study investigated the relationships between the autogenous shrinkage, and the hydration reaction of cement, and the structural change of hardened cement paste by hermetically curing the cement pastes of normal portland cement, type B - blastfurnace slag cement and 10% silicafume blended cement prepared at w/c of 0.5 and 0.25 and continuously measuring the humidity changes and the shrinkage strains of hardened cement pastes to obtain the basic data for elucidating the mechanism of autogenous shrinkage. Relationships between autogenous shrinkage and kind of hydration products were also discussed.

Autogenous shrinkage is caused by the self drying at a relative humidity (RH) from 100 to 80%. The autogenous shrinkage occurred mainly during a period from 8 hours to 4 days and slightly increased after that.

The autogenous shrinkage of type B - blastfurnace slag cement paste was approximately $1,800\mu$ which was about 1.8 times that of normal portland cement paste. Since type B - blastfurnace slag cement paste produces more C-S-H than normal portland cement paste, thereby causing remarkable autogenous shrinkage. The autogenous shrinkage of 10% silicafume blended cement paste is about $1,000\mu$ almost same as that of normal portland cement paste, since the pozzolanic reaction is suppressed in its paste prepared at w/c of 0.25 up to 4 days.
Keywords: Autogenous shrinkage, blastfurnace slag cement, gel pore, humidity change, hydration product, microstructure, self dessication, silicafume blended cement.

Autogenous Shrinkage of Concrete, edited by Ei-ichi Tazawa. Published in 1999 by E & FN Spon, 11 New Fetter Lane, London EC4P 4EE, UK. ISBN: 0 419 23890 5

1 Introduction

High-strength concrete with superior workability has recently been used for construction works to achieve reduction in construction time and labor. The autogenous shrinkage of high-strength, high fluid concrete during hardening is larger at early ages than that of ordinary concrete [1][2], particularly the autogenous shrinkage of concrete prepared at low w/c using blast furnace slag or silicafume if the blending component is large [1]. Although the autogenous shrinkage has been studied about for a long time [3], there are few papers on its mechanism.

This paper examines the relationship between the humidity change and the shrinkage strain and investigates the relationships between the autogenous shrinkage, and the hydration reaction of cement and the structural change of hardened cement paste by hermetically curing the cement paste prepared at various w/c from normal portland cement, blast furnace slag cement and 10% silicafume blended cement and continuously measuring the humidity change and shrinkage strain of hardened cement paste to obtain the basic data for elucidating the mechanism of autogenous shrinkage.

2 Experimental methods

2.1 Sample and preparation of cement paste
Normal portland cement (NPC), type B - blast furnace slag cement (BB) and 10% silicafume blended cement (SF) were used for the experiment and their compositions and surface areas are listed in Table 1. Cement pastes of those three types of cement were prepared at w/c of 0.5 and 0.25. A ß-naphthalene sulfonic acid-based high-range water-reducing admixture in amount as much as 2% of the mass of cement was added to the cement paste prepared at w/c of 0.25. Mixing of silicafume was done by dispersing the silicafume in the mixing water and then mixing with cement.

2.2 Measurement of strain
The strain of cement paste was measured with a strain gauge. The thin strain gauge with 60 mm long covered with water-proof silicone sealant, was horizontally set in the center of a 500 ml plastic vessel keeping the temperature at 20 °C as shown in Fig. 1 and the cement paste just after mixing with water was poured into the vessel. The strain was continuously measured from just after pouring the cement paste.

2.3 Measurement of humidity
Cement paste was poured into a tray so as to be 2 cm deep and the tray was tightly sealed. After curing that at 20 °C for 12 hours, the hardened cement paste was removed from the tray. The hardened cement paste was crushed with a hammer into approximately 20-mm cubes. A plastic vessel shown in Fig.2 was filled with those cubes and tightly sealed. A ceramic humidity sensor provided with a refreshing device was set in the center of the sample pieces in the vessel to measure the humidity change inside the plastic vessel continuously. The gas in the vessel was circulated to make the humidity inside the vessel uniform and the gas generated from the hardened cement paste was removed by passing the circulating gas through an activated carbon layer. The calibration of the device was performed according to JIS B 7920 (Hydrometers - Test method).

Table 1. Chemical composition and surface area of cements and silica-fume

		Chemical composition (%)							BI' (cm²/g)
		SiO_2 Al_2O_3	Fe_2O_3	CaO	MgO	SO_3	Na_2O	K_2O	
Normal portland cement	NPC	21.6 5.3	3.3	64.9	1.1	2.3	0.45	0.35	3,310
Type B - blastfurnace slag cement	BB	25.9 8.6	1.8	55.9	4.1	2.2	0.27	0.36	3,710
Silicafume		84.5 1.5	2.2	1.4	0.9	0.2	0.7	0.9	18.2 m²/g*

BI': Blaine surface area, *: BET surface area

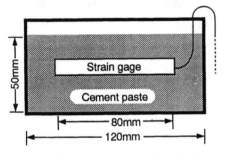

Fig. 1 Measuring method of strain of   Fig. 2 Apparatus for measuring the inside
      hardened cement paste   humidity of hardened cement paste

2.4 Analysis of hydration reaction of cement and hardened cement paste

The test pieces were prepared as follows: 1) cement pastes prepared at w/c of 0.5 and 0.25 were cured at 20 °C for 8 hours and 4 days; 2) each hardened cement paste cured for specified term was cut with a cutter into approximately 5-mm cubes; 3) these cubes were immersed in acetone to terminate the hydration; and 4) the cubes were dried by D-drying or in the atmosphere with the RH of 11%. The curing term of the sample was decided by reference to the measurements of shrinkage strain. The content of combined water and the surface area were measured by the ignition loss method and the BET method, respectively. The identification and quantitative analysis of hydrates were done by powder X-ray diffractometry (XRD), differential scanning calorimetry (DSC) and scanning electron microscopy (SEM). The microstructure of hardened cement paste was observed with secondary electron images and back-scattered electron images and its pore structure was determined by the mercury-intrusion method. The hydration reaction from just after the mixing with water was traced using a conduction calorimeter.

3. Results

3.1 Autogenous shrinkage strain

The measurements of the autogenous shrinkage strain of cement paste are illustrated in Fig.3. The cement pastes prepared at w/c of 0.5 from NPC, BB, and SF expand just after mixing with water and the expansion strains of them are approximately 200μ at the age of 7 days, while those prepared at w/c of 0.25 slowly expand for approximately three hours after mixing with water and after that shrink to zero of the strain at the age 6 hours followed by rapid shrinkage. The shrinkage strains of the cement pastes prepared from NPC, BB, and SF become constant at the ages of 2 days, 6 days and 2 days, respectively. The shrinkages of the cement pastes prepared from NPC, BB and SF are 970μ, $1,810\mu$, and $1,080\mu$, respectively. Those three cement pastes showed large autogenous shrinkages at the ages of 8 hours to 4 days.

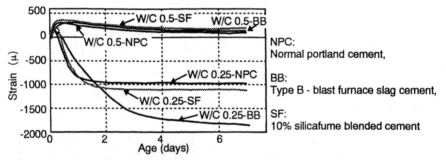

Fig. 3 Time dependency of autogenous shrinkage of hardened cement paste

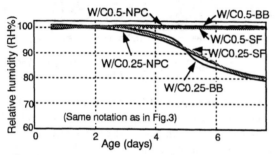

Fig. 4   Change of relative humidity inside the hardened cement paste with time (20 °C)

3.2 Humidity change in hardened cement paste

The measurements of the humidity change of hardened cement paste are illustrated in Fig.4. The humidity of the hardened cement paste prepared at w/c of 0.5 from any of NPC, BB, and SF hardly decreases till the age of 8 days, while that of the hardened cement paste prepared at w/c of 0.25 from any of them begins decreasing two to three days after the mixing, reaching the RH of 80% at the age of 7 days. The humidity reduction of the hardened cement paste prepared from NPC begins earlier than that from BB, while the reduction rate of the latter is slightly higher than the former. This trend agrees with the result of the previous study [4]. The trend of the humidity reduction for SF is nearly equal to that for NPC.   Although the autogenous shrinkage of hardened cement paste corresponds to the humidity reduction, the shrinkage takes place followed by the humidity change. It was described that the humidity change of the hardened cement paste prepared at low w/c from silicafume blended cement was larger [4]. In the present experiment, however, the humidity change of the hardened cement paste prepared from silicafume blended cement was almost the same as those prepared at the same w/c from NPC and BB.

Table 2. Combined water of hydration at various relative humidities (25°C)

	RH 100% 3170Pa		RH 11% 360Pa		D-drying 6.7×10^{-2} Pa	
	Combined water		Combined water		Combined water	
	n	Mass (%)	n	Mass (%)	n	Mass (%)
C-S-H ($1.7CaO\cdot SiO_2\cdot nH_2O$)	4	46.4	2.1	24.3	1.3-1.5	15.1-17.4
AFt ($C_3A\cdot 3CaSO_4\cdot nH_2O$)	32	85.0	29.4	78.1	15.5	41.7
AFm ($C_3A\cdot 3CaSO_4\cdot nH_2O$)	12	53.2	11.7	51.7	7.6	33.7
$Ca(OH)_2\cdot nH_2O$	1	32.2	1	32.2	1	32.2

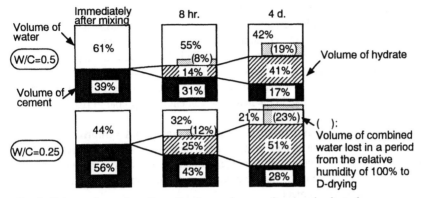

Fig. 5 Schematic explanation of volume change of water, hydrated,
and unhydrated cement prepared at various w/c

In order to elucidate the phenomenon of the humidity reduction of hardened cement paste prepared at low w/c, the content of free water in the hardened cement paste is measured to determine from the hydration rate of cement calculated according to reference [5] using the content of combined water mentioned later. Table 2 lists the changes of combined water of C-S-H, ettringite (AFt), monosulfate hydrate (AFm) and $Ca(OH)_2$ composing the hardened cement paste [5][6] under various relative humidities and drying conditions. In the hydrates including C-S-H ($1.7CaO \cdot SiO_2 \cdot nH_2O$, hydrated at room temperature), the content of combined water (value of n) varies according to the relative humidity of the atmosphere surrounding it. Although the value of n is 4 at the relative humidity of 100%, it is decreased to 1.3 to 1.5 by D-drying. For instance, assuming that the content of combined water by D-drying is 1% , those of C-S-H, AFt, AFm and $Ca(OH)_2$ at the relative humidity of 100% are 2.71, 2.03, 1.58, and 1%, respectively.

The volume ratios of unhydrated cement, water, and hydrate in the cement pastes just after mixing with water, at the ages of 8 hours and 4 days prepared at w/c of 0.5 and 0.25 from NPC are calculated and the result is illustrated in Fig.5. The volume of combined water in the hydrate evaporated from the relative humidity of 100% to D-drying is also given as the item of the volume ratio of water in this figure. The water content in hardened cement paste can be calculated from the change of the water contents in those hydrates shown in Table 2 [5] from the relative humidity of 100% to D-drying considering the kinds and quantities of hydrates coexisting with C-S-H on reference to the data given in Fig.7. For instance, assuming that the hardened cement paste prepared at w/c of 0.5 is composed of 70% of C-S-H, 10% of AFm, and 20% of $Ca(OH)_2$ and that prepared at w/c of 0.25 is composed of 70% of C-S-H, 10% of AFt, and 20% of $Ca(OH)_2$ and that the water content after D-drying is 1%, the combined water in both hardened cement pastes at the relative humidity of 100% is 2.06 and 2.10%, respectively. Accordingly, 1.06 and 1.10% of combined water are lost from the relative humidity of 100% to D-drying. The volume ratio of water in the hardened cement paste prepared at w/c of 0.5 is 42% at the age of 4 days and that of free water at the relative humidity of 100% is 23%(42-19). Since the volume ratio of water in hardened cement paste prepared at w/c of 0.25 is 21% and calculated combined water at the relative humidity of 100% is approximately 23%, the free water is short by approximately 2%. The free water content in hardened cement paste is, therefore, reduced by decreasing w/c. The content of combined water in the hardened cement paste prepared at w/c of 0.25 from NPC hardly increases from the age of 3 days on.   Maybe this is because free water required to participate in the reaction hardly remains, thereby accelerating the shrinkage caused by the self dessication.   The change of the combined water content reveals that the more the C-S-H content and the lower the content of $Ca(OH)_2$ in hardened cement paste, the more the reduction of free water is.   Since $Ca(OH)_2$ is consumed to produce C-S-H accompanied with the hydration reaction in the hardened cement paste prepared from BB, it is more short of free water than in the hardened cement paste prepared from NPC.

3.3 Hydration reaction and microstructure of hardened cement paste
The relationships between the change in strain of cement paste, and the hydration reaction of cement and the microstructure of hardened cement paste were examined to aim at acquiring the basic data for elucidating the mechanism of autogenous shrinkage.
3.3.1 Hydration reaction of cement
The rate of heat evolution in the hydration of cement from just after the addition of water to the age of 7 days is illustrated in Fig.6.   The first peak corresponding to the hydration of the interstitial phase in the cement paste prepared at higher w/c from NPC is higher and earlier, while the rate of heat evolution of the cement paste prepared at w/c of 0.25 is higher 10 minutes or more after mixing with water.   The second peak corresponding to the hydration of alite in the cement paste prepared at w/c of 0.5 from NPC is higher, begins rising earlier and is positioned on the shorter time side than that of cement paste prepared at w/c of 0.25.   The exothermic peak for the cement paste prepared from BB is almost the same as that for NPC except that the peaks are smaller because it contains 50% of slag and the second peak is broader because of proceeding of the slag hydration reaction.   The quantity of cumulative heat in the cement paste of BB till the age of 4 days is equal to that of NPC.   Although the exotherm rate from the pozzolanic reaction from SF is almost the same as that of NPC, the heat evolution peak is smaller.   Maybe this is because of the effect of the dilution by the mixing of silicafume.   The second peak is also generated at the same time as that for NPC.

The result of powder X-ray diffractometry indicates that the cement hydrates contained in the hardened cement paste prepared at w/c of 0.25 from NPC, BB, and SF include C-S-H, AFt, and $Ca(OH)_2$ at the age of 8 hours to 4 days and those prepared at w/c of 0.5 include AFm as well as those three compounds at the age of 4 days.   The conversion of ettringite (AFt) produced by the hydration of the interstitial phase to monosulfate hydrate (AFm) progresses with increasing w/c.   The result agrees with that described in a paper [7].

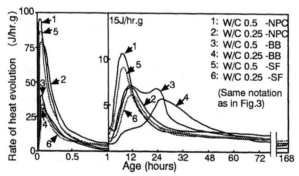

Fig. 6 Rate of Heat evolution of cement pastes (20 °C)

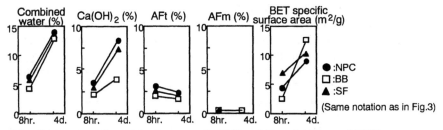

Fig. 7 Amount of combined water, Ca(OH)₂, AFt and AFm, and BET specific surface area of hardened cement paste (w/c=0.25, 20 °C)

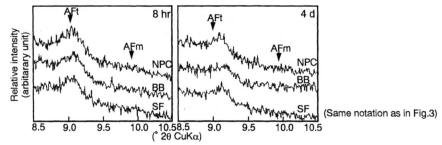

Fig. 8 Powder X-ray diffraction pattern of cement paste prepared from NPC, BB and SF (w/c=0.25, 20 °C)

The amount of combined water, Ca(OH)2, AFt, AFm, C-S-H and BET surface area of hardened cement paste prepared at w/c of 0.25 showing large autogenous shrinkage strain are illustrated in Fig. 7. Increments of the contents of combined water in the hardened cement paste prepared from NPC, BB, and SF from the age of 8 hours to 4 days are 7.4, 8.2, and 7.2%, respectively. The increment of it for BB is therefore, larger than those for NPC and SF. The powder X-ray diffraction pattern of AFt and AFm in hardened cement paste prepared at w/c of 0.25 is illustrated in Fig.8. The production of AFt in the hardened cement pastes prepared from NPC, BB and SFC at the age of 8 hours were 2.8, 2.5 and 2.2% respectively. At the age of 4 days, the production of AFt in the hardened cement pastes prepared from NPC, BB and SFC decreased to 2.5, 2.3 and 2.0% respectively. This result gave good agreement with powder X-ray diffraction pattern illustrated in Fig. 8. The production of AFm were little from 8 hours to 4 days. Amount of AFm determined by DSC coincide with the result determined by XRD. Production of AFm was not recognized in paste from 8 hours to 4 days. The production of AFt decreased slightly from 8 hours to 4 days. This result suggests that AFt decomposed partly, but the production of AFm conversion from AFt were a little in any cement pastes. Although, it is reported that partly decomposition of AFt and conversion of AFt to AFm being the increase of autogenous shrinkage [8], production and phase conversion of aluminate hydrate products hardly related to the increase of autogenous shrinkage in this study. The production of Ca(OH)2 in the cement pastes prepared from NPC and BB increases by 5.5 and 1.6% respectively. The production of Ca(OH)2 based on the content of portland cement in BB (50%) is 2.8%. Accordingly, 1.2% of Ca(OH)2 corresponding to the difference between 2.8% and 1.6% is considered to be consumed in the hydration reaction with slag. Since the increment of Ca(OH)2 in the hardened cement paste prepared from SF during same period of time is 4.7%. This value is approximately 90% of NPC. It suggests that the pozzolanic reaction with silicafume does not proceed.

The surface areas measured by the BET method of cement paste prepared from NPC at the age of 8 hours and 4 days are 4.2 and $8.7m^2/g$, respectively, hence the increment during a period of time is $4.5m^2/g$, while those of cement paste prepared from BB are 2.6 and $12.6m^2/g$, hence the increment during that period of time is $10m^2/g$. This suggests that C-S-H is more vigorously produced in the cement paste of BB during a period of time when the autogenous shrinkage takes place than in that of NPC. The surface areas of the cement paste from SF at the ages of 8 hours and 4 days are 6.9 and $10.4m^2/g$, respectively. The smaller value of increment of surface area of SF from 8 hours to 4 days of $3.5m^2/g$ than that of BB and approximate value to that of NPC also suggests that the pozzolanic reaction with silicafume does not proceed during that period of time.

3.3.2 Texture and structure of hardened cement paste

The secondary electron images and back-scattered electron images of the hardened cement pastes prepared at w/c of 0.25 from NPC, BB, and SF are shown in Fig. 9 and 10. The quantity of C-S-H produced on the surface of cement particles of NPC at the age of 8 hours is larger than that produced on the surface of BB cement particles. C-S-H is much produced at the age of 4 days and large-sized crystals of $Ca(OH)_2$ as well as C-S-H are observed in the hardened cement paste of NPC, while $Ca(OH)_2$ is less produced and small-sized crystals of C-S-H predominate in the hardened cement paste of BB. The texture of hardened cement paste of SF by secondary electron image is rather similar to those of NPC at the age of 8 hours and 4 days. This observation agrees with those of the measurements of the surface area and the quantitative analysis of the hydrates. The back-scattered electron images reveal that dark wide spaces filled with free water are observed between a lot of unhydrated cement particles in the cement paste at the age of 8 hours. At the age of 4 days, however, the free water space is greatly reduced because it is filled with the hydrates including C-S-H. The free water space in the cement paste prepared from NPC at the age of 4 days is wider than that in the cement paste prepared from BB. The unreacted silicafume in the free water space is observed in the cement paste prepared from SF at the age of 8 hours and 4 days.

The pore size distributions of the hardened cement paste prepared at w/c of 0.25 are illustrated in Fig. 11. Although the pore size distribution in the hardened cement paste prepared from SF is similar to that in the cement paste from NPC, there are more gel pores as much as 3 to 6nm in the former than in the latter. In all hardened cement pastes, the total pore volume and the volume of large capillary pores decreases with age, while the volumes of small capillary pores and gel pores corresponding to the interlayer distance of C-S-H increase instead.

Although the volumes of gel pores in the hardened cement pastes prepared from NPC, BB, and SF are small at the age of 8 hours, the total pore volume for BB is larger than those for SF and NPC, and the largest capillary pore diameter of the former is $2\mu m$ and those of the latter are 120 and 200nm, respectively. The volume of coarse capillary pores of 100nm or more in diameter predominates in the hardened cement paste of BB.

At the age of 4 days, the total pore volume of hardened cement paste prepared from NPC, BB and SF decreases and the volumes of gel pores and capillary pores 20nm or less in diameter sharply increase, hence the largest diameter of capillary pores decreases to approximately 100nm, 80nm, and 80nm in hardened NPC, BB, and SF paste, respectively. Most of the pore diameters in the hardened cement paste prepared from BB range from 3 to 30nm and the volume of gel pores is double as much as those in the cement paste of NPC and SF, because the production of C-S-H markedly proceeds.

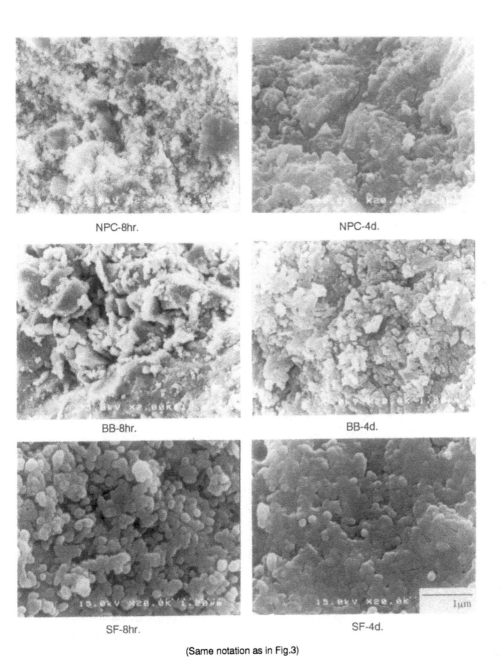

NPC-8hr.

NPC-4d.

BB-8hr.

BB-4d.

SF-8hr.

SF-4d.

(Same notation as in Fig.3)

Fig. 9   Secondary electron images of hardened cement pastes (w/c=0.25)

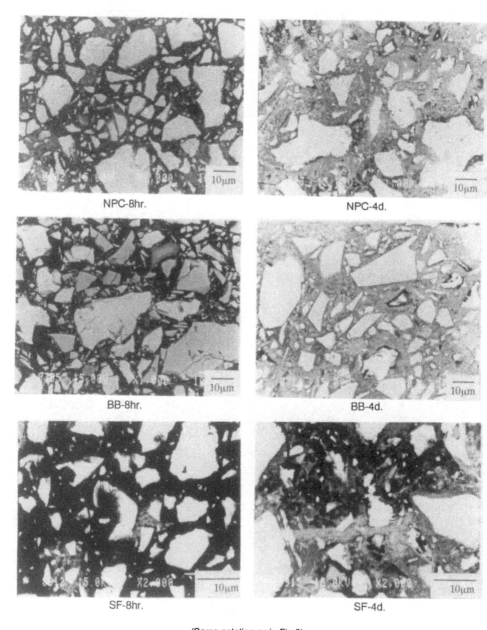

NPC-8hr.

NPC-4d.

BB-8hr.

BB-4d.

SF-8hr.

SF-4d.

(Same notation as in Fig.3)

Fig. 10 Back-scattered electron images of hardened cement pastes (w/c=0.25)

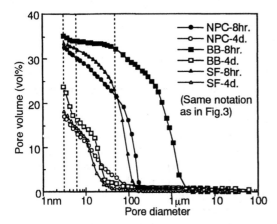

Fig. 11 Pore size distribution of various hardened cement pastes(w/c=0.25, 20 °C)

3.4 Effect of Structural Change of Hardened Cement Paste on Autogenous Shrinkage

Since the volume of hydrates is smaller than the sum of volumes of water and unhydrated cement consumed, the pore space is made by hydration. The result of this experiment indicates that cement paste prepared at w/c of 0.25 which reduces relative humidity and shrinks, though that prepared at w/c of 0.5 which does not reduces relative humidity in a hermetically sealed vessel does not shrink. It is considered, therefore, from this that the autogenous shrinkage is caused by the self dessication. Cement paste shrink even at a relative humidity as high as approximately 80%. It is, therefore, considered that cement paste shrinks by stress caused by loss of water contained in the pores [9][10]. There is a time lag between the occurrence of shrinkage strain and the reduction of relative humidity. Maybe this is because some time is required until the humidity locally changed by the self dessication in cement paste is reflected to that of atmosphere being kept in equilibrium with the specimen.

The drying shrinkage is caused by the evaporation of free water in order of size of pores according to the outer humidity [5], while the autogenous shrinkage is caused by the self dessication of the hardened structure through the proceeding of hydration reaction composing the unhydrates and the hydrates including ettringite (Aft) and C-S-H. BB paste produces more C-S-H than NPC paste and accelerates the pozzolanic reaction earlier than the pozzolanic reaction of the SF paste. It is, therefore, considered that since BB paste produces more volume of C-S-H and gel pores so that it causes more easily the self dessication of hardened paste and shows large autogenous shrinkage.

4. Conclusions

(1) The autogenous shrinkage of hardened cement paste was closely related to its humidity reduction. Although there was a time lag between the autogenous shrinkage

and the humidity change of hardened cement paste, no autogenous shrinkage took place in the cement paste prepared at w/c of 0.5 showing no humidity reduction. The autogenous shrinkage observed in the cement paste prepared at w/c of 0.25, therefore, is considered to be caused by the self drying at a relative humidity (RH) from 100 to 80%.
(2) The autogenous shrinkage was clearly recognized in the hardened cement paste prepared at w/c of 0.25 during a period of age from 8 hours to 4 days and slightly increased afterward. The autogenous shrinkages of normal portland cement paste and 10% silicafume blended cement paste were 970μ and $1,080\mu$, respectively, while that of type B - blast furnace slag cement paste was $1,810\mu$.
(3) It is considered that the autogenous shrinkage takes place because the free water contained in pore particularly in fine gel pore formed by producing a large quantity of C-S-H is consumed by the hydration reaction and the humidity in the hardened cement paste is reduced. Production and phase conversion of aluminate hydrate products hardly related to the increase of autogenous shrinkage in this study.
(4) The proceeding of the pozzolanic reaction was hardly observed in 10% silicafume blended cement paste till the age of 4 days. The hardened structure of this cement paste was similar to that of normal portland cement paste except co-existence of unhydrated silicafume. As a result, the autogenous shrinkage of silicafume blended cement paste was similar to that of hardened normal portland cement paste.

5. References

1. Japan Concrete Institute (1994) *JCI Report (II) of the Technical Reseach Committee on Super Workable Concrete.*
2. Tazawa, E., Miyazawa, S. and Shigekawa, K. (1991) Macroscopic Shrinkage of Hardening Cement Paste due to Hydration. *CAJ Proceeding of Cement & Concrete,* No.45, pp.122-127.
3. Powers, T.C. and Brownyard, T.L. (1947) Studies of the Physical Properties of Hardened Portland Cement Paste. *Proc. ACI,* Vol.43, pp.971-992.
4. Paillere, A.M., Buil, M. and Serrano, J.J. (1990) Effect of Fiber Addition on the Autogeneous Shrinkage of Silica-fume Concrete, Discussion, Authors' Closure. *ACI Material,* 1990.1/2, pp.81-83.
5. Uchikawa, H. (1993) Test and Evaluation Method of Concrete (in Japanese). *Gijutsu Shoin,* Tokyo, pp.12.
6. Taylor, H.F.W. (1990) Cement Chemistry, *Academic Press Limited,* London, pp.130-131.
7. Uchikawa, H. (1990) Characterization and Material Design of High-Strength Concrete with Superior Workability. *Am. Ceram. Soc., Ceramic Transactions.* 40, pp.143-186.
8. Takahashi, T., Nakata, H., Yoshida, K. and Goto, S. (1996) Influence of Hydration on Autogenous shrinkage of Cement Pastes. *Concrete Research and Technology,* Vol. 7, No. 2, pp.137-142.
9. Uchikawa, H., Hanehara, S. and Sawaki, D. (1991) Structural Change of Hardened Mortar by Drying. *3rd NCB International Seminar.* New Dehli, India, Vol.4, pp.VIII-1 -VIII-12.
10. Young, J.F., Berger R.L. and Bentur, A. (1978) Shrinkage of Tricalcium Silicate Pastes - Superposition of Several Mechanisms. *Il Cemento.* Vol.75, pp.391-398.

4 SHRINKAGE OF HIGH-PERFORMANCE CONCRETE

B.S.M. Persson
Lund Institute of Technology, Div. Building Materials,
Lund University, Lund, Sweden

Abstract
This paper outlines an experimental and numerical study on the effect of w/c, silica fume and age on shrinkage of High-Performance Concrete (HPC). Cylinders of 8 types of HPC were cast. Carbonation and water-losses were studied parallel. Internal relative humidity (RH) and strength were investigated on cubes from the same batches of HPC that was used in the studies of shrinkage. The results indicate well correlated relationships between shrinkage, w/c and RH. The type and amount of silica fume affected shrinkage. The study was performed in 1992-1998.
Keywords: Carbonation, High-Performance Concrete, internal relative humidity, shrinkage

1 Introduction

Sealed, carbonation and unsealed shrinkage were investigated. Sealed shrinkage is caused by the self-desiccation, which in turn occurs owing to chemical shrinkage during hydration [1]. When RH decreased, the depression in the pore water also decreased, causing compression in the aggregate and the cement paste [2]. Chemical shrinkage also occurs in Normal Strength Concrete (NSC), though, since the pores are larger, hardly affects RH in NSC at all [3].

2 Previous research on shrinkage of HPC

Roy and Larrard [4] studied shrinkage of HPC and NSC at different w/c with different amounts of silica fume. They found that the sealed shrinkage increased at lower w/c. At w/c ≈ 0.3 the sealed shrinkage after 400 days was about 220 millionths (μm/m). They observed the total shrinkage to be larger at higher w/c; for example about 600 millionths (μm/m) at w/c ≈ 0.6. Fig. 1 shows shrinkage versus w/c with different amounts of silica fume [4]. The results in Fig. 1

Autogenous Shrinkage of Concrete, edited by Ei-ichi Tazawa. Published in 1999 by E & FN Spon, 11 New Fetter Lane, London EC4P 4EE, UK. ISBN: 0 419 23890 5

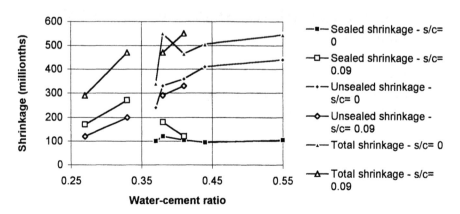

Fig. 1. Shrinkage versus w/c. s/c = silica fume to cement content ratio. Raw data from [4].

were obtained for specimens 1 m long and 0.16 m in diameter with strength varying between 46 and 101 MPa. The measurements started at 2 days' age. Other French experiments were performed on cylinders 0.12 m in diameter with a length of 0.24 m [5]. In this case the water moisture losses were 1% by the weight of the HPC per year, which was quite much. The measurements of shrinkage started at an age of 1.2 days. In Fig. 2 the result of the sealed shrinkage is shown versus time. Tazawa and Miyazawa [6] studied shrinkage at early ages by cast-in strain gauges in specimens 0.1 x 0.1 x 1.2 m, that were cast in vinyl polymer plastic moulds allowing for early movements. The measurements started at ≈ 0.2 days' age. Fig. 3 shows the shrinkage versus time. The shrinkage increased with lower w/c and higher content of silica fume. The difference between sealed and unsealed shrinkage was small at low w/c.

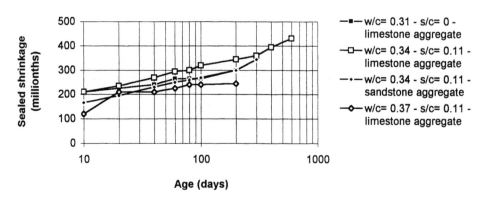

Fig. 2. Sealed shrinkage versus time. s/c = silica fume content. Raw data from [5].

3   Experimental method and results

Descriptions of materials are given in Tables 1-2. Nine-teen specimens (sealed shrinkage) and 26 (unsealed shrinkage) were cast of 8 HPCs, Table 3. After demoulding and sealing by butyl rubber clothing 6 steel screws were fixed to cast-in items in the cylinder 25 mm from each end.

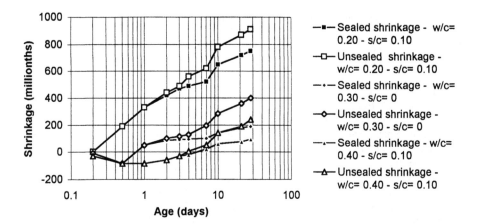

Fig. 3. Shrinkage versus time at different w/c and content of silica fume, s/c. Raw data from [6].

Table 1 - Characteristics of the aggregates.

Material/characteristics	Elastic modulus	Compressive strength	Split tensile strength	Ignition losses
Quartzite sandstone	60 GPa	330 MPa	15 MPa	0.3%
Granite, Norrköping	61 GPa	150 MPa	10 MPa	1.7%
Crushed sand, Bålsta	59 GPa	230 MPa	14 MPa	2%

Table 2 - Chemical composition and main characteristics of the cements.

X-ray fluorescence analysis (%):	
CaO	64.9
SiO_2	22.2
Al_2O_3	3.36
Fe_2O_3	4.78
MgO	0.91
ICP-analysis (%):	
K_2O	0.56
Na_2O	0.04
LECO apparatus (%):	
Ignition losses at 950 °C	0.63
SO_3	2.00
Physical properties:	
Specific surface according to Blaine (m^2/kg)	302
Density (kg/m^3)	3220
Setting time:	
Vicat (minutes)	135
Water (%)	26.0
Standard test (prisms 40x40x160 mm, MPa):	
1 day	11.0
2 days	20.2
7 days	35.8
28 days	52.6

Table 3 - Mix proportions, etceteras (kg/m³ dry material, and so on)

Material /Mix number	1	2	3	4	5	6	7	8
Quartzite, 8-11 mm	460							
Quartzite, 11-16 mm	460	965	910		1010	985		1065
Sand, Åstorp 0-8 mm	800	820	790		750	755		690
Granite, Norrköping 11-16							1030	
Gravel, Toresta 8-16 mm				1095				
Natural sand, Bålsta 0-8				780			780	
Cement, Degerhamn Std	430	440	445	455	495	530	490	545
Granulated silica fume	21	44	45		50	51		55
Silica fume slurry				23			49	
Air-entraining agent	0.02		0.02					
Superplasticiser	2.6	4.5	3.8	5.1	4.6	7.6	8.6	10.8
Water-cement ratio	0.38	0.37	0.37	0.33	0.31	0.30	0.30	0.25
Air-content (% by volume)	4.8	1.1	4.0	0.9	1.1	1.2	1.0	1.3
Aggregate content	0.74	0.73	0.72	0.75	0.71	0.70	0.72	0.70
Aggregate/cement ratio	4	4.1	3.8	4.1	3.6	3.3	3.7	3.2
Density (kg/m³)	2335	2440	2360	2510	2465	2480	2500	2490
Slump (mm)	140	160	170	45	200	130	45	45
28-day drying strength (MPa)	69	85	69	89	99	106	112	114
28-day sealed strength (MPa)	89	105	95	101	121	126	122	129

Measurements were taken on three sides of the cylinder on a length of 250 mm within 1 h of demoulding [7]. The specimens were placed in a 20 °C climate chamber with RH = 55%. Mechanical devices performed the measuring. The temperature at start of the measurement (varying between 17 °C and 24 °C) was obtained by a cast-in thermo couple. Compensation for temperature deformations was done by the coefficient 0.01 [mm/(m · °C)]. RH was measured by dew-point meters on fragment from the strength test. ASTM E 104-85 calibration was performed. Fig. 4 shows a summary of the measured shrinkage of mix 6 versus age, Fig. 5 the relative loss of weight of the specimens (the ratio of weight loss, w_e, to the mixing water of the specimen, w) versus time and Fig. 6 the measured shrinkage versus the relative loss of weight.

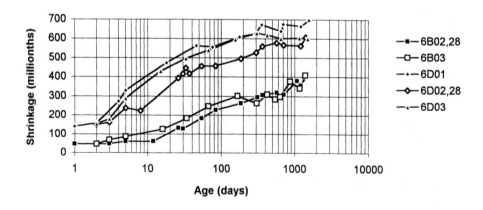

Fig. 4. Measured shrinkage of HPC mix 6. B= sealed curing, D= unsealed. 28 = batch no.

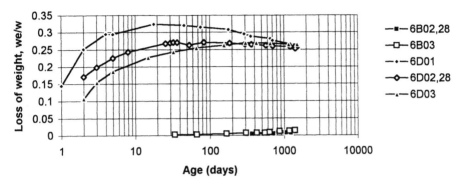

Fig. 5. Loss of weight of HPC mix 6 versus age. B= sealed curing, D= unsealed.

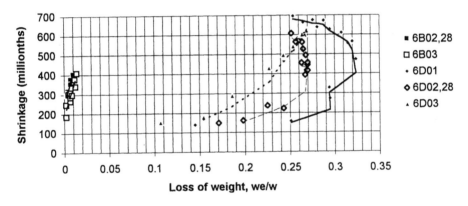

Fig. 6. Shrinkage of mix 6 versus relative loss of weight, w_e/w. w= mixing water.

4   Sources of error and accuracy

The specimens were weighed to detect the weight losses. Absorption of water in the butyl rubber sealing existed at high RH only. Effects of hydration heat in particular were avoided by compensating the first measurement to a temperature of 20 °C [a fault of ± 1 °C reduced the fault to ≈ ± 10 millionths(μm/m)]. Mechanical devices were calibrated with a micrometer. The accuracy device was within 0.004 mm, which was comparable to a fault of 16 millionths. The total fault of ± 1°C and the mechanical device was ≈ ± 16 ± 10 ≈ ± 26 millionths (μm/m) .

5   Analysis

5.1   Sealed shrinkage

From a previous work on basic creep of HPC [8] it was observed that small loss of weight affected the measured shrinkage. Linear regressions were performed to obtain the measured shrinkage at no loss of weight (sealed shrinkage). Fig. 7 shows the sealed shrinkage after 4 years versus w/c. The type and amount of silica fume affected the sealed shrinkage, Fig. 8, also noticed by others [9]. Fig. 9 shows the sealed shrinkage versus RH (in HPC fragment) [10-11].

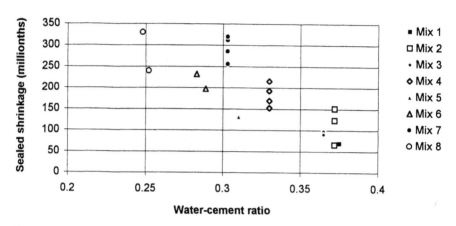

Fig. 7. Sealed shrinkage versus w/c. The HPC mix is given, Table 3.

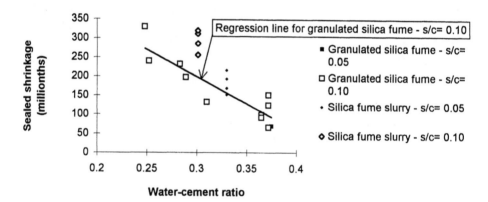

Fig. 8. Sealed shrinkage versus w/c. Type and amount of silica fume.

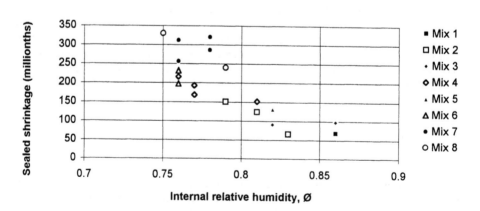

Fig. 9. Sealed shrinkage versus RH, Ø. HPC mix, Table 3.

The type and amount of silica fume affects sealed shrinkage, Fig. 10. Silica fume slurry had a larger fineness than granulated silica fume (22.5 and 17.5 m^2/g), which caused a larger shrinkage. The following correlations were obtained for the sealed shrinkage, ε_B [per mil (‰)]:

$$\varepsilon_B = k_s \cdot k_5 \cdot 1.42 \cdot [0.44-(w/c)] \tag{1}$$

$$\varepsilon_B = k_{s\varnothing} \cdot 1.75 \cdot (1-1.13 \cdot \varnothing) \tag{2}$$

k_s	= 1.5 for silica fume slurry; k_s = 1 for HPC with granulated silica fume
$k_{s\varnothing}$	= 1.3 for silica fume slurry; $k_{s\varnothing}$ = 1 for HPC with granulated silica fume
k_5	= 0.78 for 5% silica fume; k_5 = 1 for HPC with 10% silica fume
$\varnothing$	denotes RH$\{0.70 < \varnothing < 0.90\}$

5.2 Unsealed shrinkage of mature HPC
Small loss of weight was observed in the sealed specimens for reasons mentioned above. Table 4 shows the loss of weight recorded during a period of 3-4 years from 1.8-kg HPC specimen. Small loss of weight over a long period gave a simulation of the shrinkage of a large structure of mature HPC. On the left hand side of Fig. 6 the shrinkage of mature HPC is indicated versus the moisture losses, w_e/w, the inclinations of which gave the specific shrinkage of mature HPC.

Table 4 - Weight losses from 1.8-kg specimen (g). HPC mix: Table 3. Symbols: ....01: batch no.

HPC mix	01	02	03	28
1	-	3.9	-	-
2	2.6	2.0	1.1	-
3	2.7	2.3	-	-
4[1]	5.3	2.4	4.2	3.0
5	1.7	1.6	-	-
6	0.7	1.3	-	-
7[1]	3.9	2.5	2.9	5.2
8	0.9	1.0	-	-

1) 4 years

5.3 Unsealed shrinkage of young HPC
The RH, $\varnothing$, of the cylinder used for the measurements of shrinkage was obtained in the experiments. Between 5 and 28 days' age $\varnothing$ became less than 0.7, which ceased the effect of hydration and thus the sealed shrinkage in HPC [13]. However, at 28 days' age RH of the specimen was still about 5% larger than the ambient RH. Another year of unsealed conditions was required to obtain equilibrium between the internal and the ambient RH [8]. Some moisture was required for the carbonation. Due to carbonation of the calcium hydroxide of the HPC, the loss of weight ceased at an age of the HPC that was dependent on w/c, Fig. 11. Sufficient silica fume to consume all the calcium hydroxide by the pozzolanic reaction was estimated, in NSC about 16% silica fume [12]. At low w/c the required amount of silica fume, s_{rq}, to consume all the calcium hydroxide by the pozzolanic reaction was smaller due to lower hydration [12-13]:

$$3\ Ca(OH)_2 + 2\ SiO_2 = 3\ CaO.2\ SiO_2.3\ H_2O \tag{3}$$

$$s_{rq} \approx [(w/c)/0.39] \cdot 0.16 \approx 0.4 \cdot (w/c) \tag{4}$$

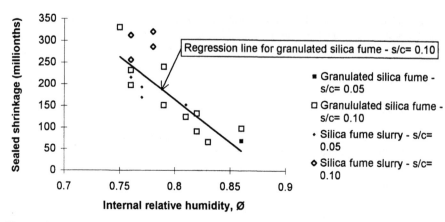

Fig. 10. Sealed shrinkage versus Ø. Amount and type of silica fume.

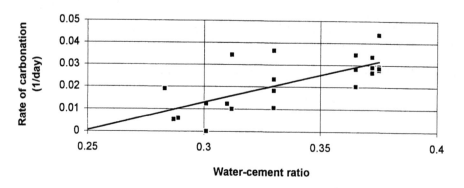

Fig. 11 - Rate of carbonation versus w/c.

According to equation (4), carbonation does not occur at w/c= 0.25 and s/c= 0.10. This was confirmed by the experiments, Fig. 11. At w/c= 0.3 some of the HPCs did not carbonate, which indicated the required amount of silica fume to be slightly lower than estimated in equation (4).

5.4   Carbonation shrinkage of mature HPC

At an age varying between 1 and 20 months the HPC began to carbonate, which was recorded by weighing. Once the carbonation started, no decline of RH was recorded. Fig. 11 shows the carbonation rate versus w/c. It was feasible to describe the carbonation rate [(kg/kg)/day]:

$$d(w_c/w)/dt= 0.25 \ (w/c-0.25)/t \qquad (5)$$

t	denotes age (days)
w	denotes the content of mixing water (kg/m^3)
w_c	denotes carbonated weight (kg/m^3) {0.2< w_c/w<0.35, 0.25< w/c<0.40}

5.5   Total shrinkage

The studies were performed for at least 3 years (4 years for HPCs 4 and 7), Fig. 12:

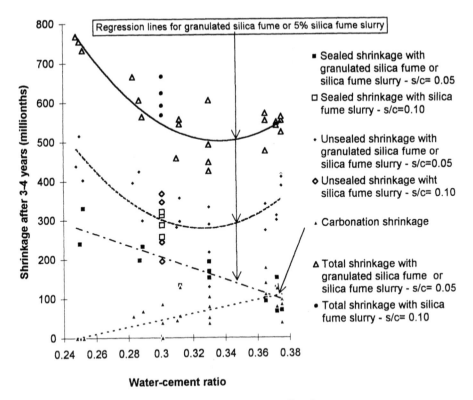

Fig. 12. Shrinkage versus the w/c. c= cement content; s= silica fume content.

$$\varepsilon = k \cdot 34 \cdot [(w/c)^2 - 0.68 \cdot (w/c) + 0.13] \tag{6}$$

$$\varepsilon_B = k_B \cdot 1.5 \cdot [0.43 - (w/c)] \tag{7}$$

$$\varepsilon_C = 0.85 \cdot [(w/c) - 0.25] \tag{8}$$

$$\varepsilon_D = 33 \cdot [(w/c)^2 - 0.654 \cdot (w/c) + 0.115] \tag{9}$$

k	=1.1 with 10% silica fume slurry; k=1 with 10% granulated silica fume or 5% slurry
k	=1.5 with 10% silica fume slurry; k_B=1 with 10% granulated silica fume or 5% slurry
ε, ε_B, ε_C, ε_D	denote total, basic (sealed), carbonation and unsealed shrinkage [per mil (‰)]

5.6   Comparison with other results

Table 5 shows a comparison of the measured shrinkage with results according to other studies. The specimens in [5] and [8] exhibited small losses of weight over time. The moisture losses were small, about 1% respectively 2 grams from a 2-kg specimen, but the sealed shrinkage seems to have been affected. It was suitable to correlate the sealed shrinkage of the HPC specimen to the loss of weight. The sealed shrinkage consequently was defined as the shrinkage at no loss of weight. A linear regression between sealed shrinkage and loss of weight was performed. Fig. 13 shows the sealed shrinkage versus w/c of 50 specimens without loss of weight. Another equation was obtained between sealed shrinkage, ε_{Bn}, and w/c [per mil (‰)]:

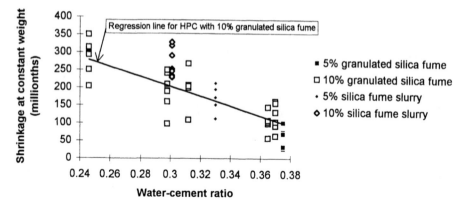

Fig. 13. Long-term sealed shrinkage of HPC versus w/c at no weight loss.

Table 5. Results of measured shrinkage for w/c= 0.3 and s/c= 0.1 [millionths (μm/m)].

Study	[4]	[5]	[6,14]	[8]	[8]	This study	This study
Silica fume	Granu -lated	Granu- lated	Granu- lated	Granu- lated	Slurry	Granulated	Slurry
Age (days)	400	600	40	1000	1000	1200	1500
Sealed	220	430	200	320	380	195	295
Unsealed	160	-	200	-	-	300	285
Carbonation	-	-	-	-	-	45	30
Total	380	-	400	-	-	540	610
No weight losses, n						200	270

$$\varepsilon_{Bn}= k_{sn} \cdot k_{5n} \cdot 1.38 \cdot [0.45-(w/c)] \qquad (14)$$

ε_{Bn} denotes shrinkage in HPC with no loss of weight [per mil (‰)]
k_{sn} = 1.33 for HPC with silica fume slurry; k_{sn} = 1 with granulated silica fume
k_{5n} = 0.69 for HPC with 5% silica fume; k_{5n}= 1 for HPC with 10% silica fume

5.7   Carbonation depth and shrinkage

No carbonation shrinkage and thus no carbonation of HPC was observed in HPC with w/c sufficiently low in combination with 10% granulated silica fume. Probably all calcium hydroxide was consumed in the pozzolanic interaction, giving no chemical component in HPC for the carbon dioxide in the air to carbonate. This observation was of course important for the durability of the HPC and a way to diminish the cover layer of the reinforcement. The way to verify the hypothesis concerning the carbonation of HPC mentioned above was to cut the specimen identical to the one used for measurements of sealed shrinkage. For this purpose cylinders formerly used in the studies of quasi-instantaneous deformations were investigated. These cylinders originated from the same batches as the specimens used for the studies of carbonation shrinkage, which were saved for future studies. The cylinders were cut in two halves, 150 mm in length. A solution of phenolphthalein was applied directly on the cut surface. The depth of carbonation was directly measured by microscope at 4 places on each cut part of the cylinder, in all 8 measurements. The difference in the measured carbonation depth of each HPC batch was small, within 1 mm. The average 5-year carbonation depth of the 8 measurements on each cylinder versus w/c is shown in Fig. 14. The carbonation became astonishingly deep in HPC with w/c= 0.37. However, for HPC with w/c < 0.30 combined with

10% granulated silica fume no carbonation appeared in HPC even after 5 years. For HPC with silica fume slurry the w/c-limit for carbonation seemed to be larger. Fig. 14 shows the carbonation shrinkage versus the carbonation depth of the cylinder. The carbonation shrinkage also depended on the size of the cylinder. A small cylinder was more affected by carbonation shrinkage than a large one. Tensile stress at the surface was compensated by compression at the inner of the cylinder. Alternatively perhaps cracking occur in the surface owing to the carbonation shrinkage, which also affects the durability of the HPC structure. From Fig. 15 the following equation was obtained between carbonation depth (in 55-mm cylinder of HPC with 10% granulated silica fume) and carbonation shrinkage ε_C [per mil (‰)]:

$$\varepsilon_C = 0.005 \cdot d_c \cdot (1 + 0.14 \cdot d_c) \tag{15}$$

d_c denotes the depth of carbonation for 55-mm cylinder (mm)
ε_C denotes carbonation shrinkage in 55-mm HPC cylinder with 10% granulated silica fume

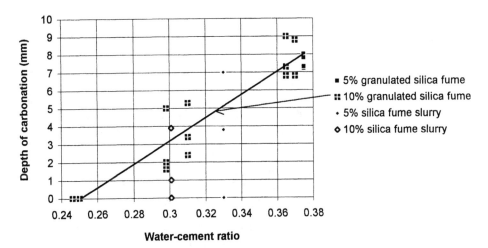

Fig. 14. Five-year carbonation depth versus w/c. Average of the 8 measurements.

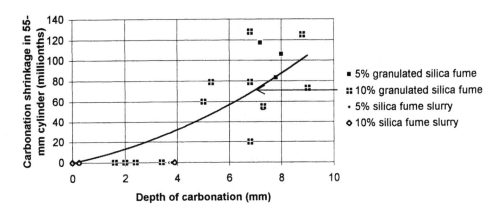

Fig. 15. Carbonation shrinkage versus the carbonation depth of the cylinder at 5 years' age.

6 Conclusions

- Sealed shrinkage was dependent on age, w/c, type and amount of silica fume.
- Sealed shrinkage was related to the decline of RH.
- The conditions for carbonation were related to w/c and content of silica fume.
- Carbonation shrinkage of HPC was related to age, w/c and depth of carbonation.
- Long-term shrinkage was related to age, w/c and type and content of silica fume.

7 Acknowledgement

The main part of the research was financed by the Consortium for Research of HPC (BFR, Cementa, Elkem, Euroc Beton, NCC Bygg, NUTEK, SKANSKA and Strängbetong) which I hereby gratefully acknowledge. I am also thankful to Professor G. Fagerlund for his review.

References

1. Persson, B. (1997). *Long-term shrinkage of HPC*. Proceedings of the 10th International Congress on the Chemistry of Cement. 2ii073. Gothenburg. 1997. Ed. by H Justnes. 9 pp.
2. Persson, B. (1996). *(Early) basic creep of HPC*. 4th International Symposium on the Utilisation of HPC. Paris. 1996. Ed. by F Larrard and R Lacroix. Pp 405-414.
3. Persson, B., Fagerlund, G. (1997). *Self-desiccation and Its Importance in Concrete Technology*. Report TVBM-3075. Lund Institute of Technology. Lund. 1997. 255 pp.
4. Roy, R.le, Larrard, F.de (1993). *Creep and shrinkage of High-Strength Concrete*. Proceedings of the Fifth International RILEM Symposium on Creep and Shrinkage in Barcelona, CONCREEP 5. RILEM 1993. E & FN Spon. London. Pp 500-508.
5. Sicard, V. (1993). *Origènes et Propriétés des Déformations de Retrait et de Fluage de Bétons à Hautes Performances à Partir de 28 heures de Durcissement*. Materiaux et Durabilité de Constructions. INSA-UPS no. 201. Toulouse. 1993. Pp 55-81.
6. Tazawa, E., Miyazawa, S. (1993). *Autogenous Shrinkage of HPC and Its Importance in Concrete Technology*. CONCREEP 5. Barcelona. RILEM. 1993. Pp 159-168.
7. Bazant, Z.P., Carol, I. (1993). *Preliminary Guidelines and Recommendations for Characterising Creep and Shrinkage in Structural Design Codes*. Proceedings of CONCREEP 5. RILEM 1993. E & FN Spon. London. Pp 805-829.
8. Persson, B. (1995). *Basic Creep of High Performance Concrete*. Report M6:14. Lund Institute of Technology. Division Building Materials. Lund. 1995. 292 pp.
9. Jensen, O.M., Hansen, P.F. (1995). *Autogenous Relative Humidity Change in Silica Fume-modified Cement Paste*. Advances in Cement Research. 1995, 7. No. 25. Pp 33-38.
10. Persson, B. (1996). *Self-desiccation and Its Importance in Concrete Technology*. Materials and Structures. Vo 30. RILEM. 1996. Pp 293-305.
11. ASTM E 104-85 (1985). *Standard Practice for Maintaining Constant Relative Humidity by Means of Aqueous Solutions*. ASTM. Philadelphia. 1985. Pp 33, 637.
12. Peterson, O. (1976). *Interaction between Silica Fume and Standard Portland Cement in Mortar and Concrete*. Cementa Ltd. Malmö. 1976. 8 pp.
13. Powers, T.C., Brownyard, T.L. (1946-1948). *Studies of Physical Properties of Hardened Portland Cement Paste*. Research Laboratories. PCA. No 22. 194 Pp 473-488, 845-864.
14. Tazawa, E., Miyazawa, S. (1997). *Influence of Cement Composition on Autogenous Shrinkage of Concrete*. 10th Int. Congr. on the Chemistry of Cement. 2ii071. 1997. 8 pp.

Discussions

The influence of cement characteristics on chemical shrinkage
H. Justnes, SINTEF, Norway
E.J.Sellevold, The Norwegian University of Science and Technology, Norway
B. Reyniers, D. Van Loo, A. Van Gemert, F. Verboven and D. Van Gemert,
Katholieke Universiteit Leuven, Belgium

F.Tomosawa	Rotating method for preparing cement paste specimens has not affected the formation of microstructure of hardening cement paste?
H.Justnes	There was a detailed description in the proceedings of 10th ICCC. There is a plastic bag or condom inside the tube, so it is actually contained in a plastic tube which is rotating. So there is no impact on the soft structure of the paste itself. It is a special technique. It is described in the text or I can send you a copy.
K. van Breugel	In your paper you present shrinkage as function of time. Did you try to plot shrinkage against progress of the hydration process?
H.Justnes	I'm not sure if it is possible. But, in the case of alite, we can determine the exact degree of hydration by NMR for instance, and then also measure the chemically bonded water, and actually calculate the true density of the C-S-H gel which can vary quite a bit we think. In this way, we try to get a better grip on contribution of each mineral.

Experimental assessment of chemical shrinkage of hydrating cement pastes
S. Boivin, P. Acker, S. Rigaud and B. Clavaud, LAFARGE Laboratoire Central de Recherche, France

B.S.M.Persson	Did you also measure the chemically bonded water in combination with the chemical shrinkage?
L.Barcelo	No, we didn't. We measured the heat flux but not chemically bonded water.
B.S.M.Persson	Because I think if you also measure the hydrated water, you can then avoid the size affect. Because, it is a question of how much water has hydrated. If you know the value then you have the specific volume of the hydrated water. It could be a way to avoid the size effect.

L.Barcelo   I'm not sure.

M.T.Leivo   Here in your paper you yourself say that you must find out how the dissolving ions affect your results, have you made any conclusions? Because you are now using distilled water.

L.Barcelo   Yes I have. I was told by a specialist that the problem is that when you use free lime saturated water at the surface of sample, when K_2SO_4 and Na_2SO_4 dissolve in the water, the solubility of the lime decreases rapidly, and afterwards the portlandite may precipitate at the top of the sample in a very dense layer, but I have no results more than that.

Relationships between autogenous shrinkage, and the microstructure and humidity changes at inner part of hardened cement paste in early age
S. Hanehara, H. Hirao and H. Uchikawa, Chichibu Onoda Cement Corporation, Japan

B.S.M.Persson   How did you calibrate the relative humidity?

H.Hirao   It is corrected by saturated solution of sodium chloride and lithium chloride according to the Japan industrial standard.

E.Tazawa   In the last conclusion, you showed that the conversion of aluminum phase has no relation with the autogenous shrinkage deformation, I appreciate that result, because somebody in Japan has presented a paper saying that there might be some correlation between autogenous shrinkage and the conversion, I personally felt at first that this would be one reason why the autogenous shrinkage is increased with the calcium aluminate content. But the reason is not solved yet. Dr. Goto, how do you think about this, you have some opinion, I think.

S.Goto   I think that the conversion of the ettringite to monosulfate is very important factor which influence to the shrinkage before the texture is produced.

M.T.Leivo   Because you have time lag between drop in the relative humidity and the shrinkage generation and it is quite big, did you try to crush the cement paste into smaller pieces? Does it affect the time lag?

H.Hirao   The reduction of relative humidity of inside of hardened cement paste and the increment of shrinkage strain occurred simultaneously. However, in this study, we measure the relative humidity in whole plastic vessel. The plastic vessel have more space and air than interstitial space of hardened cement paste. Therefore, the reduction of relative humidity at inside of the plastic vessel needs more time than that in hardened cement paste structure.

D.Van Gemert

NPC, BB and SF pastes with w/c 0.5 all show expansion from point 0 in time. I presume that all these mixes at w/c 0.5 were all very fluid. How can a strain gauge measure horizontal strains in a liquid? Can you give information about the method, procedure and intensity of mixing. What was the "fluidity" or "consistency" of the pastes?

For the other pastes, with w/c=0.25; how was the strain gauge introduced in the paste? How was it kept in horizontal position? How was the paste placed and compacted around the strain gauge?

H.Hirao

The strain gauge was fixed in the center of the plastic vessel horizontally by fine thread when cement paste was poured into the plastic vessel. Soon after pouring the cement paste, the thread was cut. By this method, the strain gauge was kept in horizontal position still the cement paste is not hardened.

The hardness of fresh cement paste just after mixing with water was smaller than that of strain gauge. However, the strain gauge was covered with silicon sealant. The hardness of silicon is as small as that of cement paste just after mixing with water. Therefore, the strain gauge covered with silicon sealant can measure the strain change of cement paste just after mixing with water.

Shrinkage of high-performance concrete
B.S.M. Persson, Lund Institute of Technology, Sweden

E.Tazawa

In your first conclusion you stated that the drying shrinkage increases with decreasing w/c ratio. So by saying that, you are saying that the knowledge until now or in the text book is not correct for the drying shrinkage.

B.S.M.Persson

This is according to our measurements. We have found that if you seal the specimen very quickly and you seal it without any moisture losses in the beginning, then we can measure drying effect even on very low w/c ratios.

S.Goto

I think that Prof.Tazawa said that shrinkage vs time, but Dr.Persson said that it is shrinkage vs. weight loss

B.S.M.Persson   Yes, maybe it is a misunderstanding.   I have plotted shrinkage vs weight loss.   I have not shown here shrinkage vs time.   So maybe it is a misunderstanding.

PART THREE

AUTOGENOUS SHRINKAGE AND MEASUREMENT

5   EVALUATION AND COMPARISON OF SEALED AND NON-SEALED SHRINKAGE DEFORMATION MEASUREMENTS OF CONCRETE

H. Hedlund and G. Westman
Luleå University of Technology, Luleå, Sweden

Abstract
In this study measurements have been performed on cylindrical and prismatic specimens surrounded by water or air. The specimens have been either sealed or non-sealed. It is shown that it is of great importance to perform shrinkage measurements on sealed specimens in order to obtain material related information about the moisture movements.

The shrinkage as a function of time for sealed concrete is mainly a consequence of the self-desiccation, which is more pronounced for concrete mixtures with low water binder ratios and high binder contents, i.e. circumstances typical for High Performance Concrete (HPC). Time dependence of the shrinkage is chosen to be expressed in a similar way as may be used in describing the degree of reaction for hardening concrete.
Keywords: Sealed deformation, non-sealed deformation, isothermal measurements, hardening concrete, autogenous shrinkage, HPC design handbook

1   Introduction

For high strength concrete/high performance concrete (HSC/HPC) both creep and shrinkage differs from what is common knowledge about the behaviour of normal strength concrete (NSC). Measuring autogenous deformations in concretes with an open pore system (NSC) may give large scatter and/or totally incorrect result if non-sealed. If the surrounding environment consists of water in direct contact with the concrete surface, a swelling can occur due to water penetration into the relatively open pore system. The stresses - caused by the moisture distribution - will be quite uniform over the cross section as the water have possibilities to effect the whole cross section of small specimens. In a study on mortar made by [1] it is shown that different curing

Autogenous Shrinkage of Concrete, edited by Ei-ichi Tazawa. Published in 1999 by E & FN Spon, 11 New Fetter Lane, London EC4P 4EE, UK. ISBN: 0 419 23890 5

conditions generates differential moisture distribution over cross section. The curing condition strongly affects the measured length change. However, in HPC with a more dense pore system and a relatively stable internal structure water penetration will only affect the surface layer, i.e. causing a stress distribution over the cross section, which will give the deformation strain for the system. Therefore, it is of great importance to measure the autogenous deformation on HPC as well as NSC on sealed specimens.

2   Influences of curing conditions on measured autogenous deformations

Measurement on one concrete mix have been performed according to:
- cylinder and prisms, sealed with plastic foil stored in water (Type 1 in Fig. 2)
- prism, sealed with bitumen - aluminium foil stored in air (Type 2 in Fig. 2)
- cylinder, sealed with plastic foil stored in a aluminium box placed in water (Type 3 in Fig. 2)
- cylinder, non-sealed stored in water, see Fig. 4.

Data have been obtained by using a mechanical measuring gauge or two strain gauges of type LVDT (Schaevitz type 010 MRH) mounted on composite bars made of invar and graphite. The gauges are symmetrically mounted on the cylinder specimen border, see Fig. 1. With this type of gauges the deformation can be measured with an accuracy of about 0.03μm to be compared with about 5μm for the mechanical device (STAEGER). Depending on the concrete it is possible to demould and apply the gauges onto the specimen about 6 - 10 hours after casting.

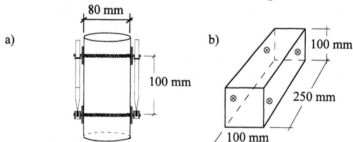

Fig. 1.   a) Inductive strain gauge mounted on the border of the cylindrical test specimen. b) Prismatic specimens where ⊗ = measuring points.

The studied concrete is used in floor castings (in housing) where there is short time between cast and coverage of the floor. Low-w/c concrete reduces as well the problem with moisture during the time of construction. Mix composition is presented in Table 1 where the binder content, b, is defined as cement + silica fume, kg/m^3 ($b = c + s$). The used cement is CEM Type I according to ASTM and the compressive strength are tested on 100 mm cubes. Until the measurements of the autogenous deformation starts the test specimens have been stored at 20 °C in a water tank inside its mould. As the mould have been kept around the specimens the water penetration can be neglected.

Table 1.   Mix composition of HPC T

Concrete	Cement kg/m³	s/c %	Aggregate content fine / coarse	w/c -	w/b -	fcc,28 MPa
HPC T	450	3	59 % / 41 %	0.39	0.38	87

All the tests are performed under isothermal conditions at 20 °C. The early temperature rise due to the hydration process was controlled by thermo-couple cast into the specimen. The maximum temperature was 26 °C in the concrete specimen. Compensation was made for this difference in temperature by using the thermal dilatation coefficient $10 \cdot 10^{-6}$ /°C. The choice of water storage has mainly been used in order to maintain a stable temperature. Air storage of the sealed specimens have been used in a temperature (20 °C) stable room for comparison to the water stored specimens. When a fully acceptable sealing have been achieved the autogenous shrinkage should show deformations of the same magnitude, see Fig. 2. In the figure presented test results have been presented from 11 hours after casting. The obtained test result show about 30 per cent scatter in measured shrinkage strain five days after casting. The type 3 sealing is probably a good carrier of the average autogenous deformation for concrete mix HPC T. The same shrinkage development is obtained with both measuring techniques for all the tested curing methods.

Some initial differences of the measured shrinkage level for non-sealed specimens stored less then 18 hours before the tests starts are shown in Fig. 3. In the figure it can also be seen that the rate of shrinkage for all tested start times after casting are about the same.

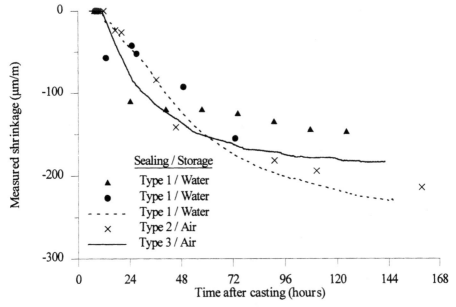

Fig. 2.   Measured deformations under sealed conditions using three types of sealing materials (Type 1 = plastic foil, Type 2 = bitumen and aluminium foil and

Type 3 = aluminium box). The sealed specimens have been surrounded either by water or air. Start time are 11 hours after casting for all tests.

Results obtained from tests of non-sealed specimens shows substantially less shrinkage deformation than sealed specimens, see Fig. 4. A comparison shows that the measured deformations at sealed conditions is twice as large as the measured non-sealed deformation one week after casting for HPC T.

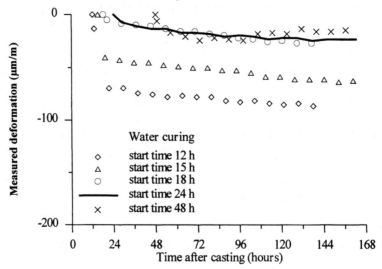

Fig. 3. Measured deformation of non-sealed specimens stored in water. Tests have been performed from 12 hours after casting and the measuring period have been about one week.

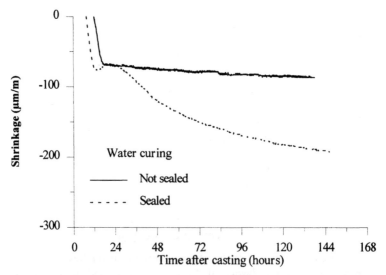

Fig. 4. Measured shrinkage at sealed and non-sealed specimens stored in water.

The measured deformation is an average volume change summoned over the cross section influenced by both swelling of the outer parts of the test specimen and a contraction of the inner parts (core). Therefore, measuring autogenous deformations in hardening concrete at early ages demands high accuracy of sealing material due to the influence of the surrounding environment.

As can be seen in Fig. 4 both the sealed and non-sealed specimen have an initial shrinkage in the level of -80 μm/m which is stabilised at about one day after casting. Thereafter, the non-sealed specimen show very little shrinkage development, due to the non-uniform stress distribution over the cross section. The sealed cylinder is continuously decreasing in volume as the hydration process prolongs. Measured strain shows autogenous shrinkage in the same level as in the study of different sealing material presented in Fig. 2.

3 Shrinkage in HPC

The basic sequence in the chosen equations is in line with CEB/FIP Model Code '90 for NSC [2], which here is adjusted in a way to fulfil the conditions of HSC/HPC. In general, creep and shrinkage tests on HSC/HPC are definitively not yet as numerous as for NSC, a fact which limits the range of applicability at present. Therefore, the presented model should be regarded as approximate with respect to effects that may be expected to play a significant role but which are not investigated so far. Such effects may be varying temperature, loading and unloading stress history sequences, tensile stresses, combined effects of varying stress and varying environmental conditions, high stress levels, types of cement, types of aggregate.

The formulas presented below are a short version of what is presented in [3] and [4]. The model is based on data from the three laboratories LCPC (Paris, France), LTH (Lund, Sweden) and LTU (Luleå, Sweden), see [5], [6], [7], [8], [9] and [10].

All HPC contain low amount of mixing water, $w/c \leq 0.40$, which results in a decrease of the pore humidity (RH < 100%) and a shrinkage even without any exchange of humidity with the environment. This shrinkage under sealed conditions, autogenous shrinkage, must be included in shrinkage models for HPC.

As the rate of drying is significantly lower in HPC compared with NSC [11], the drying shrinkage for large HPC elements only comprises the outer shell. A simple model to consider this is to calculate autogenous shrinkage for the whole body and drying shrinkage for an apparent surface layer zone, which is expressed by the total external shrinkage strain, $\varepsilon_{cs}(t)$, of a concrete member as

$$\varepsilon_{cs}(t) = \varepsilon_{cso}(t) + \varepsilon_{csd}(t) \tag{1}$$

where t = time after casting of concrete (days); $\varepsilon_{cso}(t)$ = shrinkage strain at sealed conditions; $\varepsilon_{csd}(t)$ = additional strain due to drying / wetting caused by humidity exchange with the environment.

3.1 Modelling of shrinkage at sealed conditions

The starting time for shrinkage measurements used here have varied between 9 and 24 hours. All data have been normalized to zero at 24 maturity hours (= time at 20 °C curing) after casting, and the autogenous shrinkage, $\varepsilon_{cs0}(t)$, is described by

$$\varepsilon_{cs0}(t) = \varepsilon_{cs\infty} \cdot \beta_{s0}(t) \tag{2}$$

where t = age of concrete (days); $\varepsilon_{s0\infty}$ = final value of autogenous shrinkage; $\beta_{s0}(t)$ = time distribution of autogenous shrinkage.

The main cause of autogenous shrinkage is the chemical shrinkage which make the time distribution of the autogenous shrinkage closely related to the degree of reaction. In [12] both data and a model is presented for the relation between the degree of reaction and autogenous shrinkage. In a simplified model the desorption isotherm is regarded as a straight line and the relevant part of the shrinkage as approximately linear versus the decrease of relative humidity. Then the time distribution of the degree of reaction can at the same time be valid for the autogenous shrinkage development over time, see [8].

The final value of the autogenous shrinkage is assumed to be approximately 100000 hours (11.4 years) after casting. Also, the abbreviation to level the shrinkage at zero at 24 h after casting is used. This gives - regarding data from [12] - the relative time distribution of the degree of reaction, see Fig. 5 and has been employed as the time development of autogenous shrinkage expressed by

$$\beta_{s0}(t) = 1.14 \cdot \exp(-\left[\frac{t_{so}}{t - t_{start}}\right]^{0.3}) \tag{3}$$

where t_{start} = 1 day = (chosen) start time of shrinkage at sealed conditions; t_{s0} = 5 days = time parameter regarded as constant for all HPC.

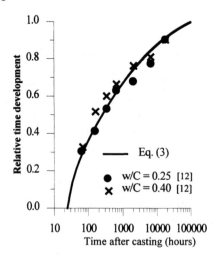

Fig. 5.   Time development of the autogenous shrinkage according to Eq. (3). Data from [12] concerning self-desiccation.

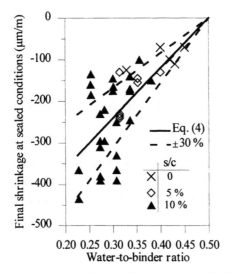

Fig. 6.   Final shrinkage at sealed conditions. Data from [5], [6], [7], and [8].

The final shrinkage at sealed conditions for HPC mixes, reflecting the main trend in the behaviour, see Fig. 6, is described with the following expression

$$\varepsilon_{s0\infty} = (-0.6 + 1.2 \cdot \frac{w}{b}) \cdot 10^{-3} \qquad (4)$$

where the binder content is defined as $b = c + s$.

Evaluated data on final shrinkage at sealed conditions presented in the figure show a rather big scatter with a coefficient of variation of approximately 30 per cent in accordance with the use of Eq. (4).

3.2   Modelling of shrinkage at variable humidity

Based on results from measurements made on columns standing in water for more than 10 years outside Bergen, Norway, see [11] and limited amount of data from [6] and [8], implies that HPC can take part of the moisture exchange with the environment into a depth of a few centimetres beyond the surface. Therefore, external humidity exchange is only regarded for the outer shell of the structural member and here denoted "surface layer drying". For a cross section of the structure perpendicular to the main flow of the moisture movement an apparent final portion of the cross-section area, $\alpha_{sd\infty}$, concerning surface layer drying is expressed by

$$\alpha_{sd\infty} = \frac{u\, l_{sd}}{A_c} \le 1 \qquad (5)$$

where u = the perimeter (m) of the cross-section in contact with the environmental humidity conditions; A_c = studied cross-section (m²) perpendicular to the water flow direction(s); l_{sd} = representative length from the surface for external water exchange expressed by

$$l_{sd} = \frac{l_{sd,ref}}{0.5 - \dfrac{w}{b}} \qquad (6)$$

where $l_{sd,ref} = 0.0045\text{m} =$ the reference depth of surface layer drying for HPC.

The additional humidity strain concerning moisture exchange (shrinkage at drying and swelling at wetting, respectively) with the environment may be calculated from

$$\varepsilon_{csd}(t) = \varepsilon_{sd\infty}\, \beta_{sd}(t)\, \beta_{sd,RH} \qquad (7)$$

where $\varepsilon_{sd\infty} =$ final reference drying/wetting shrinkage for a studied concrete member; $\beta_{sd,RH} =$ coefficient depending on the environmental humidity; $\beta_{sd}(t) =$ coefficient describing the development of humidity strain expressed as

$$\beta_{sd}(t) = \left[\frac{t - t_s}{t_{sd} + t - t_s}\right]^{0.5} \qquad (8)$$

where $t - t_s =$ time (days) after start of drying/wetting; $t_s =$ the age $(\geq 1$ day) of concrete at start of drying/wetting; $t_{sd} = 200$ days $=$ time reflecting the typical rate of humidity exchange for HPC.

A force equilibrium between an inner part of the cross section only subjected to autogenous shrinkage and an apparent surface layer zone with additional shrinkage due to external moisture exchange, see Fig. 7, results in the final reference drying shrinkage for the whole cross-section expressed by

$$\varepsilon_{sd\infty} = \alpha_{sd\infty}\, \varepsilon_{sd,tot} \qquad (9)$$

where $\varepsilon_{sd,tot} =$ final humidity strain for a cross-section where the total area is in equilibrium with the external humidity, which is expressed by

$$\varepsilon_{sd,tot} = (-0.1 - 0.8 \cdot \frac{w}{b}) \cdot 10^{-3} \qquad (10)$$

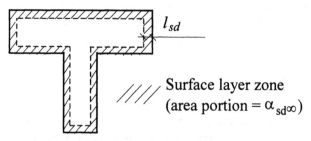

Fig. 7.   Illustration of the surface layer zone for a cross-section of a T-beam.

The dependency of the environmental humidity conditions is described by

$$\beta_{sd,RH} = \frac{RH_0 - RH}{\left| RH_{ref} - RH_0 \right|}$$ (11)

where RH = environmental relative humidity (%)
        RH_0 = relative humidity (%) representing equilibrium at sealed conditions (%)
        RH_{ref} = relative humidity (%) at chosen reference conditions

For description of HPC ($w/b \leq 0.4$) together with the use of Eq. (11) the following values are adopted as typical

$$\text{For } 60 \% \leq RH \leq 100 \%: \begin{cases} RH_0 & = 80 \% \\ RH_{ref} & = 60 \% \end{cases}$$ (12)

and

$$\beta_{sd,RH} = 1 \quad \text{for } RH < 60 \%$$ (13)

For application of Eq. (11) the following environmental humidities may be used as rough guidelines

$$RH = \begin{cases} 100 \% & \text{in contact with water or at very moist conditions} \\ 80 \% & \text{at typical outdoor athmospheric conditions} \\ 60 \% & \text{at typical indoor conditions} \end{cases}$$ (14)

For description of sealed conditions using Eq. (11) for HPC ($w/b \leq 0.4$) the formal environmental humidity $RH = 80 \%$ may be used.

Example of test results and calculated values for autogenous shrinkage and external drying shrinkage of HPC according to the presented model is presented in Fig. 8. Subfigure 8a show that there is good agreement between measured and calculated data. In subfigure 8b it is seen that the measured shrinkage at sealed conditions (filled symbols) probably show effects of some leakage of water during the test period. Still the data can be used as Eq. (3) shows the expected time development of autogenous shrinkage. Subfigure 8b also show another feature typical for drying shrinkage measurements (non-filled symbols), the rather big scatter although the recipies are exactly the same for all of these measurements.

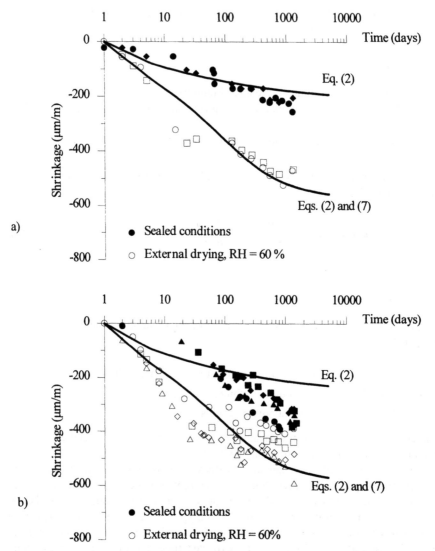

Fig. 8   a) w/c = 0.37 and s/c = 0.10 and b) w/c = 0.33 and s/c = 0.05. Measured [5] autogenous shrinkage and drying shrinkage for two concrete mixes in comparison with the presented model.

4   Conclusions

Measuring autogenous deformations in hardening concrete at early ages demands high accuracy of sealing material due to the influence of the surrounding environment. Also due to the fact that even small losses of moisture affects the internal relative humidity significantly. A new code type model for the prediction of shrinkage deformations in high strength/high performance concretes (HSC/HPC) recipes (cube strength $(f_K) > 80$ MPa and water-to-binder ratios $(w / b) \le 0.4$) is presented. Based on the fact that HSC/HPC tests not yet are as numerous as for normal strength concretes (NSC), the presented model should be regarded as approximate with respect to effects that are not investigated so far. Such effects may be varying temperature, loading and unloading stress history sequences, tensile stresses, combined effects of varying stress and varying environmental conditions, high stress levels, types of cement, types of aggregate.

The structure of the presented equations is built in such a way that the behaviour typical for HSC/HPC is gradually decreased up to the water-to-binder ratio of 0.5. This is done to allow the model to have a transitional stage to the behaviour of NSC.

5   Acknowledgement

The work presented in this paper has been supported financially within the national research program for High Performance Concrete and within the research program of Crack Control. The following funds and firms have financially supported one or both of these programs: the Swedish Council for Building Research, the Swedish National Board for Technical Development, Cementa AB, Elkem A/S, Betongindustri, NCC, Skanska, Strängbetong, Swedish National Rail Administration, Foundation for Swedish Concrete Research, and the Development Fund of the Swedish Construction Industry. Special acknowledgements are directed to professor Lennart Elfgren, head of the Division of Structural Engineering at Luleå University of Technology.

6   References

1. Tazawa, E.I, and Miyazawa, S. (1997) *Effect of Self-dessication on Volume Change and Fluxural Strength of Cement Paste and Mortar*, in Proceedings of an International Research Seminar in Lund, June 10, on "Self-Desiccation and its Importance in Concrete Technology", Report TVBM-3075, Lund Institute of Technology, Div. of Building Materials, Edited by Persson, B. and Fagerlund, G., pp 8 - 14.

2. *CEB-FIP Model Code 1990*. Comité Euro-International du Beton, Thomas Telford, London 1993, 437 pp (ISBN 0-7277-1696-4).

3 Jonasson, J.-E., and Persson, B. (1998) Chapters 2.4.6 *Shrinkage* and 2.4.7 *Creep* in *High Performance Concrete Structures - Design Handbook*. Division of Structural Engineering, Luleå University of Technology, Under preparation February 1998.

4 Jonasson, J.-E., and Persson, B. (1998) Chapter 10 *Deformations* in *High Performance Concrete - Material Handbook* (in Swedish). Division of Structural Engineering, Luleå University of Technology. Under preparation February 1998.

5. Le Roy, R. (1996) *Déformations instantanées et Différées des Bétons à Hautes Performances*. Série Ouvrages d'art - OA22, ISBN 2-7208-2520-4, Laboratoire Central des Ponts et Chaussées (LCPC), Paris.

6. Persson, B. (1998) *Quasi-instantaneous and Long-term Deformations of High-Performance Concrete with some Related Properties*. Div. of Building Materials, Lund Institute of Technology, Doctoral Thesis, Report TVBM-1016, to be published.

7. Jonasson, J.-E., Westman G., and Hedlund, H. (1998) Unpublished material, Div. of Structural Engineering, Luleå University of Technology, Luleå.

8. Hedlund, H. (1996*) Stresses in High Performance Concrete due to Temperature and Moisture Variations at Early Ages*, Division of Structural Engineering, Luleå University of Technology, Licentiate Thesis 1996:38L, 238 p.

9. Westman, G. (1995) *Thermal Cracking in High Performance Concrete*, Division of Structural Engineering, Luleå University of Technology, Licentiate Thesis 1995:27L, 123 p.

10. Persson, B. (1997) *Creep and Shrinkage in Concrete Designed for Bridge Elements*, Task Report No. U97.16, Division of Building Materials, Lund Institute of Technology, Lund, 9 pp.

11. Nilsson, L.-O. et al (1998) Chapter 11 *Moisture in High Performance Concrete - Material Handbook* (in Swedish), Chalmers University of Technology, Göteborg, to be published.

12. Norling Mjörnell, K. (1997) *A Model on Self-Desiccation in High Performance Concrete*, in Proceedings of an International Research Seminar in Lund, June 10, on "Self-Desiccation and its Importance in Concrete Technology", Report TVBM-3075, Lund Institute of Technology, Div. of Building Materials, Edited by Persson, B. and Fagerlund, G., pp 141-157.

6   AUTOGENOUS SHRINKAGE AT VERY EARLY AGES

E.E. Holt
*Department of Civil Engineering, University of Washington,
Seattle, Washington, USA*
M.T. Leivo
*VTT Building Technology, Technical Research Centre of Finland,
Espoo, Finland*

Abstract
Early age autogenous shrinkage measurements provide a challenge, due to the difficulty in making accurate measurements of the concrete prior to demoulding. The shrinkage must be measured immediately after casting in a mould which permits constant readings without disturbing the concrete. Within the first hours after casting, the concrete is the most sensitive to internal stresses. Even if cracks are internal and microscopic, drying at a later age will merely open the existing deformations and cause problems. This paper aims at demonstrating how capillary pressure induces concrete shrinkage at very early ages.

Research was carried out at the Technical Research Centre of Finland. The plastic shrinkage test specimen dimensions was 270*270*100 mm. Shrinkage was measured with respect to settlement, setting time, and the development of internal capillary pressure. Autogenous deformations were found to be a significant contributor to the total concrete shrinkage measured in early and later ages. In both the drying and moist conditions, the immediate shrinkage should not be overlooked. The major part of autogenous shrinkage was found to take place during the first 12 hours when the concrete's tensile strain capacity is at the minimum.

Keywords: autogenous shrinkage, capillary pressure, very early-age, volume changes.

Autogenous Shrinkage of Concrete, edited by Ei-ichi Tazawa. Published in 1999 by E & FN Spon,
11 New Fetter Lane, London EC4P 4EE, UK. ISBN: 0 419 23890 5

1 Background

The influence of cracking of concrete is becoming more of a concern as the building industry demands higher strength and better performing materials. This cracking is highly dependent on the concrete shrinkage due to both drying and autogenous changes at early and late ages. Autogenous shrinkage is merely due to the internal chemical reactions and structural changes when the specimen is not allowed to transfer water with the surrounding environment. The magnitude of autogenous shrinkage increases severely with the higher strength concrete [1] and is uncontrollable by field handling and placing techniques. It is these autogenous volume changes which should be identified to understand what changes can be made in the concrete mixture designs to ensure durable concrete.

At very early ages the fresh concrete goes through a phase of possible bleeding and segregation while the internal skeleton is forming. After this structure is achieved the concrete will begin to set. During these early reactions (< 12 hours) the tensile strain capacity reaches the minimum prior to the concrete hardening. [2] It is at this point which the concrete is most susceptible to cracking due to the shrinkage. This region often coincides with the time of maximum early age autogenous shrinkage. Earlier hours when the concrete is still fluid allows for higher tensile strain capacity because of the plasticity of the material. But even during this stage the concrete is at risk of being deformed and creating a less homogeneous body. The shrinkage can cause the transition zone at the aggregate to paste interface to be much weaker. This aids the later age drying shrinkage which is deforming in the same direction as the very early age autogenous reactions.

A second concern in the very early ages of the concrete strength development is the role of internal capillary pressure. Radocea [3] has shown the development of this under-pressure induces stresses in the concrete pores. This pressure at very early ages is related to the relative humidity of the capillary pores in the concrete. As the relative humidity decreases the magnitude of autogenous shrinkage directly increases, as shown in Figure 1. [4]

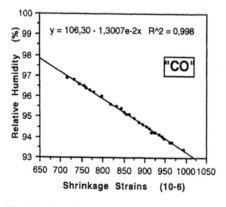

Fig. 1. Correlation between free autogenous shrinkage strains and internal RH of the materials, measured at T=21°C, from 28 days. [4]

In general, early age autogenous shrinkage measurements provide a challenge due to the difficulty in making accurate measurements of the concrete prior to demoulding. The shrinkage must be measured immediately after casting in a mould which permits constant readings without disturbing the concrete. The remainder of this paper will focus on some of the measuring methods used for evaluating early age autogenous shrinkage. The same factors influence the early age autogenous shrinkage as compared to the long term shrinkage as described elsewhere [4-7]. It must also be noted that many of the contributions of autogenous and drying shrinkage at early ages are still unknown.

2 Horizontal Shrinkage

The horizontal displacement of the concrete starts soon after the settlement is complete and internal pressure begins to develop. It is often measured as some average value over the whole concrete depth.   The exact method of measuring is varied by researchers using a variety of methods as detailed below with accompanying sketches (Figures 2 through 6).

2.1 Embedded Gauges
Original research at VTT [5] used two vertical metal supports positioned on the bottom of the mould to which LVDTs where attached. (Fig. 2) The problem with these gauges is that they risk measuring movements associated with the settling of the fresh concrete. As the concrete undergoes vertical deformation in the first hour after casting, the dead weight of the concrete exerts a pressure on the vertical mould walls and these supports. It is also impossible to distinguish at which point the gauges are measuring, i.e.: is the shrinkage magnitude which is recorded representative of the average value or a value somewhere closer to the surface?

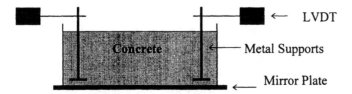

Fig. 2. Original VTT measuring method, with gauges imbedded from base.

2.2 Suspended Gauges
The next option to modify the above method of embedded gauges is to suspend the vertical metal supports from above the concrete specimen. (Figure 3) The metal strips would be hung from a wire or bar with pivots and horizontal movement again measured with LVDTs. This method would eliminate the forces associated with the internal friction angles because the supports would not extend through the whole depth of the concrete. The problem with this method is that the settling concrete will still exert some vertical force on the metal strip right at the concrete surface due to the metal friction.

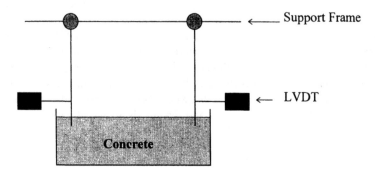

Fig. 3. Optional measuring method at VTT, with gauges suspended from above.

2.3 Gauges Through Walls

Research in Norway and Sweden [8-10] uses horizontal shrinkage gauges which are placed through the center of the vertical mould walls. (Figure 4) These gauges permit measurements similar to the above two methods without the restraint problems at the surface. The uncertainty exists at the wall and concrete interface where the gauges are attached. As the concrete settles, it is possible that the gauges would have forces exerted vertically on them which would prevent a perfectly horizontal alignment. It may also be difficult to place the concrete in the mould without clogging the hole in the mould wall. Measurements at very early ages (in the first 2 hours?) document a combination of concrete movements not necessarily due to shrinkage.

Fig. 4. Norwegian and Swedish measuring method, with gauges through mould wall.

2.4 Non-Contact Along Walls

A slight variation to the above method would be to use non-contact transducers along the mould walls. (Figure 5). This method uses the reflection of electronic pulses against any metal chip placed in the concrete. This would eliminate the problems of concrete interfering with the gauges but would require the mould be made of a non-interfering material such as PVC plastic.

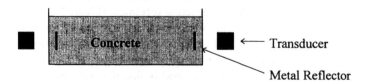

Fig. 5. Non-contact transducer option to measure along mould wall.

2.5 Lasers On Surface

Recent adaptations to the VTT test arrangement have included converting the horizontal shrinkage devices to a more accurate method of placing lightweight sensor on the concrete surface to detect laser movement. (Figure 6) This method seems the most simple (by not having forces imposed on embedded gauges by settling) as long as the sensor remains level and on the top of the concrete.

All of these equipment devices have parameters which must be satisfied prior to testing. A major concern is each test method with its own possible errors yields different results. This makes comparing shrinkage data from various researchers very difficult. It is imperative to design an equipment arrangement which produces the most accurate and precise measurements of the phenomenon of concrete at early ages.

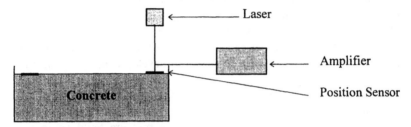

Fig. 6. Present VTT measuring method, with laser and position sensing device.

3 Capillary Pressure

The capillary pressure begins to develop after the basic concrete skeleton has developed. As the internal pores lose water the under pressure develops, as well understood and accepted in the case of plastic drying shrinkage. [11] It is also the case that the capillary pressure contributes to the onset of autogenous shrinkage at very early ages. As shown in Figure 7, the autogenous shrinkage follows the development pattern of the capillary pressure. This pressure is actually a negative capillary pressure, or suction. If the concrete has already reached its final set point then the shrinkage will not follow the rise in capillary pressure.

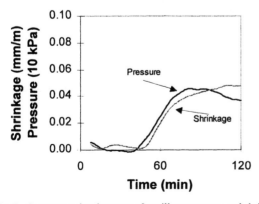

Fig. 7. Concurrent development of capillary pressure and shrinkage.

The primary cause of autogenous shrinkage at very early ages is the development of the capillary suction. This suction is a result of chemical and structural changes. The initial volume change is due to the chemical hydration reactions of the cement paste. The second factor is the development of the microstructure, which makes the body more rigid and prevents any additional plastic settlement. Finally the pressure will develop within the concrete pores due to the previous two factors and induce shrinkage.

Combining the pressure with other measured properties, the shrinkage mechanisms become more evident. Figures 8 and 9 are the results of two different autogenous shrinkage tests in the original VTT test arrangement. The variation in shrinkage magnitude for these two sample mixes is due to varying paste content, setting time, and the time period of pressure development. The test is zeroed after one hour of measuring though the computer log begins approximately 30 minutes after mixing. Double pressure transducers are placed in the slab at depths of 10 and 60 mm. Vertical shrinkage, or settlement, is measured with an LVDT and occurs immediately after placing. The horizontal shrinkage is shown to develop concurrently to the pressure increases.

The transducer which is deeper in the concrete slab (60 mm) builds up pressure faster because the concrete is impermeable enough to isolate it from any water source. The bleeding water at the surface prevents the pressure build up in the shallow transducer (10 mm). When the concrete gains stiffness the bleed water can no longer penetrate the concrete so the pressure gains at the surface as well.

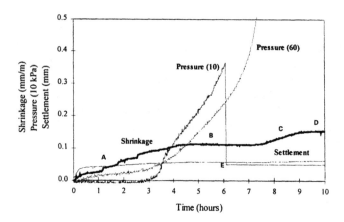

Fig. 8. Autogenous shrinkage test; w/c = 0.40 with 400 kg/m³ white cement and 1.5% superplasticizer.

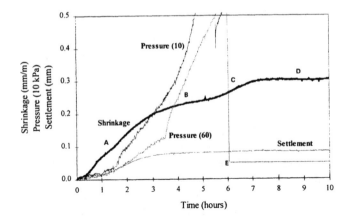

Fig. 9. Autogenous shrinkage test; w/c =0.40 with 425 kg/m³ white cement and 1.25% superplasticizer.

In the early hours immediately after casting even a small pressure increase will cause shrinkage, as shown at point A in Figures 8 and 9. But as the concrete gains strength, the shrinkage will level out (Point B) and the pressure does not induce as much shrinkage. As the pressure continues to rise the concrete strength capacity will be overcome and the shrinkage will continue, at Point C. With age the concrete again has enough strength to sustain the pressure so there is no volume changes. (Point D) Eventually the pressure will drop (though not necessarily the breakthrough pressure we see) and the shrinkage will cease. At later ages the microscopic pores control any volume changes due to long-term autogenous shrinkage.

At point E, when the pressure drops, it is merely due to a breakthrough in the transducer or concrete (due air penetration). After this point the internal capillary pressure may still be building though we can no longer measure the movement.

4 Conclusions

The measurement of very early age shrinkage can be measured in a variety of ways. All of these testing arrangements have their own problems but no standard test method exists at this time. The various arrangements can yield different results which should be taken into consideration when comparing and drawing conclusions.

It is also helpful when measuring the shrinkage to document the development of internal capillary suction. Though this is accepted in early age drying shrinkage measurements, it also a key factor in the autogenous shrinkage case. Observing this pressure build up provides information to better understand the mechanisms causing the concrete volume changes.

The capillary suction is a result of chemical and structural changes. The initial volume change is due to chemical hydration which is followed by development of the microstructure. This makes the concrete less permeable and instigates the pressure development within the concrete pores, which finally induces shrinkage.

5 References

1. Tazawa, E., and S. Miyazawa, "Experimental Study on Mechanisms of Autogenous Shrinkage of Concrete," *Cement and Concrete Research*, Vol. 25, No. 8, 1995, pp. 1633-1638.
2. Kasai, Y., Yokoyama, K., and I. Matsui, "Tensile Properties of Early Age Concrete," *Mechanical Behavior of Materials*, Society of Materials Science, Vol. 4, Japan, 1972, pp. 288-299.
3. Radocea, Adrian, A Study on the Mechanisms of Plastic Shrinkage of Cement-Based Materials, Ph.D. thesis, CTH Göteborg, 1992.
4. Baroghel Bouny, V., "Texture and Moisture Properties of Ordinary and High-Performance Cementitious Materials," *Proceedings of Seminaire RILEM 'Benton: du Materiau a la Structure'*, September 1996, Arles, France.
5. Leivo, Markku, and Erika Holt, "Autogenous Volume Changes at Early Ages," *Self-Desiccation and Its Importance in Concrete*, B. Persson and G. Fagerlund editors, Lund University, Lund Institute of Technology, Report TVBM-3075, Lund 1997, pp. 88-98.
6. Paulini, P., "Outlines of Hydraulic Hardening – an Energetic Approach," Workshop NTNU/SINTEF - Trondheim, Norway, Early Volume Changes and Reactions in Paste - Mortar - Concrete. November 28-29, 1996.
7. Tazawa, E., and S. Miyazawa, "Influences of Cement and Admixtures on Autogenous Shrinkage of Cement Paste," *Cement and Concrete Research*, Vol. 25, No. 2, 1995, pp. 281-287.
8. Hammer, Tor Arne, SINTEF Research Scientist, Trondheim, Norway, personal conversations, November 1996 through present.
9. Justnes, H., Van Gemert, A., Verboven, F., and E.J. Sellevold, "Total and External Chemical Shrinkage of Low W/C Ratio Cement Pastes," *Advances in Cement Research*, Vol. 8, No. 31, July 1996, pp. 121-126.
10. Radocea, Adrian, "Autogenous Volume Change of Concrete at Early Ages Model and Experimental Data," *Self-Desiccation and Its Importance in Concrete*, B. Persson and G. Fagerlund editors, Lund University, Lund Institute of Technology, Report TVBM-3075, Lund, Sweden, 1997, pp. 56-71.
11. Kronlöf, Anna, Leivo, Markku, and Pekka Sipari, "Experimental Study on the Basic Phenomena of Shrinkage and Cracking of Fresh Mortar," *Cement and Concrete Research*, Vol. 25, No. 8, 1995, pp. 1747-1754.

7 TEST METHODS FOR LINEAR MEASUREMENT OF AUTOGENOUS SHRINKAGE BEFORE SETTING

T.A. Hammer
SINTEF Civil and Environmental Engineering,
Cement and Concrete, Trondheim, Norway

Abstract
The relations between chemical shrinkage, bleeding, settlement, autogenous volumetric and linear shrinkage in the initial phase (before and during setting) of paste-mortar-concrete, has been discussed in order to arrive at a meaningful and reliable method to measure the linear autogenous shrinkage of concrete. The paper is based on own measurements as well as data and test method descriptions found in the literature.

The present work reveals that there is an inconsistent relation between the measured "true driving force", i.e. the volumetric autogenous shrinkage of the paste, and the measured linear autogenous shrinkage of the equivalent concrete. The re-absorption of bleeding water, taking place roughly around setting, seems to constitute an important uncertainty: In linear measurements it will give a reduced rate of shrinkage or even a swelling. Therefore, from a crack risk point of view the re-absorption is helpful. At a building site, however, one can not count with this help as the bleeding it self is a product of many non-material factors and, as the re-absorption takes place close to the top surface, the expansion may be undetectable in cases with rather thick slabs/decks/floors in particular. Accordingly, the linear autogenous shrinkage should be measured **without** the influence of bleeding water, i.e. it has to be removed during the test.
Keywords: Autogenous shrinkage, paste, mortar, concrete, before setting, test methods

1 Introduction

The use of low w/c concretes, commonly referred to as high strength or high performance concrete (HSC or HPC), has revealed that it is susceptible to cracking in the early ages. This holds for large horizontal surfaces (e.g. bridge decks) in the **initial phase** (before and during setting) in particular [1].

Autogenous Shrinkage of Concrete, edited by Ei-ichi Tazawa. Published in 1999 by E & FN Spon, 11 New Fetter Lane, London EC4P 4EE, UK. ISBN: 0 419 23890 5

The cracking of horizontal surfaces due to plastic shrinkage (caused by insufficient protection against external drying) is the traditional problem. In this respect HSC/HPC is more vulnerable than normal concrete due to minimal bleeding, i.e. extra care has to be taken to protect against the drying. However, the experience with the use of low w/c concretes in e.g. bridge decks have revealed that severe cracking may occur in spite of proper protection (curing membrane, plastic sheets, water spray, etc). Some years ago NTNU/SINTEF and the Norwegian Directory of Roads started to investigate the issue. From the work we have established that a considerable volumetric autogenous shrinkage (also called external chemical shrinkage) is taking place in the initial phase, increasing with decreasing w/c [2]. Thus, there is an "internal" driving force that may cause cracking. The tensile strain capacity of the concrete and the degree of restraint will determine whether it will crack or not. When the concrete is stiffening (e.g. seen as workability loss), the tensile strain capacity decreases considerably and will go through a minimum point [3]. Compared to normal concrete, HSC/HPC typically have more rapid workability losses (i.e. faster stiffening) but longer setting times. Our hypotheses is then that higher autogenous shrinkage in a longer period with minimum strain capacity, is the main reason for the higher crack risk of low w/c concrete.

In order to investigate the hypotheses, it is necessary to measure the **"linear" autogenous shrinkage** development (the one-dimensional horizontal component), and the development of the tensile strain capacity. The latter will not be discussed here. (At the time being, we are constructing the test rig for measurement of tensile strength and strain capacity).

The present work at NTNU/SINTEF reveals that there is an inconsistent relation between what we consider as the "true driving force", namely the volumetric autogenous shrinkage measured on pastes with rotation (see section 3.2), and the measured linear autogenous shrinkage of the concrete. This is of course not acceptable when the goal is to determine reliable values for the linear autogenous shrinkage. Furthermore, the work shows that the linear measurement is complicated for several reasons connected to the dramatic alteration of the paste-mortar-concrete reology in the initial phase, shifting from a liquid to a solid material behaviour. The paper discusses possible reasons for the inconsistency, based on own measurements and previous measurements presented in the literature, ending up with some recommendations for a method for reliable measuring of the linear autogenous shrinkage.

2   Relations between early age chemical shrinkage, bleeding, volumetric and linear autogenous deformations

2.1 General

Provided no external moisture exchange, the chemical shrinkage due to the hydration is the fundamental driving force resulting in volume changes, stresses and possible cracking, and it develops practically from "time zero" [2, 5]. Therefore, we started to measure this **chemical shrinkage** on pastes, as the first step on the way that arrives at reliable measure of the linear autogenous shrinkage of concrete. The next step was to measure the consequence on the volumetric external contraction, i.e. the **autogenous shrinkage,** and we found a logical relation between the two, see below. The third step was to measure the **linear autogenous shrinkage**, i.e. the horizontal component and the settlement. The logical relation should give the results that two times the linear shrinkage + settlement equals the volumetric shrinkage. The fourth step is to go from linear measurements on pastes to **linear measurements on concrete.** The work shows that it is the third step that causes the main concern, regarding the inconsistency, and therefore focused on in the following.

In Fig 1, it can be seen that the autogenous shrinkage develops rather similarly to the chemical shrinkage until some point that corresponds to the time around setting. Apparently, there is a full "collapse" (liquid behaviour) of the paste in this period (i.e. the stiffness is too low to withstand the slight underpressure created by the chemical shrinkage). However, from linear measurements with a similar paste, it can be seen that the horizontal component of the shrinkage starts much earlier, less than an hour, see Fig 4. This means that the paste exhibits sufficient stiffness from this point to support its own weight (i.e. a non-liquid behaviour) corresponding to a hydrostatic water pressure (i.e. 40 mm water pile = the sample height). This contradiction between the two tests may be related to the fact that the paste samples for autogenous shrinkage measurement were submerged in water during measurement, resulting in a water pressure of approximately 200 mm wp. It follows that there is a "semi-liquid" phase after the liquid phase (with full collapse), that depends on the height of the sample and possible external load (e.g. water pressure).

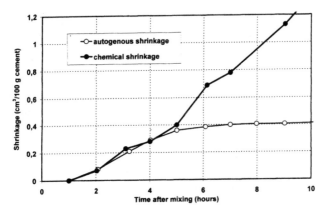

Fig 1. Chemical shrinkage and volumetric autogenous shrinkage of cement paste with
w/b = 0.4 and 5 % silica fume (based on [4]).

From the above it follows that the relation between the autogenous shrinkage and the chemical shrinkage changes rather dramatically during the initial phase. Therefore, we have divided the initial phase in two sub-phases: liquid and semi-liquid, as discussed below. Idealized relations between chemical shrinkage, volumetric autogenous shrinkage, settlement and linear autogenous shrinkage of a paste-mortar-concrete are shown in Fig 2.

2.2 The liquid phase
In the liquid phase, starting from the time of casting, the stiffness of the paste-mortar-concrete is very low and, thus, the chemical shrinkage results in an identical external volumetric contraction. Consequently, the autogenous shrinkage equals the chemical shrinkage, see Fig 2. This is also the phase where bleeding may occur. In a linear

measurement, the chemical shrinkage will appear as the settlement minus bleeding. Therefore, the settlement is higher than the chemical and autogenous shrinkage, see Fig 2. The duration of the phase is usually from "zero" to about an hour depending on the initial fluidity, material composition and height of the sample.

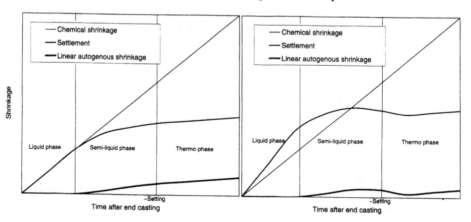

Fig 2.   Idealized relation between chemical shrinkage, volumetric autogenous shrinkage, settlement and autogenous linear shrinkage of a paste-mortar-concrete, without (left) and with (right) bleeding.

2.3 Semi-liquid phase

At the start of the semi-liquid phase a sufficiently rigid skeleton is formed to enable the sample to support its own weight. From this time a horizontal component of the shrinkage may be recorded in the given sample. The phase runs from this "self-supporting" point until setting. The stiffness development in this phase will gradually prevent the external contraction, seen as a flattening out of the autogenous volumetric shrinkage curve in Fig 2, and result in empty pore space. In cases with no bleeding the settlement plus two times the linear shrinkage equals the volumetric shrinkage, see Fig 2.

The bleeding water will be re-absorbed in the pore space, resulting in a reduced linear autogenous shrinkage rate or even a swelling, see Fig 2. The re-absorption will, however, appear, erroneously, as an **increase** of the volumetric autogenous shrinkage, i.e. the curve follows the chemical shrinkage curve up to the point where the bleeding water has been re-absorbed. According to Powers [5], the bleeding is proportional with the sample height in most of the bleeding period. Consequently, increased height gives increased bleeding and, thus, extended time with re-absorption. On the other hand, as the re-absorption takes place close to the top surface, it is likely to assume that the expansion will decrease with increasing height, and following, this particular effect of bleeding may be undetectable in a real structures, in cases with rather thick slabs/decks/floors in particular. The most important thing is, however, whether the bleeding water allows to evaporate or not.

2.4 Thermo phase

From the time about setting, the stiffness is developing very fast resulting in a much lower rate of autogenous shrinkage. The empty pore space created develops capillary forces in the pore water, which again produces external contraction. The re-absorption of bleeding water may continue in this phase too.

3 Test methods

3.1 General
The autogenous deformation may be tested either by volumetric measurements or by linear measurements, and the linear measurement may be performed vertically (i.e. settlement) or horizontally. Testing principles within these three groups are discussed below in light of the inconsistent relations discussed above. Based on own work, mainly, the following factors have been shown to influence the relation between the three:

- The re-absorption of bleeding water will reduce the shrinkage or cause a swelling of the concrete at some time after the end of bleeding. This may have a considerable effect on the total early age shrinkage, and therefore is a very important point.
- The settlement may cause a rotation of the horizontal measuring points in the paste-mortar-concrete, which may create a horizontal component giving an additional movement. Since the settlement is at least a magnitude higher than the linear shrinkage, only a small rotation will have rather great influence. This disturbs the determination of the time of "self-support" and the development in the early semi-liquid phase (i.e. in the time when the settlement is large).
- The friction between the paste-mortar-concrete and the mold will probably delay the time of "self-support". Hence, the choice of "anti-friction" solution is important. Obviously, the geometry of the specimen (i.e. height and length) have some influence on the friction, also.
- The time of self-support may increase with the height of the specimen, from "zero" on the top surface to the time of setting at the bottom of a high specimen. Consequently, both the height of the specimen and the depth of the measuring points are important factors.
- The hydration generated temperature may rise by some degrees, even in the initial phase, in pastes in particular. Thus, thermal expansion may occur.

3.2 Volumetric measurements
The total chemical shrinkage may be found as follows: A small amount of paste is filled in a glass tube, Erlenmeyer flask, etc. The rest of the volume is filled with water. The volume change is recorded as the reduction of the water level in the system [6]. The flasks are placed in a water bath for temperature control. The method has proven to be reliable, and it may be started at a short time after mixing (less than 15 minutes).

The method is based on the assumption that the pores, when created, fills with water. This is true in most cases because the thickness of the paste samples is small (i.e. maximum 10 mm), at least in the first day(s) of age. Thus, the method gives the true chemical shrinkage in the early thermo phase, also. Another set of methods is based on weighing the amount of water needed to re-fill a water filled closed case with a sealed paste-mortar sample. The sealing may be rubber membrane [7], plastic sheets [8] or oil [9].

Ziegeldorf and Hilsdorf [8] have investigated, by the use of a water filled dilatometer, the chemical shrinkage (unsealed samples) and autogenous shrinkage (sealed samples) of cement pastes with different w/c-ratios. They found an expansion (seen in the sealed samples) at some time after setting, see Fig 3, which they explained as a result of re-absorption of bleeding water. This is, however, rather illogical: In a sealed sample the re-absorption of the bleeding water should appear as a **contraction** of the total volume (= volume of solid paste + bleeding water), i.e. as a "false autogenous shrinkage" (the volume of the bleeding water must be much bigger than the swelling it causes). This is also probably the case in the period before and shortly after the point where the "sealed curve" separates from the "unsealed". Therefore, the expansion seen must occur at a time after the re-absorption, i.e. in the thermo phase.

At NTNU/SINTEF we have determined the bleeding by simply subtracting the measured settlement from the measured chemical shrinkage. Both were measured in a 50 mm cup with a sample of 40 mm in height (i.e. the same height as the samples for linear measurements, see section 3.3.2) - the settlement as the movement of a plastic mesh on the top surface, and the chemical shrinkage by filling the top of the cup with water and measure the movement of the water level (connected to a scaled thin glass tube). The results are given in Fig 6, showing that bleeding stops after about 2/3 of an hour, corresponding to the time when the pore water pressure indicates self support, see Fig 4, as shown by Radocea [10]. Apparently, all the bleeding water has been re-absorbed rather quick and the expansion continues much longer. It is, however, not yet confirmed whether this is due to measuring error or a material property, e.g. related to a redistribution of water in the paste. Expansions have been reported by others, also, e.g. Brull and Komlos [11], and Buil and Baron [12] who explained it as an effect of the gypsum content of the cement (monosulfate formation?).

Setter and Roy [7] demonstrated the "false autogenous shrinkage" due to bleeding by measuring the volume change of two similar pastes, sealed in rubber balloon, one with removing the bleeding water after the end of bleeding and the other without removing. The results is seen in Fig 5. Note that no expansion can be seen. The false shrinkage has been confirmed by Justnes et al [2, 6]. They measured the autogenous shrinkage of pastes according to the buoyancy principle, using a rubber balloon (e.g. a condom) filled with paste or mortar of which the weight in water is recorded continuously or at fixed intervals [2]. The method may allow bleeding water to lay on the surface. If so, the subsequent absorption will be observed as an extra contraction. In order to avoid the bleeding, the balloon is placed in a tube and placed on rolls to be rotated. The effect of rotation is discussed by Justnes et al [2]. The procedure has proven to give good correlation with the chemical shrinkage method [2]. Thus, the method gives the "true" chemical shrinkage in this phase. Also, they did not report any expansion. Temperature control is taken care of by the continuos immersion in water.

Another method for pastes and fine mortars based on the buoyancy principle, is presented by Paulini [13]. The sample is here placed in a plastic case with an elastic sheet on top, instead of a rubber balloon. In principle, the method measures the same as the rubber balloon method (provided that the elastic sheet with joints is water tight) in the initial phase. Thus, the absorption of possible bleeding water will be observed as an extra shrinkage. Temperature control is taken care of by the continuos immersion in water.

Ziegeldorf and Hilsdorf [8] also demonstrated that the geometry of the sample is important in this respect: The bleeding increases with increasing height and, thus, the time of re-absorption increases, too.

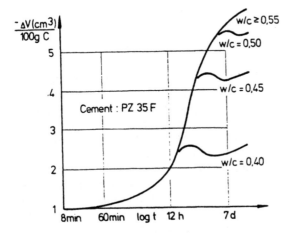

Fig 3. Shrinkage of sealed and unsealed pastes [8].

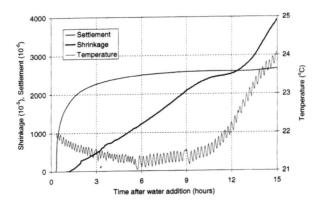

Fig .4 Linear autogenous shrinkage, settlement and pore water pressure of a cement paste with w/b = 0,4 and 5 % silica fume.

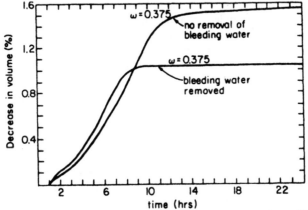

Fig. 5 Effect of removal of bleeding water on autogenous shrinkage [7]

3.3 Linear measurements

3.3.1   General

Most tests methods found in the literature are designed for measuring the autogenous and/or drying shrinkage from time after setting. A common feature for test methods designed for the initial phase and the early thermo phase, is that the specimens stay in the mold during testing. The friction between the specimen and the mold may, in combination with the low stiffness of the paste-mortar-concrete in this phase, hinder the movement of the specimen. The efforts to prevent the friction differ in the test methods: In some methods the mold surface is treated with oil, wax or asphalt, while other tests use double set of plastic with talcum in between or sheets of teflon between the specimen and the mold. The third group uses flexible tubes, i.e. corrugated tubes of metal, plastic or rubber to prevent the friction. Metal discs and oil film has also been used as under layer for pastes cast in thin plastic sheets.

3.3.2   Pastes and fine mortars

A paste dilatometer is described by Jensen [14], where the paste is poured into corrugated plastic tubes.. The method is specially design for the determination of the "thermo phase" autogenous shrinkage of pastes and fine mortars. The corrugation makes, however, the system sufficiently flexible to allow shrinkage recording in the initial phase too. The author does not trust the results in this phase due to the possible lack of contact between the transducers and the paste and some stiffness of the tube. In principle, if the contact and tube stiffness problems are solved, the method should give a reliable linear autogenous shrinkage measurement in the time after the paste-mortar is self-supporting. The deformation development starts often with an expansion, which probably is due to the hydrostatic pressure of the "liquid" paste, pressing out the transducers. The method may allow bleeding water to lay on the surface, and, thus, a swelling may be observed. The author has, however, not presented any such result. The temperature control is taken care of by placing the whole system in an oil bath.

At NTNU/SINTEF, we have measured the linear autogenous and settlement on a 40x40x160 mm paste-mortar sample in a steel mold. The shrinkage was measured as the horizontal movement of two steel plates (t/w/l = 0,5/10/50 mm) placed vertically 30 mm from both ends of sample (i.e. a measure length of 100 mm) and 35 mm down in the sample. A 4 mm thick sheet of "bubble" plastic was placed inside both ends of the mold to allow the sample to expand. The settlement, was measured as the vertical movement of the solid top surface using a plastic mesh penetrating the bleeding water. Double set of plastic sheets with talcum in between was used to prevent the friction between the sample and the steel mold. The mold was placed in water to control the temperature. An example of results is shown in Fig 4.

3.3.3   Concrete

A concrete dilatometer is presented by Jensen et al [15], based on measurement of the vertical deformation in a flexible tube via metal plates on the top. Thus, in the initial phase the autogenous shrinkage equals the settlement and the linear shrinkage in this phase can, therefore, not be found. It is not clear how the possible bleeding appears on the top, whether it occurs as a layer between the concrete and the top plate or squeezes out between the tube and the top plate. Anyway, the re-absorption may cause an expansion. Some of the results presented show expansion in a period around setting, but the authors do not discuss re-absorption of bleeding water as a reason, but suggest that it is thermal expansion (due to a temperature rise of ½ °C). They say, however, that the thermal expansion can not account for the total expansion. The dilatometer is submerged in a temperature controlled water bath.

Kasai et al [3] measured the horizontal length change of a 100x100x400 mm beam under external drying, via 30 mm anchor pins inserted in the center of both ends of the beam. The measurements started 2 or 3 hours after casting, i.e. probably rather late in the semi-liquid phase. They did not report any expansions, which probably is due to the external drying causing evaporation of the bleeding water. The temperature was controlled by the air temperature. This may have given a hydration generated temperature rise of some degrees Celsius, which in turn has caused a thermal deformation in the late semi-liquid phase and the thermo phase. Friction is taken care of by the use of paraffin paper and teflon sheets.

A similar set up, in principle, has been used by Brüll and Komloš [11 and 16]. They measured the linear shrinkage, as the horizontal length change, of cement pastes and mortars in a 40x40x160 mm [11] or in a 70x70x280 mm steel mould and concretes in a 150x150x700 mm mould [16]. It seems that these tests are influenced by the same type of temperature related uncertainties as to external temperature control and fixing of the transducers to the mold. The results did not show any deformation in the initial phase other than some thermal expansion. It is not found whether this is a material property or due to friction, fixing of the measuring points or other.

Ravina [17] used a 70x70x280 mm mold to determine the linear drying shrinkage (T = 30 °C, 50 % RH in a wind tunnel) of mortars with OPC and different amounts of fly ash (w/b ~ 0.45). He measured the movement of studs inserted in both ends of the beam, just 7 mm below the top surface. The measurements were started one hour after mixing, corresponding to the time of wind application. The results show always a very high expansion within the first hour of measuring (ie before setting), i.e. from 0.7 ‰ (without fly ash) to 1.8 ‰, dependent on the type and amount of fly ash. He explains the expansion as a "chemically-induced swelling", probably connected to the early ettringite formation. Especially the magnitude of the expansion is very different from other results as discussed above. The author says that bleeding outweighs the evaporation in this period. Hence, re-absorption may have contributed to the expansion. Some thermal expansion may have contributed, also. Still, these contributions are probably less than 10 % of the measured values.

At NTNU/SINTEF, we measure the linear autogenous and settlement on a 100x100x200 mm concrete beam in a steel mold. The shrinkage is measured as the horizontal movement of a nail placed centric in both ends of the beam, with the head 20 mm in the concrete (i.e. a measure length of 160 mm). The nails are connected to inductive displacement transducers, and the whole is fixed firmly to the mould to avoid any rotation during the settlement. A 4 mm thick sheet of "bubble" plastic is placed inside both ends of the mould to allow the sample to expand. The settlement is measured as the vertical movement of a plastic mesh placed on the top surface penetrating the bleeding water. Double set of plastic sheets with talcum in between is used to prevent the friction between the sample and the steel mould. The steel mould have sufficient heat capacity (12 mm of thickness) to keep the concrete temperature acceptably constant in the initial phase, in a temperature controlled room.

3.4 Pore water pressure measurement

Measurement of the pore water pressure, e.g. as presented by Radocea [10], is a good tool in order to obtain independent support of the results from the linear measurements and to calibrate the tests. Initially, the pressure at a given point corresponds to the hydrostatic pressure of the liquid paste-mortar-concrete given by the height and the density. The pressure decreases with time as the consequence of the creation of a "self-supporting" body (i.e. due to settlement and hydration). At the time when the concrete is able to support its own weight above this given point, the pressure corresponds to the water pressure (= the depth of the measuring point in mm water pile). At this time, the horizontal component of the shrinkage should appear. The further development is a decreasing pressure due to the tensile forces in the pore water, due to meniscus formation, caused by the chemical shrinkage and external drying (if any). Consequently, this method is suited to verify the time of "self-support". Any measured "horizontal shrinkage" in the time before this point is then erroneous (due to "false" movement of the measuring points?). Furthermore, if a horizontal shrinkage is not measured from this time, it may be a result of friction. Consequently, the method is a tool for tuning of the linear measure test method.

An example is given in Fig 4 (own results). The linear autogenous shrinkage ("shrinkage" in the graph) and settlement was measured on a 40x40x160 mm sample, see section 3.3.2. The pressure was measured in a cup with diameter 50 mm and height 70 mm, 40 mm below the top surface (corresponding to the height of the sample used in the linear measurements). The results show, as anticipated, that the settlement levels off at a pressure of 40 mm wp, corresponding to self support of the whole body. Also, it corresponds to the time of the end of bleeding, see Fig 6. The shrinkage starts, however, before this time. A repeated test show the same relation. In both cases the shrinkage starts at the time when the settlement development is no longer linear. If this early start of the shrinkage is real or not is not yet explained. It may be related to a possible increase of the time of self support with the depth of the sample (discussed earlier). If so, it should be indicated by pore water pressure measurements at smaller depths than 40 mm. This will be a part of the future work.

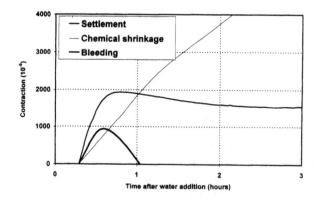

Fig. 6 Chemical shrinkage, settlement and bleeding of a cement paste with w/b = 0.4.

4   Summary and recommendations

The re-absorption of bleeding water, taking place just before, during and some time after setting, seems to be the most important point regarding measurements of volume changes, and it will be crucial in the future work. In a volumetric test it will give a "false" autogenous shrinkage in most tests (with sealed samples). It can be prevented by rotation of the samples during the bleeding period. In linear measurements it will give a reduced rate of shrinkage or even a swelling. Therefore, from a crack risk point of view of large horizontal surfaces, the re-absorption is helpful. At a building site, however, the bleeding is a product of non-material factors like protection against evaporation, time of casting, geometry, temperature, etc, which are difficult to predict. Furthermore, as the re-absorption takes place close to the top surface, it is likely to assume that the expansion will decrease with increasing height, and following, this particular effect of bleeding may be undetectable in real structures. Thus, the linear autogenous shrinkage should be measured without the influence of bleeding water, to get a "worst case" situation. The question is how to do it – either by adding agents to stabilize the paste-mortar-concrete, or by removing the bleeding water from the surface externally, continuously or at the end of the bleeding process. The first alternative needs some efforts to find the right agent to eliminate the bleeding without changing the hydration rate etc, and it has to be confirmed in some way that it is successful. Furthermore, it will alter the test conditions to be even more different from "real life. Hence, external re-absorption seems to be the best way.

The friction between the sample and the mould is another important point. In order to keep it as low as possible it is probably necessary to use some kind of sheet that allows movement between the sheet and the mould, mainly. Furthermore, the length and the height of the sample should be as small as possible, but large enough to give a homogeneous body. For concretes, a cross section of 100x100 mm and a length of 200 to 300 mm seem to be acceptable. The measuring points must not cause any additional friction, and they must not allow to rotate e.g. due the settlement.

In concrete measurements, using a 100x100 mm cross section, a steel mould may have sufficient heat capacity to keep the concrete temperature acceptably isothermal in a temperature controlled room. In paste measurements the mold should be temperature controlled by a fluid.

5   Acknowledgement

The present work is part of the Norwegian project NOR-IPACS supported by NFR (Norwegian Research Council, Selmer ASA (project leader), Elkem ASA Materials, Fesil ASA, Norcem A/S, The Directorate of Roads (The Road Laboratory) and NTNU (Norwegian University of Science and Technology).

6   References

1.   Kompen, R. (1994) High performance concrete: Field observations of cracking tendency at early ages. *Proceedings of the Int. Rilem Symposium: Thermal Cracking in Concrete at Early Ages*, Munich, Rilem Proceedings 25, Oct. 1995.

2.   Justnes, H., van Gemert, A., Verboven, F. and Sellevold, E.J. (1996) Total and external chemicla shrinkage of low w/c ratio cement pastes. *Advances in Cement Research,* 1996, 8, No. 31, July, pp. 121 - 126.

3.   Kasai, Y., Matsui, I. and Yokoyama, K. (1982) Volume change of concrete at early ages. International conference on concrete of early ages, Paris, Rilem Proceedings, April 1982, pp. 51 - 56.

4.   Ardoullie, B. and Hendrix, E. (1997) Cemical shrinkage of cementitious pastes and mortars. *Diploma theses from the Catholic University of Leuven and the Norwegian University of Science and Technology, Trondheim.* Leuven, Belgium 1997.

5.   Powers, T.C (1968) Properties of fresh concrete. *John Wiley & Sons Inc;* London 1968, pp. 533 – 600.

6.   Justnes, H., Reyniers, B. and Sellevold, E.J. (1994) An evaluation of methods for measuring chemical shrinkage of cementitious pastes. *Nordic Concrete Research,* publication no. 14. 1/94, Norwegian Concrete Association, Oslo 1994, pp. 44 – 61.

7.   Setter, N. and Roy, D.M. (1978) Mechanical features of chemical shrinkage of cement paste. *Cement and Concrete Research,* Vol 8, 1978, pp. 623 – 634.

8.   Ziegeldorf, S. and Hilsdorf, H.K. (1980) Early Autogenous Shrinkage of Cement Pastes. *7'th international conference on the chemistry of cement,* vol 4, Paris 1980, pp. 333 – 338.

9.   Slate, F.O. and Matheus, R.E. (1967) Volume changes on setting and curing of cement paste and Concrete from zero to seven days. *ACI Journal,* 1, 1967, pp. 34 – 39.

10.   Radocea, A (1992) A study on the mechanism of plastic shrinkage of cement-based materials. *Thesis for the degree of Doctor of Engineering from Chalmers Technical University.* Gothenburg, Sweden 1992.

11.   Brüll, L. and Komlös, K. (1982) Early volume change of cement pastes and cement mortars. *International conference on concrete of early ages,* Paris, Rilem Proceedings, April 1982, pp. 29 – 34.

12.   Buil, M. and Baron, J. (1982) Le retrait autogène de la pâte de ciment durcissante. *Bulletin de liaison des laboratoires des ponts et chaussée,* No 117, Jan - Feb 1982, pp. 65 – 70.

13.   Pauluni, P. (1992) A weighing method for cement hydration. *9'th international conference on the chemistry of cement,* vol 3, New Dehli 1992, pp. 248 – 254.

14.   Jensen O.M (1993) Autogen Deformation og RF-ændring (autogenous deformation and RH-development). *Thesis for the degree of Doctor of Engineering from the Technical University of Denmark,* Copenhagen, 1993.

15.   Jensen, O.M., Christensen, S.L., Dela, B.F., Hansen, J.H., Hansen P.F. and Nielsen, A. (1997) HETEK Control of early age cracking of concrete phase 2: Shrinkage and concrete. *Report No. 110 1997, Road Directorate Denmark Ministry of Transport.*

16.   Brüll, L. and Komlös, K. and Majzlan, B. (1982) Early age shrinage of cement pastes, mortars and concretes. *Materials and Structures,* Vol. 13 – No 73, Rilem 1980, pp. 41 – 45.

17.   Ravina, D. (1986) Eraly longitudinal dimensional changes of fresh fly ash mortar exposed to drying conditions. *Cement and Concrete Research,* Vol. 16, 1986, pp 902 – 910.

8 TEMPERATURE EFFECTS ON EARLY AGE AUTOGENOUS SHRINKAGE IN HIGH PERFORMANCE CONCRETES

S.L. Mak, D. Ritchie, A. Taylor and R. Diggins
CSIRO Building, Construction and Engineering,
Melbourne, Australia

Abstract

High levels of autogenous shrinkage have important implications for both cast in situ and precast concrete construction. Some of the practical issues that arise from autogenous shrinkage include dimensional incompatibility between concrete elements and cracking that is partially or wholly induced by autogenous movements. In all real structural elements, even of moderate thicknesses, the early age behaviour is characterised by a combination of moisture- and temperature-related curing effects. In this paper, results obtained from unrestrained autogenous shrinkage measurements on a range of high strength concretes with and without silica fume will be described. Measurements were made under standard conditions and also under conditions of moderate temperature development to simulate concrete elements of medium thickness. Significant unrestrained autogenous shrinkage strains in excess of 600 μs were measured within 1 day of casting. The use of 10% silica fume in a Portland cement binder is shown to significantly increase the autogenous shrinkage when compared to a plain Portland cement concrete. However, the thermal expansion that accompanies even a moderate net temperature rise of 15°C is shown to substantially reduce the net autogenous shrinkage in some concretes. A discussion on current methods of evaluating shrinkage in high performance concretes is provided in the context of the results obtained.
Keywords: Autogenous shrinkage, low water/binder, silica fume, temperature, expansion.

1 Introduction

In all real structures, a variety of different and often interrelated shrinkage-induced movements will occur. Some of these are caused by evaporative drying effects whilst others occur autogenously. Autogenous movements occur as a result of internal moisture

Autogenous Shrinkage of Concrete, edited by Ei-ichi Tazawa. Published in 1999 by E & FN Spon, 11 New Fetter Lane, London EC4P 4EE, UK. ISBN: 0 419 23890 5

depletion due to hydration and temperature-induced effects. The relative severity of each type of shrinkage depends on the type, configuration and size of the building element as well as the service environment. The level of autogenous shrinkage occurring due to hydration also depends on the binder type and the moisture content of the concrete. Recently, Aitcin *et al.* [1] advocated the adoption of a more holistic view of shrinkage within structural elements.

In trying to understand the mechanisms that govern autogenous shrinkage, we need to appreciate the real effects that exist in a structural element. This is necessary for both structures cast in situ as well as precast elements subjected to heat-accelerated curing. In most structural elements, the early age curing environment is characterised by a complex interrelationship of moisture and temperature effects. On the one hand, the bulk of concrete within an element cast in formwork does not experience a net ingress of curing moisture. Even if curing moisture is provided, the net effect of additional moist curing is likely to be restricted to near surface zones only. On the other hand, temperature development of varying degrees will occur even in moderately sized elements. This is particularly the case when high binder contents are used in either high strength or high performance concretes (HPCs). Net temperature rises in the order of 50°C are common in medium sized HPC elements [2].

There is now an increasing body of data, which links the rapid internal moisture changes in high performance concretes produced with low water/binder ratios to the often significant volume changes and associated movements at early ages. Various investigators have demonstrated the occurrence of self-desiccation in low water/binder cementitious systems, particularly those containing silica fume [3–5]. Mak and Torii [5] showed that under semi-adiabatic conditions, the relative humidity in a 100 MPa silica fume concrete reduced to less than 75% within 14 days of casting. By contrast, the relative humidity in a Portland cement concrete of similar water/binder ratio was 90% at the same age. Sellevold *et al.* [6] described the increased sensitivity to plastic cracking in HPCs subjected to high evaporation rates.

Whilst there is an increasing body of data on the autogenous behaviour of high performance concretes under standard conditions, there is currently only limited knowledge on the impact of early age temperature development. There are at least two aspects to the issue of temperature. The first is the accelerated depletion of free water due to an increase in hydration rate caused by temperature rise. The other is the thermal expansion, which serves to counteract the hydration-induced autogenous shrinkage. The net effect of both these factors may serve to significantly influence the autogenous movements in concrete under realistic in situ conditions. Tazawa *et al.* [7] demonstrated the interrelationship between autogenous and thermal strains on the level of stress generation in a range of HPCs.

In this paper, the effect of a moderate temperature rise on the net autogenous movements in high strength, low water/binder HPCs with and without silica fume will be described. Emphasis is placed on the competing effects of autogenous shrinkage and thermal expansion that occurs at and beyond the onset of setting. The influences of silica fume and water/binder ratio will be clarified. The comparison of unrestrained autogenous shrinkage movements with conventional measurements of shrinkage will be made and implications for practice will be discussed.

2 Experimental

2.1 Concrete mix design and materials

Four mixes were evaluated in this investigation, as shown in Table 1. All concrete mixes were produced with a Type GP (general purpose) cement and/or silica fume. The control mix (GP30) did not contain silica fume. For the other three mixes, silica fume was used as a 10% partial replacement by weight of Type GP cement. The water/binder ratio (w/b) of the control mix was 0.3. For the silica fume mixes, the w/b for two mixes (GSF30M and GSF30H) was 0.3. For the other silica fume mix (GSF25H) the w/b was 0.26. For mixes GSF30M and GSF30H, two initial slump values of 100 and 170 mm were obtained using different dosages of superplasticiser. The coarse aggregate was basalt while the sand was a blended concrete sand. The superplasticiser was Darex Super 20, which is used widely in the precast concrete industry in Victoria. All concrete mix designs are shown in Table 1.

2.2 Unrestrained shrinkage measurements

For each concrete mix, unrestrained shrinkage was measured under four conditions, as shown in Table 2. Under STD conditions, specimens were demoulded after 24 hours and then moist cured in a fog room at 23°C for a further 6 days before air drying in the laboratory. Under S23 conditions, specimens were demoulded after 24 hours and then sealed and conditioned in the laboratory at 50% RH and 23°C.

Table 1. Concrete mix design, slump and compressive strength

Material	GP30	GSF30M	GSF30H	GSF25H
Type GP (kg/m^3)	500	450	450	473
Silica fume (kg/m^3)		50	50	52
Water (kg/m^3)	150	150	150	131
Coarse aggregate (kg/m^3)	1235	1235	1213	1266
Sand (kg/m^3)	665	655	653	624
Superplasticiser (l)	3.7	7.0	8.4	10.5
Water/binder ratio	0.30	0.31	0.31	0.26
Slump (mm)	150	110	170	220
1-day strength (MPa)	43.0	44.0	44.5	55.5
7-day strength (MPa)	66.0	75.0	77.0	88.0
28-day strength (MPa)	75.5	92.0	90.5	108.0

Table 2. Shrinkage measurement regimes

Regime	Conditioning and measurement procedure
STD	Demoulded after 24 hours/moist cured for 6 days/air dried at standard temperature
S23	Demoulded after 24 hours/sealed and conditioned at standard temperature
AS	Measurements from time of casting in sealed conditions at standard temperature of 23°C
AT	Measurements from time of casting in insulated formwork until cooled to standard temperature of 23°C and then sealed and conditioned at standard temperature

For autogenous shrinkage measurements, $75 \times 75 \times 275$ mm prisms were cast in a false-form set-up where shrinkage was measured 1 hour after casting for periods of up to 3 days. Of these, one set was measured in uninsulated formwork whilst the other was measured in an insulated formwork, as shown in Fig. 1. The floating end studs were in contact with LVDTs at each end and the data was logged automatically at 10-minute intervals. A thermocouple was inserted into the middle of each specimen to monitor the temperature development. The temperature development characteristics obtained in the insulated test set-up closely simulated the heating conditions obtained in 300–400 mm thick concrete elements cast in steel formwork. This was verified by accompanying measurements on 300 mm diameter concrete cylinders.

2.3 Penetration resistance tests

Setting time was determined using penetration resistance tests on a series of mortar mixes, which were designed to simulate the mortar in concrete. All mortar was proportioned to provide similar initial flow values of between 180 and 200 mm. The initial and final setting times are as shown in Table 3.

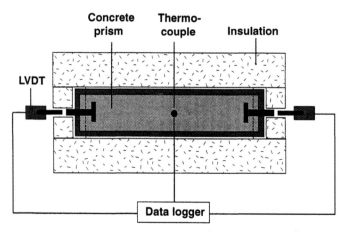

Fig. 1. Set-up for unrestrained autogenous shrinkage measurements in insulated forms

Table 3. Setting characteristics of simulated mortars

Binder	w/b	SF/b	SP/b (L/100 kg)	Flow (mm)	Initial set (hours)	Final set (hours)
GP	0.3	0.0	0.5	185	3.1	4.3
GSF	0.3	0.1	1.0	180	3.8	4.9
GSF	0.25	0.1	2.0	185	4.6	5.7

3 Results and discussion

3.1 Temperature development

The net temperature rise in the insulated (AT) prisms were approximately 3 times higher than those measured in the standard (AS) specimens. On average, the net temperature rise above the temperature of fresh concrete in the AT specimens was between 13 and 15°C as compared to 4–6°C for the standard prisms. A typical temperature development profile in the two types of prisms is shown in Fig. 2a. The temperature development results are summarised in Table 4. Under the test conditions, the differences in temperature development due to materials and mix design were not very significant, with the low w/b mix GSF25H showing slightly reduced temperature rise. For this mix, the time taken to reach the maximum temperature was also slightly longer due to the comparatively large dose of superplasticiser used.

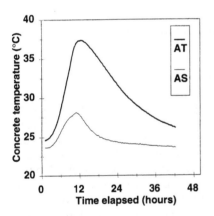

Fig. 2a. Temperature development in AS and AT specimens – GSF30M

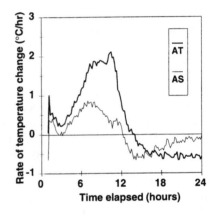

Fig. 2b. Rate of temperature change in AS and AT specimens – GSF30M

Table 4. Temperature development parameters

Mix	Regime	T_{net} (°C)	$t(T_{net})$ (hours)	R_{max} (°C/hour)	$t(R_{max})$ (hours)
GP30	AT	14.4	12.1	2.4	6.1
	AS	4.0	10.2	1.2	5.6
GSF25H	AT	12.2	15.7	2.4	11.2
	AS	4.5	14.3	1.2	9.8
GSF30M	AT	13.4	13.0	2.4	7.8
	AS	4.2	11.8	1.2	6.5
GSF30H	AT	14.3	11.0	3.9	8.8
	AS	5.8	10.3	1.8	6.7

Note: $R_{max} = (dT/dt)_{max}$

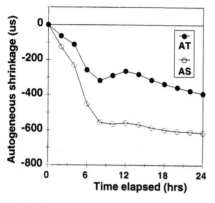

Fig. 3a. Autogenous movements up to 24 hours
– GSF30M

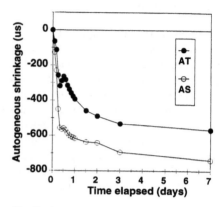

Fig. 3b. Autogenous movements up to 7 hours0
– GSF30M

3.2 Autogenous movements in relation to temperature development

3.2.1 Standard specimen

Autogenous shrinkage begins from the moment the concrete is cast, as shown in Figs 3a and 3b for measurements made in standard non-insulated prisms for mix GSF30M. However, the rate at which autogenous shrinkage proceeds varies at different stages of setting and hardening. This is evident from the plot of autogenous shrinkage rate against time in Fig. 4 for mix GSF30M. From the time the specimen was cast, autogenous shrinkage proceeded continuously, but at a decreasing rate until approximately 3–4 hours after casting, marked as point A on Fig. 4. This corresponded to the initial setting time as determined from penetration resistance tests on mortar, and also marked the stage when temperature development began to proceed significantly. This is evident from Fig. 2b where the rate of temperature development against time is plotted. Beyond this point, autogenous shrinkage began to occur at an increasing rate, reaching a maximum

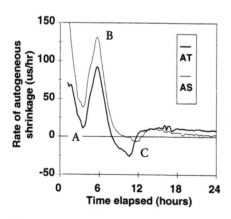

Fig. 4.   Rate of change in autogenous movements – GSF30M

rate of 130 µs/hour at about 5.5 hours for this mix. This point, marked B on Fig. 4, corresponded approximately to the time of final set as determined from penetration resistance tests on mortar. This time also corresponded to the stage when the fastest rate of temperature rise occurred, as shown in Fig. 2b. Beyond this point, the rate of autogenous shrinkage began to slow down again due to the counter-effect of thermal expansion from the temperature rise. Approximately 9 hours after mixing, the autogenous shrinkage rate had almost stopped, registering a small net expansion rate at point C marked on Fig. 4. Thereafter, the cooling down of the specimen is accompanied by a resumption of autogenous shrinkage at a slow rate.

When viewed over the first 24 hours, the results in Fig. 3a show that autogenous shrinkage proceeds at a rapid rate, reaching 450 µs in the first 6 hours. After this, the rate of autogenous shrinkage slows down substantially to reach a 24-hour value of 615 µs. The results in Table 5 show that the 7-day autogenous shrinkage for this mix was 740 µs. The development of autogenous shrinkage up to 7 days for mix GSF30M is shown in Fig. 3b. It was found that under standard temperature conditions, the development of autogenous shrinkage for the other mixes investigated followed similar qualitative trends to the one just described.

3.2.2 Insulated (AT) specimen

A moderate temperature rise in concrete had a very significant effect on the rate of autogenous shrinkage. The effects of temperature were already evident in the uninsulated (AS) specimens where the net temperature rise was only 4–5°C. In the insulated (AT) specimen, the net temperature rise was between 13 and 15°C. The measurements plotted in Figs 3a and 3b for mix GSF30M show very clearly that the autogenous shrinkage in the insulated specimens was significantly reduced. The overall patterns of shrinkage rate in relation to setting and temperature rise in the insulated specimens were similar to those of the standard specimens. However, the results in Fig. 4 clearly show that the autogenous shrinkage rate in the insulated specimens was significantly lower than that of the standard specimens. At point B, the maximum autogenous shrinkage rate in the AT specimens was 95 µs/hour or 30% lower than for the AS specimens. The changeover in autogenous shrinkage rate to a net expansion at point C occurs 2 hours earlier in the AT specimens when compared to the AS specimens. The rate of expansion in the AT specimens was also significantly higher than measured in the AS specimens. After 6 hours, the autogenous shrinkage for the AT

Table 5. Autogenous shrinkage strains for AS and AT specimens at various times up to 7 days

	AS regime				AT regime			
	GP30	GSF25H	GSF30M	GSF30H	GP30	GSF25H	GSF30M	GSF30H
4 hour	–195	20	–230	–265	–55	15	–110	–75
6 hour	–365	–20	–450	–430	–125	–15	–255	–205
12 hour	–340	–330	–560	–625	–15	–230	–265	–240
24 hour	–401	–410	–615	–675	–145	–315	–390	–365
2 day	–425	–450	–640	–695	225	–425	–490	–440
3 day	–450	–490	–695	–710	–230	–485	–530	–465
7 day	–490	–555	–740	–755	–240	–530	–565	–490

specimens was 255 μs or about 50% of that measured in the AS specimens. This early reduction in autogenous shrinkage is maintained where the 24-hour and 7-day values were 390 and 565 μs respectively, as shown in Table 5. These were approximately 35 and 25% lower than the respective autogenous shrinkage values for standard specimens at 1 and 7 days.

3.3 Influence of silica fume

For concrete of the same w/b of 0.30, the presence of silica fume led to the development of comparatively larger autogenous shrinkage strains. This was the case for both AS and AT specimens. As shown in Table 5, the 7-day autogenous shrinkage values for AS specimens for mixes GSF30M and GSF30H were in the order of 740 μs. By contrast, the corresponding 7-day autogenous shrinkage for the AS specimens for mix GP30 was 490 μs, which was 35% lower than for the silica fume mixes. Under the influence of a moderate temperature rise, the 7-day autogenous shrinkage of the control concrete was between 50 and 60% lower than for the silica fume mixes of similar 0.3 w/b.

3.4 Influence of water/binder ratio

The influence of w/b on the autogenous shrinkage in silica fume concrete varied depending on the temperature history. For the silica fume concrete with 0.26 w/b (mix GSF25H), the autogenous shrinkage in the uninsulated AS specimens was significantly lower than for the silica fume concretes with 0.3 w/b. The maximum rate of autogenous shrinkage in the AS specimens for mix GSF25H occurred after 11 hours as compared to 6 hours for mix GSF30H. This slower onset of autogenous shrinkage was also accompanied by a significantly slower maximum rate of shrinkage. For mix GSF25H, the maximum autogenous shrinkage rate under standard conditions was 85 μms/hour as compared to 120 μs/hour for mix GSF30H. After 24 hours, the autogenous shrinkage in the 0.26 w/b mix was 410 μs, which was 40% lower than for the 0.3 w/b mix. This margin reduced slightly after 7 days where the autogenous shrinkage in the 0.25 w/b mix was 25% lower than for the 0.3 w/b mix.

At relatively early ages within the first 24 hours of casting, the autogenous shrinkage of the 0.25 w/b mix was significantly reduced by a moderate temperature rise in the insulated concrete (AT) specimens. However, beyond 24 hours, the net effect of temperature rise became less significant, as shown by the results plotted in Fig. 5 comparing the autogenous shrinkage in AS with AT specimens for mix GSF25H. As a result, the autogenous shrinkage behaviour beyond 24 hours for the 0.26 w/b mix under the influence of a moderate temperature rise was quite similar to those of the silica fume mixes with 0.3 w/b. This is evident from a comparison of results shown in Table 5 between mixes GSF25H and GSF30H. The net temperature rise in the insulated specimens of mixes GSF25H and GSF30H was 12.2 and 14.5°C respectively. These occurred at 16 and 11 hours after casting for GSF25H and GSF30H respectively. The maximum autogenous shrinkage rate in the 0.26 w/b mix was 70 μs/hour or 30% lower than for the 0.3 w/b mix, resulting in similar 7-day autogenous shrinkage between the 0.26 and 0.3 w/b mixes for AT specimens.

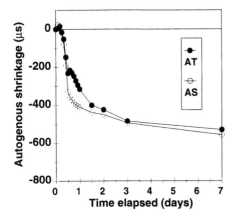

Fig. 5. Autogenous movements up to 7 days – GSF25H

3.5 Comparison of results with 'standard' measurements

Shrinkage measurements made according to 'standard' procedures essentially provide an indication of evaporative drying effects as well as residual long-term autogenous movements. Obviously, early age autogenous movements are not recorded since specimens are normally measured upon demoulding 24 hours after casting. The results in Table 5 show that net autogenous strains ranging from 400–675 µs were obtained within the first 24 hours under standard temperature conditions in the AS specimens. Even when the mitigating effects of a moderate temperature rise is considered, net 24-hour autogenous shrinkage strains ranging from 150–400 µs were measured in the concretes investigated.

The autogenous shrinkage strains obtained within the first 24 hours may be more than the subsequent shrinkage measured over 1 year using standard procedures. For instance, Figs 6a and 6b show the shrinkage of specimens obtained under STD and S23 regimes for three mixes similar to GP30, GSF30 and GSF25, produced from an earlier experimental program. For specimens measured from 7 days after an initial 6 days of moist curing under the STD regime, the 1-year shrinkage values ranged between 380 and 450 µs. When specimens were measured in a sealed condition from 1 day under the S23 regime, 1-year shrinkage values for the three mixes ranged between 340 and 380 µs. Therefore, for concretes undergoing a similar temperature history, the 1-day autogenous shrinkage for the AS specimens were between 1 and 1.5 times those of the S23 specimens.

3.6 The significance of total shrinkage measurements

There are two basic issues related to the practical impact of autogenous movements in concrete. These are dimensional compatibility between elements and autogenously induced cracking.

For some of the concretes investigated, the total 1-year shrinkage was in the order of 1000 µs if the early age autogenous components are included. This means a shortening of 1 mm in every 1 m length of concrete, which translates to a 50 mm shortening in a

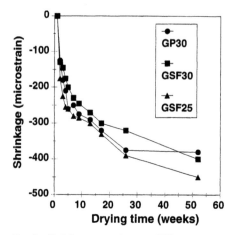

Fig. 6a. Shrinkage up to 1 year – STD

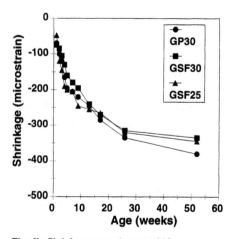

Fig. 6b. Shrinkage up to 1 year – S23

50 m beam. Drying shrinkage is directly size-dependent due to the availability of evaporative surfaces. However, autogenous shrinkage is an intrinsic material effect, which affects elements of any size to the same degree unless mitigated by temperature effects. The temperature effects are in turn controlled by size because the extent of heat retention, and thus temperature development, is a function of surface area/volume ratio. The present results show that even a moderate degree of temperature rise may reduce the early age autogenous shrinkage movements with some residual impact on medium-term shrinkage reductions. Based on the results obtained, the 7-day autogenous shrinkage was reduced by up to 50% depending on the concrete mix. Therefore, in thick elements, it is expected that the bulk autogenous shrinkage can be reduced due to temperature effects.

The next important issue is related to the potential for cracking due to stress generated in a restrained element undergoing significant autogenous shrinkage strains. In this respect, various investigators have used stress rigs to study the onset and progress of cracking due to early age volume changes [7–9]. Bjontegaard *et al.* [9] pointed out that in terms of stress generation, the relevant component of autogenous movement is that which occurs when concrete begins to harden. This is due to the fact that no stress is generated when concrete is in a plastic state. The tendency for cracking is related to not only the level of restraint and relaxation, but also to the balance between stress generation and the development of tensile strain capacity in the concrete. Mak [10] has shown that under either standard temperature or semi-adiabatic curing, the elastic modulus develops at a faster rate than compressive strength or tensile strength. This implies that stress generation occurs at a faster rate than the development of tensile strain capacity.

4 Conclusions

The dual effects of hydration-induced autogenous shrinkage and the counteracting thermally induced expansion significantly influence the early age unrestrained autogenous movements in a range of HPCs with low water/binder ratios. Even a moderate temperature rise of 15°C has a significant impact in reducing the early age autogenous shrinkage in some concretes by between 25 and 50%. The influence of binder type and mix design on net autogenous movements is not always consistent when temperature effects are included. Early age autogenous shrinkage accounts for 50–100% of the total 1-year shrinkage in the concretes investigated.

5 Acknowledgments

The authors gratefully acknowledge the assistance provided by M. El–Hassen, G. Shapiro and T. Brown.

6 References

1. Aitcin, P.–C., Neville, A.M. and Acker, P. (1997) Integrated view of shrinkage deformation. *Concrete International*, Vol. 19, No. 9, pp. 35–41.
2. Mak, S.L., Attard, M.M, Ho, D.W.S. and Darvall, P.LeP. (1994) Cross-sectional strength gradients in high strength concrete columns. *Cement and Concrete Research*, Vol. 24, No. 1, pp. 139–149.
3. Hooton, R.D. and McGrath, P. (1991) Influence of self-desiccation of non-air entrained silica fume mortars on resistance to freezing and thawing. Proc. CANMET/ACI International Workshop on the Use of Silica Fume in Concrete, Washington DC.
4. Atlassi, E. (1992) Some moisture properties of silica fume mortar. In ACI SP–132, pp. 903–919.
5. Mak, S.L. and Torii, K. (1995) Strength development of very high strength concretes with and without silica fume under semi-adiabatic temperature conditions. *Cement and Concrete Research*, Vol. 25(8), 1791–1802.

6.   Sellevold, E., Bjontegaard, O., Justnes, H. and Dahl, P.A. (199) High performance concrete: early volume change and cracking tendency. In *Thermal Cracking in Concrete at Early Ages* (ed. R. Springenschmid), E&FN Spon, London, pp. 229–236.

7.   Tazawa, E., Matsuoka, Y., Miyazawa, S. and Okamoto, S. (1994) Effect of autogenous shrinkage on self-stress in hardening concrete. In *Thermal Cracking in Concrete at Early Ages* (ed. R. Springenschmid), E&FN Spon, London, pp. 221–228.

8.   Springenschmid, R., Breitenbucher, R. and Mangold, M. (1994) Development of the cracking frame and the temperature–stress testing machine. In *Thermal Cracking in Concrete at Early Ages* (ed. R. Springenschmid), E&FN Spon, London, pp. 137–144.

9.   Bjontegaard, O., Sellevold, E.J. and Hammer, T.A. (1997) High performance concrete at early ages: self-generated stresses due to autogenous shrinkage and temperature. Proc. 3rd CANMET/ACI Int. Symp. on Advances in Concrete Technology, Auckland, Supp. Papers.

10. Mak, S.L. Thermal reactivity of slag cement binders and the response of high strength concretes to in-situ moisture and temperature conditions. Submitted to *Materials and Structures*.

Discussions

Evaluation and comparison of sealed and non-sealed shrinkage deformation measurements of concrete
H. Hedlund and G. Westman, Lulea University of Technology, Sweden

E.Tazawa	I have a question for conclusion No.4. Have you used the same type of aggregate for the concrete which was tested in different Institutes?
H.Hedlund	I am sure it was not the same type of aggregate.
E.Tazawa	What was the difference in qualities of aggregates and content of aggregate?
H.Hedlund	I don't have the exact number.
E.Tazawa	If you adjust your data, paste and aggregate volume, do you think the scatter in your curve get to closer or not?
H.Hedlund	It will probably be closer. If you would have done the adjustment.

Autogenous shrinkage at very early ages
E.E. Holt, University of Washington, USA
M.T. Leivo, VTT Building Technology, Finland

B.S.M.Persson	How did you seal your specimen? How do you prevent sealing to stop movement of the gauges?
E.E.Holt	Method we have used for autogenous shrinkage is to have a hood that is surrounding concrete specimen because we have measurement taking place on a surface. It's not easy to put piece of plastic or something over the surface. So I tried method of using oil across the top of the surface. I have also used case of using a hood with 100% relative humidity within the surrounding methlab. The most majority of our research has been concerned with early age drying shrinkage just now we are getting in full scale autogenous testing as well.
T.A.Hammer	Have you measured the temperature development during the testing?

E.E.Holt	Yes. Many of my tests last year included temperature measurement, though no result of temperature is shown today. I have temperature rise usually 4 or 5 degree Celsius and it's occurring around 3 or 4 hours. There is a platform in a capillary pressure when a capillary pressure is at platform shrinkage is basically holding still. And at the holding still moment I have a slight expansion because of temperature that's about 4 hours. I have not analyzed my results beyond 10 or 12 hours in most cases.
T.A.Hammer	Do you have a temperature increase before setting?
E.E.Holt	I believe there are still a couple of degrees that are occurring around at the initial set time. We measure initial setting as well as final set based on penetration test. We are getting slight temperature rise around at the initial set time.
T.A.Hammer	Have you tried to keep it isothermal in order to see the effect? Because we know that temperature effect is very high at initial stage
E.E.Holt	Yes I have.

Test methods for linear measurement of autogenous shrinkage before setting
T.A. Hammer, SINTEF, Norway

P.-C.Aïtcin	First of all, you said that good curing practice failed with your high-performance concrete. You should have said that good curing practices for normal strength concrete fail for high-performance concrete. Because in concrete technology we have tendency to repeat things which are familiar to us. And as very few people understand what is shrinkage, they apply curing membrane on high-performance concrete in order to cure it. Secondly they spray water but they wait for twenty-four hours when should have spray water four or five hours just after casting concrete. So I think that a curing practice can be good for normal strength concrete but very harmful to a high-performance concrete. Second point I would argue with you: why you want to take off the bleed water? I think that I don't want that concrete experience autogenous shrinkage I prefer it swell at the beginning. And we must use extra water at early beginning to have this swelling. So I know you want to make proper measurement, but why we don't use extra water to diminish autogenous shrinkage?

T.A.Hammer   I agree with your first comment.   In some of the cases I refer to, water was sprayed on the surface just one hour after finishing.   Still, cracking occurred.   May be cooling of the surface due to the spraying has an effect.   Anyway, I think autogenous shrinkage may have a major influence.

To the second point: At a building site it is difficult to count on the re-absorption because the bleeding is a product of many non-material factors: Protection against evaporation, time of casting, geometry, temperature, etc.   Therefore, in order to be on the safe side I mean that the measurements should be performed under external absorption of bleed water.

K. van Breugel   First of all I would like to say that I very much like the research that is going in Norway and I am happy to see again your challenging experimental results.   When you remove some of the bleeding water in your test in order to avoid deformations caused by suction effects, the next question is how to relate your results to the practice.   If you have a massive concrete structure, for example a large wall, you will have concrete with a water-cement ratio which varies over the height of the structure.   How do you deal with that?

T.A.Hammer   That's true.   This is a general problem.   I think those differences are much smaller than the effect of re-absorption of bleed water.   I.e., shrinkage differences due to small differences in water-cement ratio compared to the effect of re-absorption which may turn the deformations into swelling.

K. van Breugel   I am a little bit reluctant to accept it the ways of reason.   In your test, you remove quite some water, so your remaining concrete has then much lower water-cement ratio than what you have in your structure.

T.A.Hammer   That's right.   The shrinkage may be higher in the test, which is on the safe side.   I am not saying that this way of handling the re-absorption of bleeding water during testing is the only way. I am stressing the problem with the re-absorption.   I think we should come to some kind of agreement on how to proceed on that kind of testing.

S.L.Mak   I suppose I can accept that re-absorption of bleed water provides a much more conservative value since much data indicates that it decreases autogenous shrinkage.   What is your opinion on this dependence of the phenomena on size?   If I have a thin element, re-absorption of bleeding water could affect a greater portion of the specimen.   If I had a very thick slab, the small bleeding layer and its effect perhaps will be quite marginal compared to what is happening with the rest of the element.   So

I suppose there is some question on the relevance in relation to how this will relate to real performance, although it is very important concerning development of your test method. Perhaps the phenomena of bleed water re-absorption will have variable effects depending on size, thickness, and orientation.

T.A.Hammer   That's right. We have to take care of geometrical effect. We have to remove the bleeding water during testing in order to simulate better what is happening in real life.

E.Tazawa   May I ask size of the specimen which you used and method of measuring if you don't mind?

T.A.Hammer   Size for the paste measurement 40x40x160 mm. For concrete measurement, we used 100x100x300 mm.

E.Tazawa   To which direction have you measured the expansion strain, is it lateral or vertical?

T.A.Hammer   Both. The test method is similar to your proposal except for the length of the specimen.

E.Tazawa   Was it same to the horizontal and vertical direction?

T.A.Hammer   We have got higher expansion in the settlement measurement than in longitudinal measurement for concrete due to the size, suggesting that the expansion is highest close to the top surface.

E.Tazawa   Same time we have expansion due to destruction of meniscuses. When we measure length change in between lightweight aggregates, we measure expansion with increasing weight loss. Also for cement paste, in some occasion we have the same situation, we measure expansion with increasing weight loss. In your measurement, when you lost the bleeding into the paste level, then same kind of destruction of meniscuses which located near top surface may be occurring. How do you think?

T.A.Hammer   I think that the mechanism of re-absorption of bleeding water is the same as the mechanism of any external water supply.

Temperature effects on early age autogenous shrinkage in high performance concretes
S.L. Mak, D. Ritchie, A. Taylor and R. Diggins, CSIRO, Australia

P.-C. Aïtcin	You said that temperature increases with the cement content in the concrete. In high-performance concrete the amount of water can be a limitation to the temperature increase. So what is true for a normal strength concrete that contains more cement is not always true for high-performance concrete.
S.L. Mak	I agree with that. I did try to make that point. Water does play an important role in limiting temperature rise. So even though you increase the cement content for instance, form 450 to 550 kg/m³ this is also normally accompanied by a reduction in water/binder ratio. So some results that I showed indicated an increase in temperature development with increasing binder content up to a certain point beyond which temperature development decreased with increasing binder. Actually, the mitigating effect is the self-desiccation that also retards continued hydration in spite of the cement content. So you have reduced degree of hydration and a large amount of un-hydrated clinker. The reduction in chemical shrinkage also means that autogenous deformation are somewhat limited by these low water/binder ratios used. But it depends on how do you design your mix as well.
Ø. Bjøntegaard	I find some of your results quite surprising: You found a reduction of autogenous shrinkage when the water-to-binder ratio was reduced, and you also found that autogenous shrinkage was reduced at increased temperatures.
S.L. Mak	When you put together three or four factors that are occurring at that time, you can understand why. First of all, of course, you get more expansion effects with increasing temperature. But autogenous shrinkage is also mitigated because you have a limit on degree of hydration that can occur per unit mass of cement because of the limited amount of water that you have in there. We observe the same phenomenon for shrinkage. For instance, you can increase binder content and decrease water binder ratio up until a certain point. Due to an increase in paste volume, shrinkage increases. However beyond that point, shrinkage starts to decrease. Simply because you do not have enough pore volume in there although the paste volume itself is pretty big. But the amount of residual space between cement grain is so small. You actually also have limited water left for cement hydration. So there is actually a limit. This happens with either shrinkage or temperature rise.

Ø.Bjøntegaard	Largest difference was within the first day. The system is kind of soft in that period?
S.L.Mak	Yes.
Ø.Bjøntegaard	Average distance between grains has great importance?
S.L.Mak	The effect of temperature on a material is not always consistent. The difference begins to show between two to three hours and up to 24 hours. But do remember that in our case, it is not held isothermally and the temperature goes up and then down. If I express the results in terms of maturity the effect of temperature may be clearer. We are doing some mathematical modeling of this phenomenon.
F.Tomosawa	You have said that temperature goes up autogenous shrinkage is smaller for standard cured specimen?
S.L.Mak	Because of two or three effect due to temperature in this case.
F.Tomosawa	We have measured autogenous shrinkage of 80 MPa concrete by 100x100x400 mm specimen under 20 ℃ and under a temperature condition which was simulated from an actual column. And autogenous shrinkage of the two specimen are quite different. Under 20 ℃, autogenous shrinkage goes to 600 or 700 micro-strain. When you cure under the increasing temperature condition, autogenous shrinkage stops at 300 or 400 micro-strain, it is very small. You can see the result in the proceedings on self-desiccation held in Lund University last year.
S.L.Mak	Similar findings, thank you very much. We have a set of similar findings. The point is that if we measured autogenous behavior under standard temperature conditions only, we may not be able to accurately estimate actual deformations. But in real structure, you have temperature effects which may mitigate some autogenous deformations. However if you look at a column which is hot in the center and cold in the outside, you may have a mitigation of autogenous deformations in the center but you also get fairly large deformations on the outside. That actually compounds the effects of thermal gradients. That is another effect that we have to consider.

EFFECT OF MATERIALS ON AUTOGENOUS SHRINKAGE

9 EFFECT OF CHEMICAL COMPOSITION AND PARTICLE SIZE OF FLY ASH ON AUTOGENOUS SHRINKAGE OF PASTE

S. Tangtermsirikul
Department of Civil Engineering, Sirindhorn International Institute of Technology,
Thammasat University, Patumthani 12121, Thailand

Abstract
The aim of this paper is to study the effect of fly ashes with various chemical composition, particle sizes and replacement percentages on autogenous shrinkage of the pastes with fly ashes. It was found in the experiment that for the effect of chemical composition, fly ash with higher SO_3 content resulted in lower autogenous shrinkage. For the effect of particle size, paste with fly ash having smaller average size than cement exhibited larger autogenous shrinkage whereas pastes with fly ashes having bigger average size than cement showed smaller autogenous shrinkage than that of the cement paste. For the effect of fly ash content, non-classified fly ash and classified fly ashes which have larger average size than cement showed the same tendency i.e. larger autogenous shrinkage in 20% fly ash paste than in 50% fly ash paste. On the other hand, smaller autogenous shrinkage in 20% fly ash paste than in 50% fly ash paste was found in case of pastes with classified fly ash having smaller average size than cement. It could be concluded based on the results of the study that not only chemical composition which affects rate of hardening and volume change of pastes with fly ashes but also particle size which affects the pore structure of the pastes, has to be considered for modeling autogenous shrinkage of paste with fly ash.
Key words: autogenous shrinkage, fly ash, chemical composition, particle size

1. Introduction

The current development on some types of high performance concrete tends toward reducing water to binder ratio and increasing the paste content. Two obvious examples are high strength concrete and self-compacting concrete. As it is well known that the

Autogenous Shrinkage of Concrete, edited by Ei-ichi Tazawa. Published in 1999 by E & FN Spon, 11 New Fetter Lane, London EC4P 4EE, UK. ISBN: 0 419 23890 5

autogenous shrinkage is a kind of shrinkage which is dominant in the pastes having low permeability, it is essential to clarify the mechanisms and factors which have effect on the autogenous shrinkage so that simulation of the shrinkage leading to the appropriate mix proportioning of concrete regarding the autogenous shrinkage can be derived.

Fly ash is one of the pozzolans confirmed to be effective for improving various properties of concrete. Due to the inconsistency of the chemical and physical properties of fly ashes in Thailand, an attempt has been made to study the properties of concrete utilizing fly ashes with different chemical composition [1,2], and different size and fineness [3]. For an appropriate use of fly ash and to be able to design the fly ash concrete with the required performance, it is necessary to study the effect of chemical composition and physical properties of the fly ashes on properties of concrete added with the fly ashes.

This paper summarizes the test results on autogenous shrinkage of pastes which were mixed with various types of fly ash. Those fly ashes had different chemical composition and different average particle size. The average particle size of fly ash was varied by classifying the fly ash into various size groups with an air classifier machine. To reduce the specimen volume, paste specimens were selected for the test. Also, it is realized that autogenous shrinkage occurs only in the paste phase. The results of this study will be useful for the modeling of autogenous shrinkage of the paste phase in the concrete and will then be a part of the two-phase model for predicting shrinkage of concrete by considering restraint from the aggregate phase which had already been proposed by the author [4]. This paper will be divided into two parts. Part 1 concerns the effect of chemical composition and part 2 involves the effect of particle size.

2. Effect of Chemical Composition of Fly Ash on Autogenous Shrinkage

2.1 Materials Used and Mix Proportions

The materials used for studying the effect of chemical composition of fly ash on autogenous shrinkage were ordinary Portland cement Type 1, a sample of ASTM Class F (low-lime) fly ash imported from Hong Kong, and two samples of Class C fly ash obtained from Mae-Moh power plant in Lampang province, Thailand. Physical properties and chemical composition of the materials are shown in Table 1 and Table 2, respectively. According to the physical properties of the fly ash samples in Table 1, it was assumed that the average particle size of the three fly ashes was not much different so that the effect of chemical composition of the fly ash was more dominant. The results of the chemical analysis in Table 2 verify that fly ashes from Mae-Moh are of Class C and fly ash from Hong Kong are of Class F, in accordance with the ASTM Test Designation C-618.

The mixture proportions of the paste specimens which were tested in this study are listed in Table 3. The value of the ratio between water and total binders was selected to be 0.3 for the whole study whereas the ratio of fly ash to total binders was varied from 0, 0.30 to 0.5.

Table 1 Physical Properties of Cement and Fly Ashes

Physical Properties	Portland Cement Type I	Types of Fly Ash		
		Class C Mae-Moh (FM1)	Class C Mae-Moh (FM2)	Class F Hong Kong (FHK)
Specific Gravity	3.15	2.41	2.64	2.26
Blaine Fineness (cm^2/g)	3467	2,739	2,845	3,926
Bulk Density (g/l)	1.02	1.06	1.16	0.85
Moisture Content (%)	0.06	0.06	0.11	0.18
Loss on Ignition (%)	1.03	0.27	0.11	2.59
Pozzolanic Activity Index	-	86	91	73

Table 2 Chemical Composition of Cement and Fly Ashes

Chemical Composition (%)	Portland Cement Type I	Types of Fly Ash			
		Class C Mae-Moh (FM1)	Class C Mae-Moh (FM2)	Class C Mae-Moh (FM3)	Class F Hong Kong (FHK)
Silicon Dioxide (SiO_2)	21.14	33.62	29.61	33.44	45.48
Aluminum Oxide (Al_2O_3)	5.52	15.08	15.03	18.69	30.36
Ferric Oxide (Fe_2O_3)	3.25	11.27	15.08	13.08	5.07
Calcium Oxide (CaO)	65.93	23.96	20.55	13.19	6.41
Magnesium Oxide (MgO)	1.41	3.52	3.17	2.48	1.01
Sulfur Trioxide (SO_3)	2.48	1.64	4.87	1.86	0.34
Sodium Oxide (Na_2O)	0.1	1.12	0.93	0.62	0.20
Potassium Oxide (K_2O)	0.37	1.93	1.88	2.32	0.73
Manganese Oxide (MnO)	0.0	0.12	0.11	0.09	0.05
Loss on Ignition	1.03	0.27	0.11	0.65	2.59

Table 3 Mixture Condition of Tested Cement Paste Specimens

Mix No.	w/(c+f)	f/(c+f)	Designations
1	0.3	0	CTA
2	0.3	0.3	FHK
3	0.3	0.5	FHK
4	0.3	0.3	FM1
5	0.3	0.5	FM1
6	0.3	0.3	FM2
7	0.3	0.5	FM2

w : water, c : cement, f : fly ash
(1) : sealed curing   (2) : submerged curing
CTA   : cement pastes for testing autogenous shrinkage
FHK : Class F fly ash from Hong Kong (see Table 1 and Table 2)
FM1, FM2 : Class C fly ashes from Mae-Moh (see Table 1 and Table 2)

2.2 Specimen Preparation and Test Procedure

Paste specimens with dimensions 15x40x160mm were prepared for measuring length change due to autogenous shrinkage. One average test result was obtained from the average of results of three specimens.

Specimens were subjected to two kinds of curing conditions, sealed curing and submerged curing. Sealed curing was done by wrapping the specimen with vinyl sheet so that no moisture exchange could occur between the specimen and its environment. The sealed curing condition, in which there was no moisture exchange between the specimens and their environment, was considered proper for observing autogenous shrinkage. Submerged curing was performed by submerging the specimens in water so that water would be sufficiently supplied to the specimen to study the effect of water curing period. Both curing conditions were conducted under a room temperature of 25 ±1°C. For drying shrinkage tests after the submerged curing period, the specimens were put in the room with constant temperature and humidity of 25±1°C and 60± 2%RH, respectively.

The parameters considered for autogenous shrinkage test were type of fly ash (two samples of Class C fly ash from Mae-Moh and a sample of Class F fly ash), SO_3 content in Mae-Moh fly ashes (1.64% and 4.87%), fly ash content (0%, 30% and 50% by weight of the total cementitious material content) and submerged curing period (0, 3 and 7 days).

2.3 Results and Discussion

2.3.1 Autogenous Shrinkage

Fig.1 and Fig.2 show the comparison of length change of paste specimens containing 30% and 50% fly ash, respectively, with that of cement. It is shown that all specimens with fly ash exhibit smaller autogenous shrinkage than the cement paste did. Specimens containing the Mae-Moh fly ashes expressed a smaller autogenous shrinkage than those containing the Class F fly ash, especially when the replacement percentage was large and when the SO_3 content in fly ash was high (compare mixtures containing FM1 with those containing FM2). It can be seen from these figures that the higher the fly ash content, the more the autogenous shrinkage is reduced. However, increasing the content of the Class F fly ash did not affect the amount of autogenous shrinkage. This can be explained based on two mechanisms of autogenous shrinkage reduction. The first, which occurs in both the Class C and Class F fly ashes, was the result of particle shape of fly ash. Fly ash particles were spherical, as verified by the electron microscope photo. Therefore they retained less water than cement particles which were irregular. This resulted in a larger free water content in the mixture with fly ash than in mixture without fly ash when prepared with the same w/c. As autogenous shrinkage is the result of water consumption in the hydration process, larger free water content can, therefore, reduce the shrinkage. The second, which is the special characteristics of high SO_3 fly ash from Mae-Moh, is the chemical expansion which compensates the autogenous shrinkage. This expansion will be discussed in the next section. As a consequence, the tested high SO_3 fly ashes are considered more effective than the Class F fly ash for the autogenous shrinkage reduction.

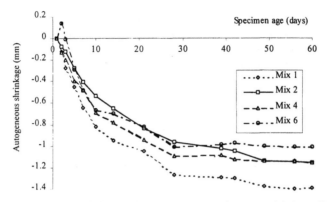

Fig.1 Autogenous shrinkage of cement paste and pastes with 30% fly ash

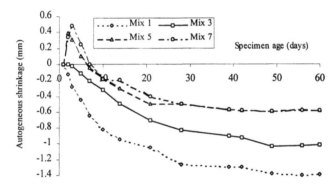

Fig.2 Autogenous shrinkage of cement paste and pastes with 50% fly ash

2.3.2 Expansion

Expansion can be considered to consist of two components, namely, swelling due to water absorption and chemical expansion. In this study, only the chemical expansion, which is the major component of expansion, will be concentrated on, since swelling is a recoverable process which is considered not effective for autogenous shrinkage reduction. The submerged curing condition was designed for the study of expansion because it was found that the chemical expansion requires water [5]. It was found that chemical expansion occurred in the pastes with Mae-Moh fly ashes because the fly ashes contained high SO_3 content. This was supported by the author's results which showed higher chemical expansion in specimens containing higher SO_3 fly ash (FM2) than in the specimen with lower SO_3 fly ash (FM1) when they were compared by the same fly ash replacement percentage, as shown in Fig.3. The results of chemical expansion in Fig.3 were obtained by subtracting the expansion of the specimens with Class F fly ash, which was assumed to be only the swelling expansion, from the total expansion of the specimens with the Mae-Moh fly ashes by assuming that swelling of specimens with the Class F fly ash were similar, even not exactly equivalent, to those with the specimens with Mae-Moh fly ashes. This was assumed because both swelling and chemical expansion occurred in pastes with Mae-Moh fly ash, and there was no method for measuring swelling by prohibiting chemical expansion of the pastes with Mae-Moh fly ash. It might be possible to measure swelling of the Mae-Moh fly ash if Mae-Moh fly ash with very small SO_3 content could be obtained, however, it was not possible to obtain this from the Mae-Moh source. Nevertheless, the assumption is considered reasonable in the light that the value of swelling is small when compared with that of the chemical expansion. It can be seen from Fig.3 that not only the SO_3 content in the Mae-Moh fly ash but also the fly ash replacement percentage have effect on chemical expansion. The more the Mae-Moh fly ash in the mixture, the larger the chemical expansion was obtained.

2.3.3 Effect of Curing Period

Demonstrated in Fig.4 are the results of the effect of submerged curing period on autogenous shrinkage of specimens that have the same mix proportions with Mix 6 in Table 3 (containing 30% Mae-Moh fly ash replacement). It can be seen that the specimens with 7-day submerged curing have lower autogenous shrinkage than those with 3-day submerged curing. In addition to the known effect of longer curing period which reduces the subsequent autogenous shrinkage, larger chemical expansion when submerged curing period is longer, especially during the first week, helps reduce the autogenous shrinkage too. The water supplied into the specimens during the curing period might have some effect but was considered much smaller than the effect of chemical expansion.

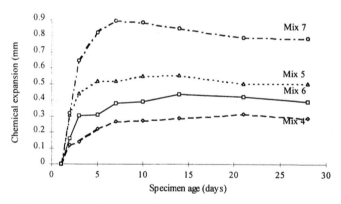

Fig.3 Chemical expansion of paste specimens with Mae-Moh fly ashes from Mae-Moh

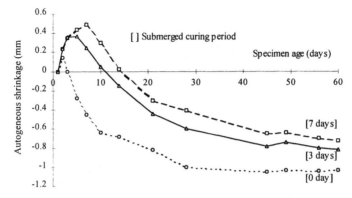

Fig.4 Effect of submerged curing period on autogenous shrinkage of paste specimens
for Mix 6 (30% Mae-Moh fly ash, FM2)

3. Effect of Particle Size of Fly Ash on Autogenous Shrinkage

3.1 Materials and Mix Proportion

Similar to the first part, paste specimens were also used for testing autogenous shrinkage in this part. However, all materials used in this part were not the same as those used in the first part. The chemical compositions of ordinary Portland cement (ASTM Type 1), original nonclassified fly ash and classified fly ashes with different sizes are given in Table 4 whereas their physical properties are shown in Table 5. The mean particle size and specific surface area by Blaine fineness of the original fly ash (FA0) and four classified fly ash with varied sizes, ranging from the coarsest (CFA1) to the finest (CFA4), are shown in Table 5. The classified fly ashes, CFA1 to CFA4, were classified from the original fly ash FA0 using an air classifier. The mix proportions of the paste specimens tested for effect of particle size of classified fly ashes are shown in Table 6.

3.2 Specimen Preparation and Test Method

The paste specimens were prepared according to ASTM C305 and ASTM C157. One day after casting, the specimens were demoulded and wrapped with vinyl sheets to prevent the exchange of moisture between the specimens and the environment, and the initial length of the specimen was recorded at the same time. The specimens were then stored in the specified test environments i.e. in the room having temperature of 25±2°C and relative humidity of 60±2 %RH. After that the length of the specimens was measured once a day by a length comparator.

Table 4 Chemical composition of powder materials used in this study

Chemical	Type of powder materials					
composition(%)	FA0	CFA1	CFA2	CFA3	CFA4	Cement
SiO_2	46.26	47.08	47.04	46.56	41.52	22.2
Al_2O_3	23.32	23.22	25.04	22.60	22.30	5.4
Fe_2O_3	10.89	11.29	10.88	11.32	10.81	2.9
CaO	9.34	9.66	9.19	10.01	10.70	63.8
MgO	2.42	2.48	2.49	2.50	2.05	1.5
Na_2O	1.20	1.19	0.68	1.18	1.13	2.7
K_2O	2.9	2.89	2.38	2.80	2.74	0.3
SO_3	0.98	0.41	0.74	1.21	3.37	2.3
LOI.	1.38	0.73	1.08	1.42	3.59	2.9

Remarks data of FA0, CFA1, CFA2, CFA3 and CFA4 are obtained from the King Mongkut's Institute of Technology Thonburi [3].

Table 5 Physical properties of powder materials used in this test.

Type of material	% Retained on # 325 sieve (%)	Blaine fineness (cm.^{2}/g)	Mean particle size (micron)	Specific gravity
Cement	3.4	3386	11	3.15
FAO	10.4	3730	26	2.04
CFA1	62.8	1927	90	1.85
CFA2	61.8	2253	90	1.89
CFA3	0.4	5167	16	2.25
CFA4	0	17260	3.2	2.59

Remarks data of FA0, CFA1, CFA2, CFA3 and CFA4 are obtained from the King Mongkut's Institute of Technology Thonburi [3].

Table 6 Mix proportions of the paste specimens.

Specimen	w/(c+f)	Cement : Fly ash	Curing temperature
CEM	0.3	100 : 0	25 °C
20FA0	0.3	80 : 20	25 °C
20CFA1	0.3	80 : 20	25 °C
20CFA2	0.3	80 : 20	25 °C
20CFA3	0.3	80 : 20	25 °C
20CFA4	0.3	80 : 20	25 °C
50CFA0	0.3	50 : 50	25 °C
50CFA1	0.3	50 : 50	25 °C
50CFA2	0.3	50 : 50	25 °C
50CFA3	0.3	50 : 50	25 °C
50CFA4	0.3	50 : 50	25 °C

Remark w/(c+f) : ratio of water to total powder (cement and fly ash) by weight.

3.3 Test Results and Discussion

3.3.1 Effect of Fly Ash Content on Autogenous Shrinkage

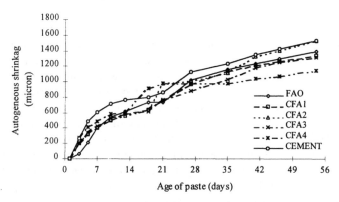

Fig.5 Autogenous shrinkage of cement paste with 20% fly ash replacement

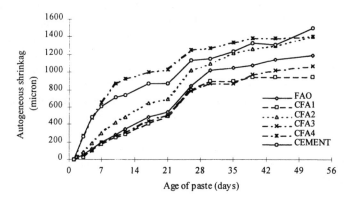

Fig.6 Autogenous shrinkage of cement paste with 50% fly ash replacement

Fig.5 and Fig.6 illustrate the results of autogenous shrinkage of all tested paste specimens cured in the environment of 25 °C and 60%RH. It can be generally concluded from the figures that for both 20% and 50% fly ash replacement, the pastes with fly ashes, either original or classified, have lower autogenous shrinkage than the cement paste. Also, for all types of fly ash except for CFA4 which has the smallest size among all tested fly ashes, a higher fly ash content causes the reduction of autogenous shrinkage (the explanation for CFA4 will be discussed in the later section concerning the effect of particle size of fly ash on autogenous shrinkage). This is because the addition of fly ash leads to less hydration, also because pozzolanic reaction can only occur when there is $Ca(OH)_2$ produced from the hydration. The spherical particle shape of fly ash also contributes to larger amount of free water in the specimens, especially when the specimens were made with the same water to binder ratio. Therefore, when replacing some part of cement with fly ash, autogenous shrinkage reduces, and the reduction is more pronounced when the ratio of replacement is higher.

3.3.2 Effect of Particle Size of Fly Ash on Autogenous Shrinkage

A definite conclusion may not be drawn from the test results for the effect of particle size of fly ash. For 50% fly ash replacement, paste with larger size of fly ash tends to have lower autogenous shrinkage, however, the trend is not clear for 20% fly ash replacement. It is known that autogenous shrinkage is related to the content and structure of the pores in the paste. Denser paste and paste having discontinuous pore structure are considered to undergo serious autogenous shrinkage. Since the content and structure of the pores vary during the hardening of the paste due to hydration and pozzolanic reactions for the cement-fly ash paste, the content and structure of the pores in the paste depends not only on the original void content in the cement-fly ash solid system but also on the change due to hydration and pozzolanic reactions. To be able to explain the results for the total time domain, there is a need to investigate into the pore structure during the hardening of all paste specimens.

Fig.7 demonstrates the comparison of autogenous shrinkage of pastes with 20% and 50% CFA4 replacement. Fly ash CFA4 is the smallest classified fly ash and was mentioned earlier to behave differently from the others regarding the effect of replacement ratio i.e. paste with 50% CFA4 replacement illustrates larger autogenous shrinkage than the paste with 20% CFA4 replacement for the total time domain. The two pastes are considered to have equal pore volume but pore structures are not the same. Paste with 50% CFA4 has smaller pore size than the paste with 20% CFA4. This is because with a constant particle concentration, larger content of small particles in the paste leads to a closer distance between particles when considering the same particle concentration, therefore bringing smaller pore size in the paste. Smaller pore size results in larger capillary pressure in the pores when water is consumed. The larger autogenous shrinkage of silica fume mixture, as known, might be explained in the similar manner.

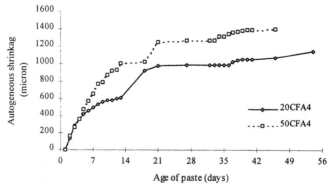

Fig.7 Autogenous shrinkage of cement paste with 20% and 50% replacement by CFA4

4. Conclusions

Based on the test results in this paper, the following conclusions can be drawn.

1. Both samples of the Mae-Moh fly ash are more effective for reducing autogenous shrinkage than the sample of Class F fly ash. The Mae-Moh fly ash with higher SO_3 content is more effective than that with lower SO_3 content. Increased fly ash content in the mixture results in smaller autogenous shrinkage.
2. The autogenous shrinkage reduction of the Mae-Moh fly ashes used in this project is due to chemical expansion. For the same Mae-Moh fly ash content, higher expansion is obtained in the mixture containing Mae-Moh fly ash with higher SO_3 content. Considering same Mae-Moh fly ash sample, higher expansion occurs in the mixture with greater fly ash content. Expansion of specimens with Mae-Moh fly ashes is much larger than that of the specimen with the Class F fly ash.
3. Longer submerged curing period leads to lower autogenous shrinkage.
4. Except for CFA4 which has the smallest mean size and is the only fly ash which is smaller than cement, the original and all classified fly ashes reduce autogenous shrinkage of the paste when compared to the cement paste especially when the replacement ratio of fly ashes becomes higher.
5. On the other hand, paste with 50% CFA4 exhibits larger autogenous shrinkage than the paste with 20% CFA4 because of smaller pore size in the paste with 50% CFA4 than in the paste with 20% CFA4.

To be able to simulate the autogenous shrinkage for the total time domain, further studies are still needed to simulate the pore structure and behavior of water in the paste during the process of autogenous shrinkage since pore content and structure, which affect the autogenous shrinkage, changes during the processes of hydration and pozzolanic reactions.

5. References

1. Tangtermsirikul, S., Sudsangium, T. and Nimityongsakul, P. (1995) Class C Fly Ash as a Shrinkage Reducer for Cement Paste. *Proceedings of the 5th CANMET/ACI International Conference on Fly Ash, Silica Fume, Slag and Natural Pozzolans in Concrete,* 4-9 June 1995, Milwaukee, Wisconsin, USA, Vol.1, pp.385-401
2. Tangtermsirikul, S., Bumrungwong, C. and Pongparit, V. (1997) Conceptual Model for Determining Optimum Proportion of Fly Ash in Sulfate Resisting Concrete. *Proceedings of the 3rd National Convention on Civil Engineering,* Engineering Institute of Thailand, Hadyai, 16-18 January 1997, pp. MAT4-1 to 4-9
3. Kiattikomol, K., Jaturapitakkul, C., et.al., (1996) Classified Fly Ash for Concrete Utilization. *Proceedings of the 1996 Annual Conference of the Engineering Institute of Thailand,* Vol.1, pp.257-269
4. Tangtermsirikul, S. and Nimityongskul, P. (1997) Simulation of Concrete Shrinkage Taking into Account Aggregate Restraint. *Structural Engineering and Mechanics,* Vol.5, No.1, January, pp.105-113

5. Raksasataya, B. (1990) Influence of Rice Husk Ash on Volume Changes and Compressive Strength of Fly Ash-Portland Cement Mortar. *A master thesis submitted to Asian Institute of Technology*, Bangkok, Thailand, Thesis No. ST-90-13.

10 INFLUENCE OF EXPANSIVE ADDITIVES ON AUTOGENOUS SHRINKAGE

A. Hori and M. Morioka
Omi Plant of Denkikagaku Kogyo Co., Ltd, Niigata, Japan
E. Sakai and M. Daimon
*Tokyo Institute of Technology, Inorganic Material Engineering Course of
Department of Engineering, Tokyo, Japan*

Abstract
The compensation effects for autogenous shrinkage of two types of high-fluidity mortar containing different binders prepared by adding with two types of expansive additives at various additive content were experimentally studied. The results indicated that any expansive additive has the compensation effect for autogenous shrinkage, especially a free-lime rich calcium sulfoaluminate based-expansive additive has larger compensation effect for autogenous shrinkage than conventional calcium sulfoaluminate based-expansive additives, and that the effect of the mortar prepared with the same additive content of the expansive additive on the expanded volume varies according to the type of binder.
Keywords: Autogenous shrinkage, Expansive additive, High-fluidity mortar

1. Introduction

Since high-fluidity concrete prepared using a high-performance AE water-reducing agent and a segregation-controlling agent (thickening agent) has very high fluidity without segregation of the components, the workability of concrete can be remarkably improved. Such a type of concrete as this is generally prepared at so low water/binder ratio that the powder contents are high. The autogenous shrinkage is, therefore, increased[1].

Meanwhile, the most effective method for reducing the autogenous shrinkage is the shrinkage compensation by adding the expansive additives. This is because the autogenous shrinkage can be extremely reduced by the autogenous expansion of expansive hydrate produced at the beginning of the hydration[2].

This paper presents the experimental study on the compensation effect for autogenous shrinkage by adding an expansive additive to various types of high-fluidity mortar.

Autogenous Shrinkage of Concrete, edited by Ei-ichi Tazawa. Published in 1999 by E & FN Spon,
11 New Fetter Lane, London EC4P 4EE, UK. ISBN: 0 419 23890 5

2 Outline of experiment

2.1 Materials used for experiment

Ordinary Portland cement and high belite cement were used as cement and blast furnace slag powder was used as the admixture. The standard sand for the compressive strength test of cement in accordance with ISO 5202 was used as the fine aggregate. A polycarboxylic acid based-high-performance AE water-reducing agent was used as the high-performance water-reducing agent. Two types of expansive additives, Expansive additives A and B, were used in this experiment. Expansive additive A is a free-lime rich free lime-Hauyne-gypsum anhydride based-expansive additive[3] manufactured from industrial materials in a rotary kiln, while Expansive additive B is a commercially available calcium sulfoalurninate based-expansive additive.

The physical properties, chemical compositions, and mineral compositions of the materials used in the experiment are listed in Tab l e 1.

Tab l e 1. Physical properties, chemical compositions, and mineral composition

Material	Sp. gravity	Sp. surface area (cm^2/g)	Ig.loss (%)	Free lime (%)
Ordinary Portland cement	3.16	3270	2.1	
High belite cement	3.22	3470	0.8	
Blast-furnace slag powder	2.90	5960	0.0	
Expanssive Additive A	3.03	2970	0.9	48.6
Expansive Additive B	2.86	3010	1.1	19.0

Material	Chemical composition (%)						
	SiO$_2$	Fe$_2$O$_3$	Al$_2$O$_3$	CaO	MgO	SO$_3$	R$_2$O
Ordinary Portland cement	21.2	3.0	5.2	64.4	0.9	1.9	0.64
High belite cement	26.4	2.5	3.0	64.1	0.4	2.3	0.16
Blast-furnace slag powder	7.0	0.4	14.2	41.2	7.0		0.39
Expanssive Additive A	1.4	0.5	8.4	68.8	1.2	17.7	0.1
Expansive Additive B	1.2	0.6	16.1	51.3	1.1	27.5	0.1

Material	Mineral composition (%)			
	C$_3$S	C$_2$S	C$_3$A	C$_4$AF
Ordinary Portland cement	29	49	5	10
High belite cement	53	22	9	9

2.2 Mix

Two types of mortar were prepared at a water/binder ratio of 35% and a sand/binder ratio of 2 using different binders and admixtures as shown in Table 2. The water-reducing agent was added so as to adjust the flow value (no tap) to 280 to 300 mm.

The test specimens were prepared by adding Expansive additives A and B to those two types of mortar within the ranges of 3 to 7% and 7 to 11%, respectively, and submitted to the experiment. The ratio of the expansive additive to the binder was calculated by the inner wt%. The test specimen of each mortar without containing the expansive additive (hereafter referred to as plain mortar) was prepared to submit to the experiment for comparison.

Table 2. Types of mortar

Type of mortar	Cement used	Admixture (additive content)
BS	Ordinary Portland cement	Blast-furnace slag powder (40%)*
HB	High belite cement	—

*:Inner wt.% to cement

2.3 Mixing method

Mortar was mixed with additives using a Hovert type mixer in accordance with ISO. The temperature of mixed material was adjusted to 20℃.

2.4 Measuring method of autogenous shrinkage

2.4.1 Before the age of 24 hours

Each test specimen of mortar was placed to a mortar-bar type frame with the bottom covered with a Teflon sheet and the inner surface covered with polyester film. Immediately after that, the surface of mortar was covered with polyester film and a wet straw mat to prevent it drying for aging. The variation of the length of the test piece was measured with a non-contact laser

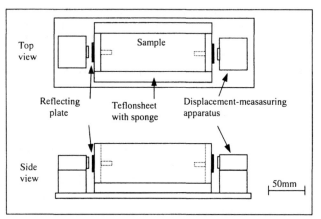

Fig. 1   Rough sketch of apparatus

displacement-measuring apparatus (manufactured by Keyence, resolution power: 2 (μ m) in a room kept at 20℃. The measuring was started at the age of four hours. The rough sketch of the measuring apparatus is illustrated in Fig. 1. The variation of the length of test piece is determined by measuring the distance between the displacement-measuring apparatus and the reflecting plate using laser. Since the measurements can be directly input to a computer, the continuous automated measurement can be made.

2.4.2   After the age of 24 hours

The autogenous shrinkage for a long period of time after the age of 24 hours was measured in accordance with a report of Autogenous Shrinkage Committee[4].   The test specimens used for this measurement were prepared separately from those used in the measurement described in 2.4.1.   The test specimens were being aged in a thermo-hygrostat kept at $20 \pm 1^{\circ}\text{C}$ and $80 \pm 5\%\text{R.H.}$ during the measuring period of time.

2.5   Measuring method of compressive strength

The compressive strength was measured using the test specimens of the age of 28 days in accordance with JIS A 1108.

3.   Experimental results

3.1   Measurements of autogenous shrinkage

The measurements of the autogenous shrinkage by types of mortar are illustrated in Figs. 2 and 3.   The axes of abscissas and ordinates in those figures represent the age and the autogenous shrinkage, respectively.   The solid line and broken line in the figures represent the measurements of the mortar with the expansive additive and of the plain mortar, respectively, and the letters pointing to the graphs with arrows represent the additive contents of expansive additive.

Properly speaking, it is necessary to discuss the autogenous shrinkage characteristics of the test specimens with different expansive additives after the hydration including the binder has been quantitatively studied and the hydration-dependent shrinkage have been elucidated.   It is, however, important to concretely show the effect of the addition of expansive additive on the design of a structure using a material containing the expansive additive.   In this experiment, therefore, the "autogenous shrinkage compensation" was defined as the evaluation method for the addition of expansive additive (eq.1).   It was only aiming at the variation of the length change of the specimen.

Autogenous shrinkage compensation =
   (Autogenous shrinkage of the specimen with expansive additive)
      − (Autogenous shrinkage of it without expansive additive)   (eq.1)

Fig. 2 illustrates the measurements of the autogenous shrinkage of blast furnace slag (BS) mortar with the expansive additive.   The autogenous shrinkage of plain mortar at the age of 91 days is as large as 500×10^{-6}.   Maybe this is because the constitution of the hardened cement is densified by using the blast-furnace slag powder with the fineness as large as 6000 cm^3/g for preparing high-fluidity mortar.   The mortar with the expansive additive autogenously expands until the age of about three days, and after that the expanded volume is reduced by the gradual proceeding of the autogenous shrinkage.   Compared with the plain mortar, however, the compensation effect for the autogenous shrinkage is clearly observed.   It is elucidated that the compensation effect for the autogenous shrinkage of the mortar even added with Expansive additive A at lower additive content than Expansive additive B is the same as each other.   This is the same result described in the previous paper[3].   Since the high-fluidity concrete contains generally much quantity of binder, the contents of the expansive additive must be increased. The autogenous shrinkage of high-fluidity concrete can be compensated by adding the same quantity of such an effective expansive additive as expansive additive A as-to normal concrete.

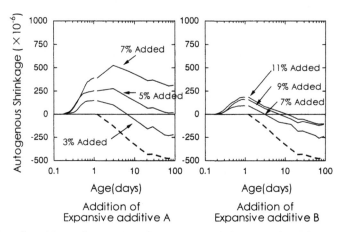

Fig. 2. Measuring results of autogenous shrinkage using BS mortar

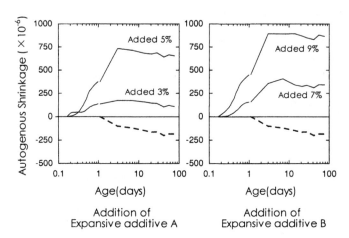

Fig. 3 Measuring results of autogenous shrinkage using HB mortar

Fig. 3 illustrates the measurements of autogenous shrinkage of HB mortar with the expansive additive. The autogenous shrinkage of plain mortar is as small as approximately 200×10^{-6} at the age of 91 days, which is remarkably smaller than that of BS mortar. HB mortar with the expansive additive causes the autogenous expansion at the early age in the same way as BS mortar. Unlike BS mortar, however, autogenously expanded volume is hardly reduced by the autogenous shrinkage. For example, BS mortar with 5% of Expansive additive A shows the difference in the autogenous shrinkage as large as approximately 250×10^{-6} between the ages of 3 and 91 days, while HB mortar with 5% of Expansive additive A shows that as small as approximately 70×10^{-6}. The autogenous expansion of HB mortar content with 5% of Expansive additive A is larger than that of BS mortar under the same conditions. Since the autogenous expansion of high-fluidity mortar prepared from high belite based-cement is increased by adding the expansive additive and moreover the expanded volume is hardly reduced, it is considered that the application of the mortar to concrete components to be chemical prestress is effective.

Fig. 4 illustrates a diagram having the axes of abscissas and ordinates showing the additive content and the autogenous shrinkage compensation at the age of 28 days, respectively. The figure reveals that the effect of the addition of expansive additive to HB mortar on the autogenous shrinkage compensation is more conspicuous than that to BS mortar. The effect of the addition of Expansive additive B to BS mortar on the autogenous shrinkage compensation is remarkably smaller than that of Expansive additive A. It is, therefore, considered that the addition of Expansive additive A to BS mortar is more effective than that of Expansive additive B. Although the addition of Expansive additive A to HB mortar is effective for the autogenous shrinkage compensation even at low rate in the same way as for BS mortar, the effect on HB mortar is so large that it is practically apprehended that a very small weighing error fatally affects the concrete skeleton. It is, therefore,

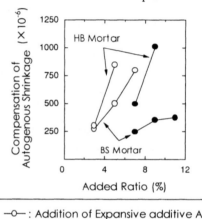

Fig. 4 Relationship between content of expansive additive and autogenous shrinkage compensation

considered that Expansive additive B with relatively little effect of the addition is appropriate as the expansive additive for HB mortar. A paper[5] reports that the neutralization of concrete containing blast furnace slag powder is inhibited by adding the concrete with a free lime rich free lime-Hauyne-gypsum anhydride based-expansive additive (the same system as Expansive additive A). It is, therefore, important to choose an expansive additive, cement or admixture according to the purpose.

The autogenous expansion of HB mortar with any of Expansive additive A and B is larger than that of BS mortar with the same quantity of any of them. It is, therefore, considered that this phenomenon does not depend upon the type of the expansive additive but the type of cement or binder. Maybe this is because either the generating of the expansion pressure coincides with the strength development of the hardened mortar or the quantity or type of the expansive hydrate is different between BS and HB mortars. This problem will be necessary to be further studied in the future after all.

The high belite cement used in this experiment is normal low-heat value cement. In addition to that, various types of high belite cement containing different contents of C_2S are commercially available. And low-general purpose property blast-furnace slag powder is used in the experiment. Then, it is necessary to examine whether or not the high belite cement with different mineral compositions and the blast furnace slag cement with different finenesses and additive contents can obtain the same result as that mentioned above.

3.2 Compressive strength

Fig. 5 illustrates the relationship in BS and HB mortars at the age of 28 days between the compressive strength and the autogenous shrinkage. The compressive strengths of both types of mortar decrease with the increase of autogenous expansion. The strength of BS mortar is, however, higher than that of HB mortar at the same autogenous expansion. It is, therefore, considered that concrete contained blast-furnace slag powder is applicable to the uses requiring high fluidity and high strength.

It is considered that concrete prepared by adding inorganic powder, for example, lime stone powder and fly ash, as well as blast-furnace powder so as to show the same compressive strength as that of high belite cement inhibits the autogenous shrinkage, it will be necessary to make an experiment from such a point of view as this in the future.

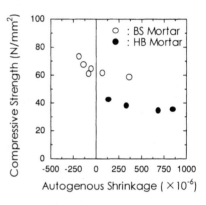

Fig. 5 Comparison of compressive strength with autogenous shrinkage

4. Conclusions

The autogenous shrinkage of various types of high-fluidity mortar with the expansive additives was investigated. The following conclusions were obtained from the investigation results:

The addition of the expansive additive is effective for the autogenous shrinkage compensation of high-fluidity mortar. A free-lime rich calcium sulfoaluminate based-expansive additive is equally effective for the autogenous shrinkage compensation even at low additive content to conventional calcium sulfoaluminate based-expansive additives.

The autogenous shrinkage compensation by the addition of expansive additive depends upon the types of binder. It was elucidated that the compensation effect for the autogenous shrinkage of mortar with a binder containing blast-furnace slag is small, while that of mortar prepared from high-belite cement is large.

Since the compressive strength of mortar prepared from blast-furnace cement is higher than that of mortar prepared from high belite cement, the former mortar is considered applicable to the uses requiring high strength and high fluidity.

Acknowledgments
We acknowledge the very special assistance received from Keyence Co., Ltd., in particular Messrs. Kawakubo and Hoshino and Mr. Fujita of JIG-Design Co.

References
1. Tazawa, E., Miyazawa, S., Sato, T., and Konishi, K., (1992), Autogenous Shrinkage of Concrete, Proceeding of Japan Concrete Institute, Vol.14, No.1, pp.561-566
2. Tazawa, E. and Miyazawa, S., (1992), Autogenous Shrinkage Caused by Self Desiccation in Cementitious Material, 9th International Congress on the Chemistry of Cement, IV, pp.712-718
3. Morioka, M., (1997), Expansive Properties of Expansive Additives in CaO-$4CaO_3Al_2O_3SO_3$-$CaSO_4$ System, Proceeding of Japan Concrete Institute, Vol.19, pp.271-276
4. The Comittee of Autogenous Shrinkage (Chairman: E.tazawa), (1996), Report of The Comittee of Autogenous Shrinkage, pp.195-198
5. Morioka, M., Nikaido, Y., Kubotoa, K., and Asaga, K., (1996), Neutralization of Hardened Blast-Furnace-Slag Cements with some Additives, Proceeding of Japan Concrete Institute, Vol.18, pp.741-746

11 FACTORS AFFECTING THE AUTOGENOUS SHRINKAGE OF SILICA FUME HIGH-STRENGTH CONCRETE

J.J. Brooks, J.G. Cabrera and M.A. Megat Johari
CEMU, School of Engineering, University of Leeds, Leeds, UK

Abstract
A laboratory investigation into the factors affecting the autogenous shrinkage of high strength silica fume concrete is presented. The parameters investigated were water/binder ratio and silica fume content. It was found that greater part of total shrinkage of the concrete is contributed by autogenous shrinkage particularly at a lower water/binder ratio and at higher silica fume content. The autogenous shrinkage was found to be a polynomial function of silica fume content. In addition, good polynomial correlation was obtained between autogenous shrinkage and porosity.
Keywords: Autogenous shrinkage, silica fume, high strength concrete, drying shrinkage.

1 Introduction

Autogenous shrinkage is a phenomenon whereby hardened concrete or cementitious materials shrink due to self-desiccation resulting from the hydration of cementitious materials. This type of shrinkage occurs under no moisture movement to or from the surrounding environment. Thus, the measurement of autogenous shrinkage is performed on sealed concrete specimen. Davis [1] found that the autogenous shrinkage of ordinary Portland cement concrete used for construction of dams was in the range of 50 to 100 x 10^{-6}. However, for high strength concrete with very low water binder ratio, the total amount of autogenous shrinkage could be more than 700 x 10^{-6} [2]. The cracking of restrained high strength silica fume concrete under sealed condition was attributed to an intense autogenous shrinkage [3]. In addition, autogenous shrinkage was also considered to be one of the causes of cracking observed in the high strength silica fume concrete used for repairing the stilling basin of the Kinzua Dam [4].

Autogenous Shrinkage of Concrete, edited by Ei-ichi Tazawa. Published in 1999 by E & FN Spon,
11 New Fetter Lane, London EC4P 4EE, UK. ISBN: 0 419 23890 5

The creep of high-strength concrete has been demonstrated to be greatly influenced by autogenous shrinkage [5]. The higher basic creep reported for high-strength concrete was due to the concomitant autogenous shrinkage. The same phenomena applies to the higher total creep of normal strength concrete, which is due to a higher concomitant drying shrinkage. The influence of autogenous shrinkage on the basic creep of high performance concrete has been reported to be particularly significant at a low stress level [6].

Previous research indicates that autogenous shrinkage is influenced by water/binder ratio [2] and dosage of silica fume inclusion [7]. Other factors, i.e. cement type and fineness, mineral composition of cements, inclusion of mineral admixtures as well as chemical admixtures and aggregate content have been identified to influence the autogenous shrinkage of cementitious materials [8]. This paper reports the early findings of an investigation into the factors affecting the autogenous shrinkage of silica fume high strength concrete. It is part of an ongoing research dealing with the deformation properties of high strength concrete containing cement replacement materials.

2 Experimental Details

2.1 Materials
The materials used in this investigation are ordinary Portland cement, silica fume, natural river sand, quartzitic gravel with a maximum size of 10 mm, and Sikament 10 superplasticiser, a new generation low alkali water soluble vinyl copolymer based on formaldehyde-free with a specific gravity of 1.11 kg/litre. Mix proportions of 1 : 1.5 : 2.5 (binder : sand : gravel) by mass were used. The control mix was cast using ordinary Portland cement, while three other mixes were prepared by replacing 5, 10 and 15 % of the cement with silica fume on mass-for-mass basis. These concrete mixes are referred to as OPC, SF/5, SF/10 and SF/15 respectively. An optimum water/cement ratio of 0.28 was obtained for the control mix using the Cabrera Vibrating Slump, a method developed by Cabrera and Lee [9]. The same water/binder ratio of 0.28 was used for the mixes containing silica fume. In addition, concrete mixes using water/binder ratios of 0.23 and 0.33 were prepared to investigate the influence of water/ binder ratio.

2.2 Test Procedure
Cylindrical specimens of 267 x 76 mm diameter were used for measuring the autogenous shrinkage as well as drying shrinkage. The specimens for autogenous shrinkage were sealed with aluminium waterproofing tape after they were demoulded. This type of sealing is very effective since the specimens showed no weight loss, which indicates no moisture movement. Measurement of total shrinkage was performed on unsealed specimens. All specimens were stored in a controlled environment of 21°C and 65 % relative humidity. The shrinkage measurement was performed using a mechanical Demec gauge of 200 mm gauge length at four circumferential positions. This measurement commenced at about 24 hours after casting, i.e. after the specimens were demoulded and sealed. Mortar samples were prepared and tested for porosity and

pore size distribution using Mercury Intrusion Porosimetry (MIP).

3 Results and Discussion

The results of the laboratory measurements of autogenous shrinkage are shown in Fig. 1 and 2 for water/binder ratios of 0.28 and 0.23, respectively. After times of between 40 and 80 days the magnitude of autogenous shrinkage was found to range from about 140×10^{-6} to more than 300×10^{-6}. Le Roy and De Larrard [10] reported long-term values of autogenous shrinkage to be between 130 and 250×10^{-6} for high performance concrete containing silica fume and between 100 to 140×10^{-6} for those without silica fume. The higher values obtained in this investigation are due to the lower water/binder ratio and higher silica fume content for SF/10 and SF/15 concrete mixes.

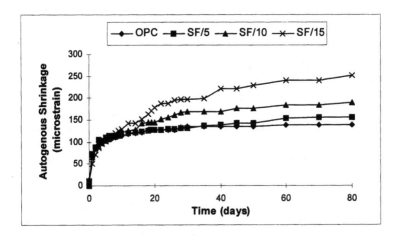

Fig. 1. Autogenous shrinkage of concrete with water/binder ratio of 0.28.

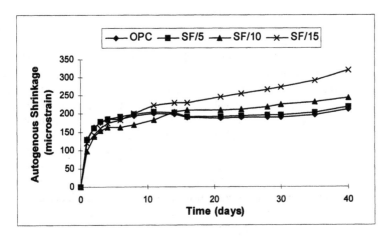

Fig. 2. Autogenous shrinkage of concrete with water/binder ratio of 0.23.

Table 1. Shrinkage of concrete with water/binder ratio of 0.28 at 80 days

Concrete Mix	OPC	SF/5	SF/10	SF/15
Total Shrinkage (10^{-6})	386	372	427	470
Autogenous Shrinkage (10^{-6})	138	155	189	252
Drying Shrinkage (10^{-6})	248	217	238	218

Table 2. Shrinkage of concrete with water/binder ratio of 0.23 at 40 days

Concrete Mixes	OPC	SF/5	SF/10	SF/15
Total Shrinkage (10^{-6})	548	448	438	496
Autogenous Shrinkage (10^{-6})	212	220	244	320
Drying Shrinkage (10^{-6})	336	228	194	176

The summary of results for total, autogenous, and drying shrinkage for concrete mixes with water/binder ratios of 0.28 and 0.23 are given in Tables 1 and 2. Drying shrinkage is taken as the difference between total shrinkage measured on unsealed specimens and autogenous shrinkage. The values of autogenous shrinkage for concrete mixes with water/binder ratios of 0.28 at 80 days represent 36, 42, 44, and 54 % of the total shrinkage for OPC, SF/5, SF/10 and SF/15 concrete respectively. While for concrete mixes with water/binder ratio of 0.23 at 40 days, autogenous shrinkage represent 39, 49, 56, and 65 % of the total shrinkage for OPC, SF/5, SF/10 and SF/15 respectively. This shows that as silica fume content increases, greater part of the total shrinkage is contributed by autogenous shrinkage particularly at the lower water/binder ratio.

From Figs. 1 and 2, it is clear that both water/binder ratio and silica fume content have significant influence on the autogenous shrinkage of concrete. Autogenous shrinkage increases with lower water/binder ratio and higher silica fume replacement. As the water/binder ratio is reduced from 0.28 to 0.23, there is a significant increase in autogenous shrinkage for all concrete mixes. According to de Larrard [11], at lower water/binder ratios, the capillary tension of the pore water increases, therefore increasing the autogenous shrinkage. However, this explanation is not entirely satisfactory, and therefore it is proposed here that the main cause of autogenous shrinkage is the large ionic concentration difference between the pore water and the adsorbed water at the hydrating front. Water moves by suction effect due to this large difference in concentration and produces a large increase in capillary tension, which then causes a large increase of the autogenous shrinkage component of shrinkage. At the testing age of 40 days, the increase in autogenous shrinkage due to the reduction in water/binder ratio from 0.28 to 0.23 was 57, 61, 44, and 61 % for OPC, SF/5, SF/10 and SF/15 concrete mixes, respectively. The inconsistency of trend for the SF/10 concrete may have been due to some variation in the time of starting the test.

The results indicate that partial replacement of cement with silica fume increases the magnitude of autogenous shrinkage particularly at higher replacement level. For concrete mixes with water/binder ratio of 0.28, the effect of silica fume content is to increase the autogenous shrinkage at 40 days by 1.5, 25, and 47 % respectively for 5,

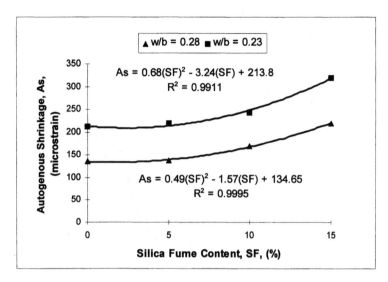

Fig. 3. Correlation between autogenous shrinkage of concrete and silica fume content at 40 days.
R^2 = coefficient of determination

10 and 15 % silica fume content. While for the concrete mix with water/binder ratio of 0.23, the increase is found to be 3.8, 15, and 51 % respectively. The general trend is that autogenous shrinkage increases as silica fume content increases. Figure 3 shows that at both water/binder ratios, there is a good polynomial correlation between autogenous shrinkage and silica fume content.

The inclusion of silica fume has been reported to refine the pore size distribution in the concrete [12]. This refinement of pores increases the capillary tension, the suction potential and the autogenous shrinkage. Furthermore, as in the case of cement hydration, the "pozzolanic" reaction of silica fume also leads to an increase in autogenous shrinkage. The pozzolanic reaction has been reported to be not as sensitive to self-desiccation as cement during hydration [7]. It has also been reported that about 24 % of silica fume by weight of cement is required to eliminate the calcium hydroxide content in the hardened silica fume-cement paste system [12]. The $Ca(OH)_2$ consumption depends on the curing temperature. Cabrera and Claisse [13] have shown that at 20°C and 100% RH, the $Ca(OH)_2$ of hydrated opc mix reacts entirely with 20 % SF at 90 days of age, however curing the OPC-SF mix at 6°C in water, the pozzolanic reaction is barely detectable. Therefore, the concrete mixes containing silica fume particularly at higher level are expected to undergo prolonged autogenous shrinkage. The trend of the curves in Fig.1 seems to agree with this expected phenomena.

The results of MIP test (Fig. 4 and 5) performed on mortar samples with water/binder ratios of 0.28 at the age of 28 days, show that as the level of silica fume replacement increases, the pore size distribution is shifted toward finer distribution values, the average pore size is reduced and the porosity decreases. The porosity was found to be a linear function of silica fume content as shown in Fig. 5. In addition, there is a good polynomial correlation between autogenous shrinkage at 28 days and porosity as shown in Fig. 6.

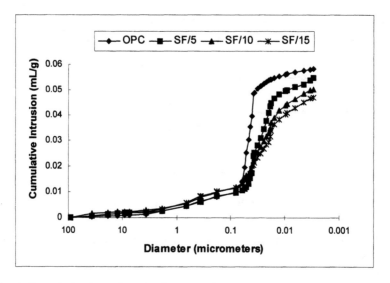

Fig. 4. Cumulative intrusion Vs diameter of pores with and without silica fume.

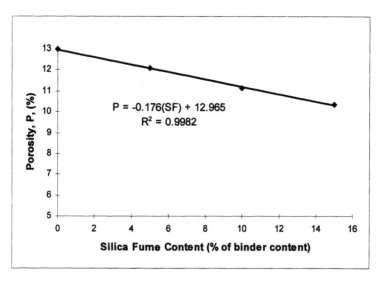

Fig. 5. Correlation between porosity and silica fume content of mortar.

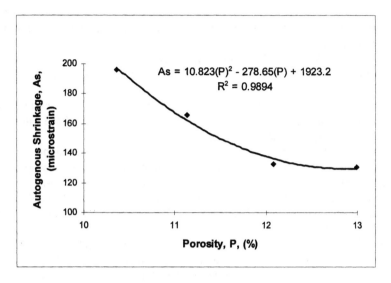

Fig. 6. Correlation between autogenous shrinkage of concrete and porosity of mortar.

4 Conclusions

From the results presented in this paper, the following conclusions are offered:
1. Both water/binder ratio and silica fume content play significant roles in the autogenous shrinkage of concrete. For the same water/binder ratio, autogenous shrinkage of concrete can be expressed as a polynomial function of silica fume content.
2. The greater part of the total shrinkage of high strength concrete is contributed by autogenous shrinkage particularly at a lower water/binder ratio and higher silica fume content.
3. There is a good correlation between autogenous shrinkage and porosity, and porosity and silica fume content. As the silica fume content increases, the pores become smaller.

5 References

1. Davis, H.E. (1940) Autogenous volume change of concrete. Proceeding ASTM, 40, pp. 1103 - 1110.
2. Tazawa, E. and Miyazawa, S. (1993) Autogenous shrinkage of concrete and its importance in concrete technology. The Fifth International SP-91, pp. 841 - 863.
3. Brooks, J.J. and Hynes, J.P. (1993) Creep and shrinkage of ultra high-strength silica fume concrete. The Fifth International RILEM Symposium on Creep and Shrinkage of Concrete. E&FN Spon, London, pp. 493 - 498.
4. Persson, B. (1997) Basic deformation of high-performance concrete. Nordic Concrete Research. Publication No. 20, pp. 59 - 74.

5. Jensen, O..M. and Hansen, P.F. (1996) Autogenous deformation and change of relative humidity in silica fume-modified cement paste. ACI Materials Journal, V. 93, No. 6, pp. 539 - 543.
6. Tazawa, E. and Miyazawa, S. (1997) Influence of constituent and composition on autogenous shrinkage of cementitious materials. Magazine of Concrete Research, 49, No. 178, pp. 15 - 22.
7. Cabrera, J.G. and Lee, R.E. (1985) A new method for the measurement of workability of high pulverised fuel ash concrete. Proceedings of the International Ash Utilization Symposium, Vol. 1, pp. 347 -360.
8. Le Roy, R. and De Larrard, F. (1993) Creep and shrinkage of high-performance concrete: the LCPC experience. The Fifth International RILEM Symposium on Creep and Shrinkage of Concrete. E&FN Spon, London, pp. 499 - 504.
9. De Larrard, F. (1991) Creep and shrinkage of high strength field concrete. CANMET/ACI International Workshop on the Use of Silica Fume in Concrete, Washington D.C. 22pp.
10. Sellevold, E.J. (1987) The function of silica fume in high strength concrete. Utilization of High Strength Concrete, Stavanger, Norway, pp. 39 - 50.
11. Cabrera, J.G. and Claisse, P.A. (1991) The effect of curing conditions on the properties of silica fume concrete. Blended Cements in Construction, Elsevier, U.K., pp. 293 - 301.

12 MEASUREMENTS OF AUTOGENOUS LENGTH CHANGES BY LASER SENSORS EQUIPPED WITH DIGITAL COMPUTER SYSTEM

M. Morioka, A. Hori and H. Hagiwara
Omi Plant of Denkikagaku Kogyo Co., Ltd, Niigata, Japan
E. Sakai and M. Daimon
Dr. of Engineering, Tokyo Institute of Technology, Inorganic Material Engineering Course of Department of Engineering, Tokyo, Japan

Abstract
A measuring method of an autogenous length change by using laser sensors equipped with a digital computer system was proposed. This method is excellent in accuracy and reproducibility, and moreover measurement can be done automatically and continuously by applying it. Simultaneously laser sensors and thermocouples in this method are used, the correction of a change in length for temperature can be carried out simply and easily. This method is effective for quality control, and on top of that, it is suitable especially for measuring an autogenous length change of special cement such as expansive cement and rapid hardening cements.
Keywords: Autogenous length change, Quality control, Laser sensor, Digital computer system, Non-contact way

1 Introduction

The real circumstances at present are that an autogenous length change of special cement or cement additives and so forth is not yet made clear. Expansive cement presents autogenous expansion at early age along with hydration [1][2]. Rapid hardening cement generates a large quantity of hydration heat from initial setting [3]. Therefore, in order to discuss an autogenous length change of these special cements, that measurement of expansion can be carried out, and moreover correction of heat are able to be executed, are required. A dial gauge methods as a measuring method of an autogenous length change [4]. However, because this method was in need of skill, it turned frequently out a failure especially in a period of 24hours from initial setting. It reasons was that this method was a contact type, and in addition measurement was performed by manual work.

Autogenous Shrinkage of Concrete, edited by Ei-ichi Tazawa. Published in 1999 by E & FN Spon,
11 New Fetter Lane, London EC4P 4EE, UK. ISBN: 0 419 23890 5

A non-contact type method by using inductive gauging sensor was proposed by Takahashi et al in 1995[5]. By this method, measurement can be executed automatically and continuously. However, input work of massive data obtained through measurement into a computer had to be done by manual work.

In addition, measuring a temperature change could not be carried out together with a length change. Accordingly, we have contrived a measuring devise and mold for an autogenous length change (not only shrinkage, but also expansion), and furthermore have devised so that correction of a change in length caused by heat could be done simply and easily. In this paper, a measuring method of an autogenous length change by using laser sensors equipped with a digital computer system is investigated.

2 Measuring device

2.1 Features of the laser sensor
Laser sensors (Type:LB-02, produced by Keyence corp.) are used. Features of a laser sensor are shown in table-1 in comparison with those of an inductive gauging sensor used by Takahashi et al.. For a laser sensor, because a distance between a sensor-head and object (reference distance) is sufficiently long, placing a sample and object is easy compared with inductive gauging. In addition, also a measuring range is broad enough. Therefore, measuring a large sample, like concrete is possible. An object is not limited only to metal. (For example, ceramics, paper, plastics, glasses, and so on). An angle between a sensor-head and object is not limited. When we use inductive gauging sensors, it is necessary for that an angle between a sensor-head and object is vertical, and otherwise, measuring accuracy can not be maintained.

As mentioned above, laser sensors are in good accord with a demand for measuring an autogenous length change for the purpose of doing quality control, and in consequence they are more useful than inductive gauging sensors.

Table-1 Features of laser sensor and inductive gauging sensor

Kind of sensor	Laser	Inductive gauging
Reference distance	40mm	1mm
Measuring range	± 10mm	± 1mm(0 ~ 2mm)
Objects	Not limited	Metal
Resolution	2 μ m	0.03% of full scale
Linearity	1.0% of full scale	0.3% of full scale
Output(analogue)	± 4v(0.4v/mm)	0 ~ 5v(2.5v/mm)
Angle	Not limited	Vertical

2.2 Outline of device
Fig.1 shows an outline of laser sensors equipped with a digital computer system. Analogue data (mv), which are output from laser sensors and amplifiers, are converted to digital data by PC-card. This PC-card has large memory capacity (256k word), and moreover has high resolution (14bit). Upon this, digital data are taken directly in cells of table calculation software for note type personal computer by data logger software. Also analogue data (mv), which are output from thermocouples (K-

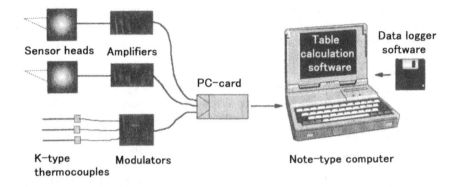

Sensor heads   Amplifiers

K-type   Modulators
thermocouples

PC-card

Table
calculation
software

Data logger
software

Note-type computer

Fig. 1 Outline of laser sensors with digital computer system

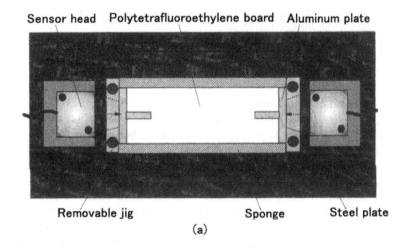

Sensor head   Polytetrafluoroethylene board   Aluminum plate

Removable jig   Sponge   Steel plate

(a)

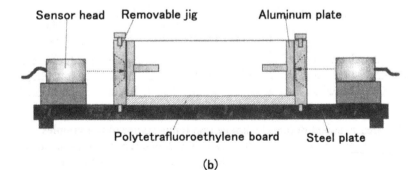

Sensor head   Removable jig   Aluminum plate

Polytetrafluoroethylene board   Steel plate

(b)

Fig. 2 Outline of the mold (a): over view, (b): side view

type Thermocouples) and modulators, are similarly converted to digital data by a PC-card. Accordingly, measuring an autogenous length change and temperature change can be carried out at the same time.

2.3   Mold

Fig.2 illustrates outline of the mold. The mold is so designed that size of a sample becomes to be 40 x 40 x 160 mm. In order to measure a sample with the length of 160 mm from the both sides in a longitudinal direction, 2 sensor-heads were installed for 1 mold. These sensor-heads are perfectly fixed on a steel plate. Hereupon, a position of sensor-heads is kept at 40mm, which is a reference distance, apart from an object. Therefore, ±10mm , which was a full scale of measuring range could be secured. In order to reduce friction resistance from a mold [5], a polytetrafluoroethylene (Teflon) board was laid at a base of the mold. 2 of sponges with Teflon sheet were arranged, so as not obstruct free expansion. Jigs were so designed that a hole is opened in the middle of jigs at measuring face, and moreover the jigs can be removed if necessary. Incidentally, the jigs are not required to be removing when autogenous shrinkage is measured. In the case of measuring autogenous expansion, measurement is possible by removing the jigs at initial setting.

3   Experiments

3.1   Material used and mixproportion

Calciumsulphoaluminate-type expansive additive (CSA) in order to evaluate measurement of autogenous expansion was used. Furthermore, in order to evaluate correction of hydration heat, rapid hardening cement (RHC) was used. Chemical compositions and physical properties of materials are shown in table-2. RHC contains calcium aluminate glass and anhydrite, and it forms great deal of Ettringite [6]. CSA was evaluated by using CSA mortar and expansive cement (EC) mortar with a water/binder ratio = 50 % and binder / sand (ISO 679) ratio = 1/3. Binder in CSA mortar is CSA only. And binder in EC mortar was prepared by 10 wt% of CSA with OPC of 90wt%. RHC was evaluated by using a paste with a water/binder ratio = 35 %.

Table-2   Chemical compositions and phisycal properties of materials used

Kind of binder	Chemical composition							Blaine (cm^2/g)	Gravity (g/cm^3)
	SiO_2	Al_2O_3	Fe_2O_3	CaO	MgO	TiO_2	SO_3		
CSA	1.2	16.1	0.6	51.3	1.1	0.2	27.5	3010	2.86
RHC	15.5	9.4	2.1	58.0	0.6	0.5	10.0	4730	3.02
OPC	20.9	5.4	2.9	64.7	0.9	0.1	1.8	3310	3.14

3.2   Measurement of length change and temperature of the samples

Interval of measurement was set at 5minutes. Measurement for 24 hours after start of initial setting was carried out. Sampling of data was performed by data logger software automatically and continuously. Accordingly, manual work is only placing of a sample. Initial setting of CSA mortar was 45minutes after placing, and initial setting of EC mortar was 5 hours and 45minutes. Initial setting of RHC paste was 1 hour. Because

CSA mortar was prepared only by using an expansive additive (CSA) and sand , in order to observe further larger expansion, start of it's initial setting was substantially fast. Since a retarder was added in RHC paste, sufficient handling time was acquired. In order to prevent drying, a sample was wrapped by polyethylene sheet, and furthermore wet towel was put on it while measuring. Thermocouples were installed in the middle of a sample, and subsequently temperature of sample was measured.

3.3   Calculation
All the calculation was executed by table-calculation software automatically and continuously. First of all, digital data, which were input as a difference in voltage (mv), were converted to a length change (mm) or temperature change (°C) through a computer. In the case of a length change, 0.4v corresponds to 1 mm, and on the other hand, in the case of a temperature change, 0.1v corresponds to 1 °C. If a formula is designated in advance, calculation will be made automatically and continuously.

Calculation of strain, temperature changes, and correction for thermal expansion was carried out as shown in the following.

Correction value of thermal expansion = (ΔT) x (thermal expansion coeffic.)   (eq.1)

Strain after correction =   (measuring strain) – (correction value)   (eq.2)

4   Results

4.1   Measuring precision
Fig.3 shows strain of CSA mortar and EC mortar measured by 2 sensors (sensor-A and sensor-B) respectively. As for both of CSA mortar and EC mortar, the measurement results of sensor-A and sensor-B approximate very well.

In other words, it means that equal length changes are occurring in an axial direction, due to extremely small friction resistance between sample and the mold. In the case of dial gauge method, which has been a conventional measuring method of an autogenous length change, strain of an equal amount could not been frequently observed while measuring of an axis from 2 directions, especially at early age.

It reasons are thought to be that a dial gauge method has points at issue as mentioned below.

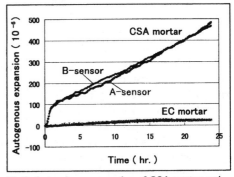

Fig.3 Autogenous expansion of CSA mortar and EC mortar measured by 2 sensors respectively

(1) Because a dial gauge method is contact type, repulsive force is generated. (2) It is rather difficult to fix a dial gauge completely. (3) Since measurement is performed by manual work in the case of a dial gauge method, there is a limit to secure linearity

between a gauge and plug. (4) Manual work produces human measuring mistakes.

On the other hands, in a method of this study, due to an anti-contact type, repulsive force is not generated. Friction resistance is remarkably small, because a Teflon board is laid at a base of the mold, and in addition, 2 sponges with a Teflon sheet are placed at the both sides of the mold. In consequence, linearity while measuring can be secured. No human measuring mistake will occur, because measurement is automatically carried out by table-calculation software of a note type personal computer. It is considered in this method that problems of a conventional method were dissolved. Fig.4 shows dispersion of measurement values when measuring autogenous expansion of CSA mortar is executed over again. Strains in Fig.4 show sum of strain measured by 2 sensors.

From these results, it is found that dispersion is remarkably small, when measuring autogenous expansion of CSA mortar is carried out in 3 times repeatedly. It is quite clear that measuring an autogenous length change at early age can be performed very well in reproducibility, when a method of this study is applied. This fact is significantly important in the case of carrying out quality control of concrete by introducing measuring results of an autogenous length change into cracking analysis.

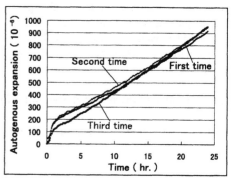

Fig.4 Dispersion of measurement values when measuring autogenous expansion of CSA mortar is executed over again

4.2 Correction of thermal expansion

Fig.5 shows an autogenous length change of RHC paste. From these results, it was turned out that an autogenous length change of RHC paste presents expansion from initial setting, then it gets up to 150×10^{-6}, but subsequently it commences to shrink at 15 hours of measuring time.

Fig.6 shows temperature change (ΔT) of RHC paste from initial setting. A temperature change of RHC paste rose drastically at early age, and then went down. Hereupon, because the temperature of a sample at initial setting (T_I) was higher than that of a sample while paste was prepared, (ΔT) had a – value. In other words, RHC paste had already been in exothermic reaction at a point of time of initial setting. An amount in a temperature change arrived at about 11 degree from the maximum value to the

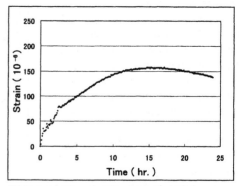

Fig.5 Autogenous length change (strain) of RHC paste

minimum value.

Therefore, measured values for an autogenous length change have to be corrected by thermal expansion. An autogenous length change, in which correction had been made about thermal expansion, was shown in Fig.7. From this figure, it is understood when correction has been made about thermal expansion, that an autogenous length change results in shrinkage just at very early age, and thereafter comes in effect to expand. With respect to stress analysis, it is very important to fractionate stress caused by an autogenous length change from stress resulted from hydration heat.

In order to calculate stress caused only by an autogenous length change, which has been corrected by ΔT, is indispensable. In this study, it was ascertained that correction of a length change caused by heat could be easily carried out, by using laser sensors and thermocouples concurrently. Therefore, it is thought that this method is quite effective especially for a large sample like concrete, which is accompanied with hydration heat. In addition, because data can be input directly into table calculation software in the case of a method of this study, also a volume change can be simply and easily calculated, if a formula has been designated beforehand. As explained above, a method of this study is quite suitable for measuring an autogenous length change of special cement such as expansive cement, rapid hardening cement and so forth. Furthermore, this method is superior in accuracy and reproducibility, and moreover measurement can be performed automatically and continuously by applying it.

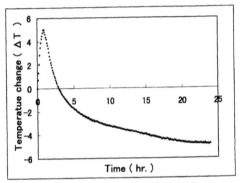

Fig.6 Temperature change (ΔT) of RHC paste from initial setting

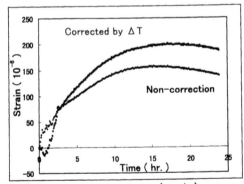

Fig.7 Autogenous length change (strain) of RHC paste corrected by ΔT.

It is confirmed that this method can be effectively applied for quality control.

5 Conclusions

A measuring method of an autogenous length change by laser sensors equipped with a digital computer system have been proposed. In consequence, by using this devise, measuring an autogenous length change of expansive additive mortar and rapid hardening cement paste was carried out.

This method is excellent in accuracy and reproducibility, and moreover measurement can be executed automatically and continuously by applying it, in the case of this method, a change in length caused by heat can be easily corrected, by using thermocouples together with laser sensors.

This method is very useful as a method for carrying out quality control of cement concrete, by introducing the measuring results of an autogenous length change into cracking analysis, and furthermore it is quite effective especially for measuring an aoutogenous length change of special cement such as expansive cement and rapid hardening cements and so forth.

Acknowledgement
Messrs. Kawakubo and Hoshino of the Keyence Corp., as well as, Mr. Fujita of the Jig-design Co. kindly contributed us by giving considerable collaboration. We wish to express here our deep indebtedness to them to their kindness.

References
1. Tazawa, E., and Miyazawa, S.,(1995) Influence of Cement and Admixture on Autogenous Shrinkage of Cement Paste, Cement and Concrete Research, Volume 25, No.2, pp.281-287
2. Hori A., Morioka M., Sakai E., and Daimon M., (1998) Influence of Expansive Additives on Autogenous Shrinkage, JCI, Autoshrink'98
3. Nikaido Y., Ito T., Sakai E., and Daimon M., (1995) Ettringite formation and micro- structure of rapid hardening cement at early hydration, Proceedings of JCI, Vol.17, No.1, pp.319-324
4. Report of the Committee of Autogenous shrinkage Japan Concrete Institute, (1996) pp.195-198
5. Takahasi T., Tanaka H., Goto S., (1996) Trial of Devise of Autogenous Shrinkage for Cement Paste, Proceedings of JCI, Vol.18, No.1, pp.621-626
6. Kondo N., Nikaido Y., Nakaya S., Handa M., Ando T., Sakai E., and Daimon M., (1997) Relationship Between Etrringite Formation and the Development of Strength for Rapid Hardening Cement, 10th International Congress on the Chemistry of Cement, Volume 2, 2ii017

13 CHEMICAL SHRINKAGE OF CEMENT PASTE, MORTAR AND CONCRETE

H. Justnes and T. Hammer
SINTEF, Norway
B. Ardoullie, E. Hendrix, D. Van Gemert and K. Overmeer
K.U. Leuven, Belgium
E.J. Sellevold
NTNU, Trondheim, Norway

Abstract
The total chemical shrinkage and the volumetric autogenous shrinkage have been determined experimentally for cement pastes incorporating different amount of the following additions: silica fume, fly ash, a silicate mineral filler fine (< 125 mm), medium (< 250 mm) and coarse (< 2 mm). The inert fillers were found to have little effect on the total chemical shrinkage; i.e. on the rate of the reaction of the cement. The effect on autogenous shrinkage is also minor; i.e. the implication is that volume change data determined on pastes are also relevant to concrete. The active additions silica fume and fly ash influence both types of volume change, as expected. Linear and volumetric measuring methods on the concrete are compared.
Keywords: chemical shrinkage, filler, mortar, paste, sand, shrinkage measuring methods, silica fume.

1 Introduction and background

High performance concrete (i.e. concrete with water-to-binder ratio below around 0.4) has been shown to be sensitive to early cracking, and this sensitivity has been postulated to relate to early volume change [1]. Many techniques have been explored to measure early volume change both in cement paste and concrete. For concrete linear measurements are most common. One feature is puzzling when comparing paste/concrete results, namely that linear measurements commonly display an early expansion phase, while this phase never have been seen in our volumetric work on cement paste. The main purpose of the present work was to start to "bridge the gap" between paste cement and concrete, by making volumetric measurements on paste as reference; and then on modified pastes

Autogenous Shrinkage of Concrete, edited by Ei-ichi Tazawa. Published in 1999 by E & FN Spon, 11 New Fetter Lane, London EC4P 4EE, UK. ISBN: 0 419 23890 5

containing various types and dosages of reactive and non-reactive fillers/sands. An additional purpose was to see if the expansive phase commonly seen in linear measurements on concrete is real (and a product of aggregate/paste interaction), or merely a consequence of the linear measuring technique. The volumetric condom method using specimen rotation in a water bath [1] is considered to give the "truth" about early external volume change.

2 Chemical and autogenous shrinkage

When cement reacts with water the volume of the reaction products is smaller than the sum of the reactants. This phenomenon is called chemical shrinkage, and it may easily be measured by established techniques [1]. Chemical shrinkage has been found to be linearly related to the degree of hydration of the cement, and numerically to correspond to about 25 % by volume of the reacted water.

Autogenous shrinkage is a commonly used term to describe linear changes in concrete dimensions when hardening under sealed conditions with regard to moisture exchange. In the "condom" method [1] the early volume change is measured volumetrically under isothermal conditions, and the results are here named external chemical shrinkage, while the chemical shrinkage is referred to as total chemical shrinkage.

The two quantities are of course closely related: initially with the paste in the liquid state the two are identical. However, during setting a self-supporting skeleton forms in the paste, allowing empty pores to form internally and consequently the external volume change becomes much smaller than the total chemical shrinkage. This transition from liquid to solid we believe to be important for cracking sensitivity and the transition may be characterised by the two types of volume change experiments.

External shrinkage causes deformation of the structure. When these deformations are restraint, additional stresses and forces will act on the structure. Restraint deformations can also lead to fissures and cracks. Fissures and stresses can be predicted and taken into account at the structural design of the construction. The effects of internal shrinkage are not so clear, but cannot be neglected. Internal shrinkage might cause micro fissures. These affect the strength of the material. This phenomenon is much more difficult to take into account in engineering design calculations. However, the material degradation by internal shrinkage induced damages affects the durability of the structure. Therefore, external as well as internal chemical shrinkage are studied.

3 Experiments

3.1 Materials
Earlier studies focussed on the chemical shrinkage of cement pastes and on the chemical shrinkage of concrete. This paper tries to fill the gap between cement and concrete. First, a reference cement paste was prepared. Second, a part of the cement was substituted with silica fume. Third, aggregates, ever more coarse, were added to the reference paste. With silica fume, substitution was preferred, because the assumption was made that silica fume

will react as a binder, in the same way as cement does. Table 1 shows the composition of the different mixes, in weight percentages of cement. A commercial Norwegian cement, Anlegg cement, is used. This is an ordinary Portland cement of type CEM I 32.5, according to the European Standard ENV 197-1.

Table 1. Composition of the different mixes in weight percentages of cement

	cement	Silica Fume	Filler (0 - 0.125 mm)	Filler (0.125 - 0.250 mm)	Sand (0 - 2 mm)	Fly Ash
ref. mix	100	0.0				
mix 1	96.4	2.6				
mix 2	94.7	5.3				
mix 3	89	11				
mix 4	80	20				
mix 5	75	25				
mix 6	100		15			
mix 7	100		15	20		
mix 8	100				38	
mix 9	100					20

The reference mix is a pure cement paste. The water-binder ratio of all mixes is 0.4. Of each mix three wide bottom glass containers, called "Erlenmeyers", were filled for the total shrinkage tests and three rubber bags (condoms) for the external shrinkage tests.

3.2 Mixing procedures
First, the cement was put into a bowl and during the first minute of mixing, the water was added. An extra minute of mixing followed. Second, three minutes of waiting passed, during which the bowl was covered with a towel. Finally, 1.5 minutes of finishing mixing followed in order to avoid false set. A vibrating table was used to keep as less air in the mix as possible. Vibrating was done for maximum five minutes and less when bleeding occurred.

3.3 Testing procedures

3.3.1 Testing procedures for total chemical shrinkage
Each slurry was placed into three 50 ml Erlenmeyer flasks, forming ≈ 10 mm layer, and then it was weighed. The flasks were then topped up with distilled water, avoiding turbulence. A silicon rubber stopper was fixed in each flask and a pipette filled with water was passed through a hole in the stopper: a graded pipette of 0.2, 0.5 or 1 ml was chosen, depending on the expected volume change. The flasks were placed in a water bath at 24 ±1 °C. The position of the meniscus of the pipette was recorded hourly for 48 h. The decrease in level of the water column (in ml) is a direct reading of the total chemical shrinkage (a drop of liquid paraffin on top prevented evaporation), expressed as ml/100 g cement or solids. The method relies on the assumption that all contraction pores are filled with water, which probably is correct at the early ages and for the thin

slurry layers investigated. The method is described and discussed in detail by Justnes et al. [1].

3.3.2 Testing procedures for external chemical shrinkage

Three elastic rubber bags (i.e. condoms) were filled with slurry while they were inserted in a 100 mm plastic tube of 50 mm inner diameter. Each condom was closed by twisting the upper part and tying it with a thin copper wire, and sealed by spraying silicon glue into the open end. The excess end part was cut off and the total mass determined.

The filled and sealed condoms in their tubes were kept in water bath of 24 ±1 °C after the tubes had been turned perpendicular to their axes in order to let all air bubbles escape. The tubes were kept on an ordinary rotating table modified to function under water (i.e. chain transfer between the motor and the rollers placed under water). During the first 8 h the condoms were weighed every hour under water. Relying on the Archimedes principle, external shrinkage will lead to a reduction in buoyancy, which will be registered as weight increase. Each condom was weighted in a basket under water hanging on a scale in a separate water bath, with no stirring to avoid turbulence. The transfer between the rotation bath and the weighing bath took place under water at all times by placing the tube in a water-filled glass under water in one bath and taking it out of the glass in the second bath. This was done to avoid trapping any air bubbles in the transfer process.

After the final weighing under water, the condoms were wiped dry and weighed in air. Finally, the condom, including copper wire and silicon glue, were stripped off and weighed in order to calculate the net weight of the slurry after subtracting the weight of the tube. The external shrinkage is determined as the mean value of three measurements and given as ml/100 g binder or cement. The method is described and discussed in detail by Justnes et al. [1].

4 Mineral admixtures

4.1 Silica Fume

The total and external chemical shrinkage of cement with silica fume replacements of 0.0, 2.6, 5.3, 11, 20 and 25 % has been plotted as ml per 100 g cement in figure 1. Plotted in this way lower, equal or higher shrinkage relative to the reference mix means that the pozzolanic silica fume reaction expands, has no volume change or shrinks, respectively. If it was plotted against 100 g cement and silica the same observations should have been interpreted as the pozzolanic reaction having less, equal or higher chemical shrinkage, respectively, than the cement hydration.

The total chemical shrinkage for the silica fume mixes in figure 1 was measured until about 7 days (168 h). The total chemical shrinkage at 7 days increases with increasing silica fume dosages, revealing that the pozzolanic reaction leads to chemical shrinkage as expected. However, at the first glance it varies in a non-systematic way with the two lowest and the two highest dosages as two equal sets. For the two lowest dosages the explanation can be that the 2.6 % silica fume has reacted completely earlier and can not contribute more until 7 days relative to the 5.3 % mix, while for the two highest dosages it is likely that about the same amount of silica fume has reacted for both the 20 and 25 % replacement but that they have a different amount unreacted at 7 days. Thus, it is the total amount silica fume reacted at a given time, and whether it is depleted before this or not, that governs the total chemical shrinkage. Secondary effects will be for high silica fume dosages the lower C/S ratio of the gel from the pozzolanic reaction, which may lead to higher shrinkage per unit silica fume reacted due to extensive polycondensation of the silicate anions [2].

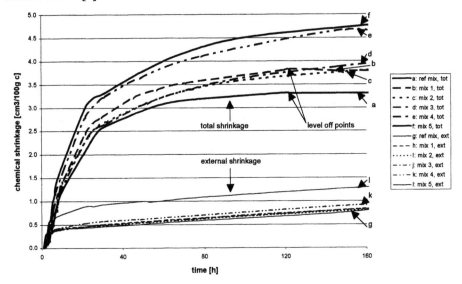

Fig. 1. Total and external shrinkage of the reference mix and the mixes 1 to 5

The external chemical shrinkage for the silica mixes in figure 1 was measured until about 7 days (168 h), and shows an equal or slightly lower flattening out level relative to the reference for the lower dosages up to 11 % replacement, while 20 and 25 % silica fume gives successively higher external shrinkage levels. The reason for this behavior can be explained by the particle packing of the system. Up to 11 % replacement, the silica fume particles (average diameter 0.15 µm) packs in the gaps formed between the irregular cement grains (average diameter 20 µm), maintaining close contact between individual cement grains. Thus, about the same degree of hydration is required to give a hydrate network strong enough to resist the contraction forces (or a little less taking into account the pozzolanic reaction, which may yield a relatively weak product at the early stage, when lime is scarce). With increasing silica fume replacements (20 and 25 %) cement grains are increasingly dispersed in excess silica fume. When further apart they need a higher degree of hydration (and thereby increased external shrinkage) before a hydration network strong enough to resist the contraction forces is formed.

In theory, water menisci can unlike for the total chemical shrinkage experiment play a role in the external chemical shrinkage measurement. When after the "knee"-point empty contraction pores are formed, water will be distributed from coarser pores to finer pores. With the higher fraction of finer pores introduced initially by the packing of finer silica fume particles, greater contraction forces created by the water menisci in pores of smaller radii could be imagined. However, since no extra shrinkage can be seen experimentally up to a silica fume replacement of 11 %, this does not seem to happen.

4.2 Fly Ash

To compare the influence of silica fume with another pozzolan, a mix is made with addition of 20 % fly ash. This fly ash is ground to cement fineness in a lab mill. In this procedure the spherical shells of the fly ash are crushed as it is common practice in Norway.

As figure 2 shows, both the total and the external chemical shrinkage of a mix with fly ash are higher then the ones of the reference mix. This shows that the pozzolanic reaction of the fly ash gives additional shrinkage. Because the total and the external chemical shrinkages of a mix with fly ash are less then the ones of the mix with silica fume, the pozzolanic reaction of fly ash has to give a lower chemical shrinkage than the silica fume slurry reaction. This may be explained by less dispersion of cement grains by the less voluminous fly ash since it is a higher density material: the fly ash is present in the shape of "egg shells" and not in the shape of hollow spheres due to grinding. The shrinkage rate of the reference mix and the mix with fly ash is about the same. Addition of silica fume accelerates the shrinkage rate. This can be seen on figure 2, when looking at the slope of the total shrinkage curves between 0 and 10 hours. Reason for the accelerated schrinkage rate is the particle size. Silica fume consists out of particles of smaller size than the cement particles. The fly ash is ground to cement fineness. That makes the fly ash such a reactive pozzolan, whereas in other countries fly ash does not react so fast.

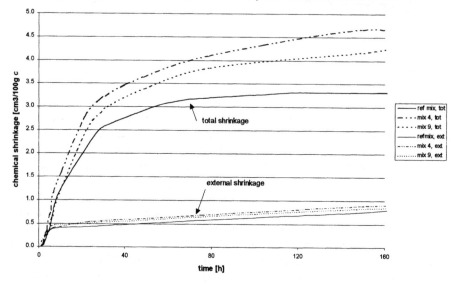

Fig. 2. Total and external shrinkage of the reference mix and the mixes 4 and 9

4.3 Fillers and sand

Fillers and sand in various amounts were added to cement paste as a first step to "bridge the gap" between the behaviour of cement paste and concrete. Filler F1 and F2 were both silicates derived from sand; their fineness and dosage are given in Table 1. The volumetric results are given in figure 3. Each curve is the mean of three parallel samples, which explains the somewhat irregular shapes. With the uncertainty this implies, the sets of curves do not show any clear effects of the fillers - except possibly somewhat increased total shrinkage at longer times, implying increased degree of hydration. For external shrinkage the position of the "knee-points" where the curves depart from the total shrinkage curves are not influenced, i.e. the formation of a self-supporting skeleton is apparently not altered by the fillers. There are no signs of expansion in any of the external shrinkage curves, i.e. the commonly observed expansion in linear measurements are not confirmed - and therefore cannot be attributed to the filler additions.

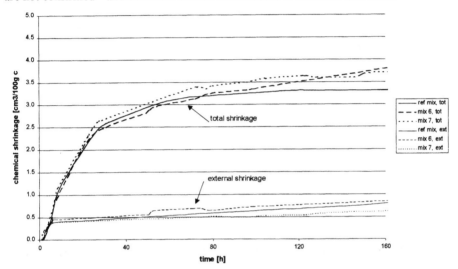

Fig. 3. Total and external shrinkage of the reference mix and the mixes 6 and 7

Figure 4 shows total shrinkage for mixes with sand additions, i.e. filler as well as particles up to 2 mm size. As for the filler, there is no effect on the total chemical shrinkage curves. External chemical shrinkage was not measured for this mortar. Reason for this are the difficulties, encountered in filling the rubber bags with the sticky mortar.

The net result of these filler/sand experiments is that additions do not influence the early reactions and volume changes of the pastes/mortars - when the comparison is made on the basis of weight of cement (which corresponds to a basis of volume of cement paste in these mixes with fixed w/c). It should be added that the measuring methods are fairly coarse and the condition therefor only applies at that level of accuracy. Normally fillers are expected to accelerate cement hydration (so called "filler-effect"). This is not found here, but it is known to be a function of filler mineralogy.

Sand with greater diameters is also added. Unfortunately, the results were not reliable, because of a very high standard deviation between mixes of the same recipe. Some practical problems occurred. The coarser sand easily damaged the rubber bags because of the sharp corners of the sand. The results of the Erlenmeyer method were also not reliable due to low accuracy. The cement layer can only be 10 mm thick to ensure the filling of all pores with water. Because of the high amount of sand in the mix, there was only a small amount of cement in this layer. Expressing the total shrinkage as cm³/100 g cement leads to high standard deviations. Other methods for measuring total and external shrinkage will be required to complete the link between cement pastes and concrete.

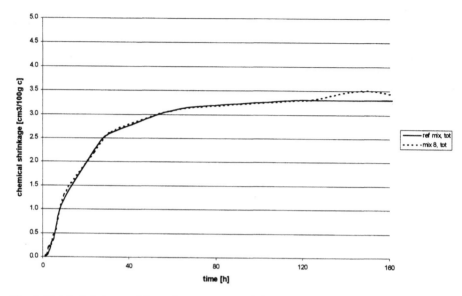

Fig. 4. Total shrinkage of the reference mix and the mix 8

5 Comparison of volumetric and linear methods for shrinkage measurements on concrete

T. A. Hammer and Ø. Bjøntegaard use a linear measurement technique for their research on the relation between thermal dilatation, external chemical shrinkage and stress generation in concrete, [3]. A disadvantage of the linear measurement technique is the fact that linear measurement of external chemical shrinkage during the liquid state is not possible. The liquid state lasts about 3 hours for the concrete used by Hammer and Bjøntegaard. The recorded shrinkage values between 0 and 4 hours are therefore questionable. Volumetric measurements are carried out directly after water cement contact.

With the linear measurement technique, an expansion is measured during the period between 6 and 11 hours after mixing. The expansion occurs between the points A and B in figure 6. Other authors also notice this expansion, [4]. A special test programme was set up to find out whether expansion also occurs when a volumetric method is used. With the rotating method [5] no expansion could be measured. An explanation for the expansion of concrete might be the growth of bridges of ettringite and hydration products in between sand and aggregate grains. The growing bridges may cause an expansion of the skeleton, leading to stresses in the constrained direction. The hydration contraction only appears in the vertical direction. Linear measurements mean horizontal measurements and no measuring of the vertical dimensional changes. A volumetric method only measures the total external volume change, as an integral of all dimensional changes.

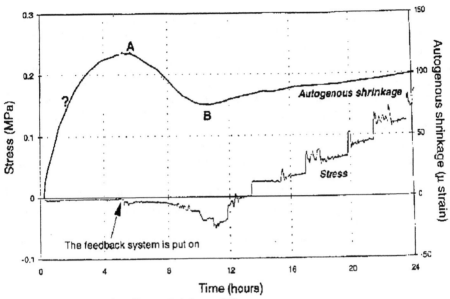

Fig. 5. Stress generating linear shrinkage, [3]

6 Conclusions

The presence of silica fume, in normally used amounts, does not influence the chemical shrinkage. Adding fly ash has less influence on both total and external chemical shrinkage than adding silica fume in the same amount. The addition of fillers reduces the external chemical shrinkage, but has no influence on the total chemical shrinkage. Therefore, the internal shrinkage increases. Adding fine sands has no influence on the total chemical shrinkage. The measurement techniques failed when coarse sand and stone were added. The rubber bags fissured and complete filling was impossible. Other bags are required to complete the study of mixes between cement paste and concrete. Linear measurement methods measured an expansion between about 6 and 11 hours after mixing, which volumetric techniques did not show.

7 References

1. Justnes, H., Reyniers, B., Van Loo D., Sellevold, E.J. (1994) An Evaluation of Methods for Measuring Chemical Shrinkage of Cementitious Past, *Nordic Concrete Research*, Vol. 14, No. 1, pp. 45-61.
2. Justnes, H., Sellevold, E.J., Lundevall, G. (1992) High Strength Concrete Binders. Part A: Reactivity and Composition of Cement Pastes with and without Condensed Silica Fume, *Proceedings of 4th International Conference on Fly Ash, Silica fume, Slag and Natural Pozzolans in Concrete*, CANMET/ACI SP 132-47, Vol. 2, pp. 873-889, Istanbul.
3. Bjøntegaard, Ø. (1996) Thermal Dilatation, Autogenous Shrinkage and Stress Generation, *Workshop "Early Volume Change and Reactions in Paste - Mortar - Concrete."*, Trondheim.
4. Takahashi, T., Nakata, H., Yoshido, K., Goto, S. (1997) Autogenous Shrinkage of Cement Paste During Hydration, *Proceedings of the 10th international congress on the chemistry of cement*, Vol. II, 2ii070.
5. Van Gemert, A., Verboven, F. (1994) *Chemical shrinkage of cement pastes*, Master Thesis, Leuven.

14 EFFECTS OF SLAG AND FLY ASH ON THE AUTOGENOUS SHRINKAGE OF HIGH PERFORMANCE CONCRETE

Y.W. Chan, C.Y. Liu and Y.S. Lu
Department of Civil Engineering, Taiwan University, Taipei, Taiwan

Abstract

This paper presents the test results on autogenous shrinkage of high performance concrete containing various amounts of pozzolanic materials, namely slag, fly ash, and their combinations. The effects of pozzolans and paste-to-aggregate ratio on autogenous shrinkage are investigated in this study. The weight ratios of cement replaced by pozzolans included 0%, 20%, and 40%. The paste-to-aggregate ratio adopted in this study were 0.44, 0.41, and 0.38. The test results indicate that the autogenous shrinkage of HPC is closely related to the content of the cementitious materials. While slag increases the autogenous shrinkage of HPC, fly ash helps to reduce the possible amount of the autogenous shrinkage due to its inert nature.
Keywords: Autogenous shrinkage, drying shrinkage, high performance concrete, slag, fly ash, paste-to-aggregate ratio.

1 Introduction

When the w/c ratio is low and no water can be further supplied during curing process, autogenous shrinkage in concrete occurs due to the "self-desiccation" in the pore structure of concrete as the moisture is consumed during the hydration process [1,2]. Autogenous shrinkage is negligible for conventional concrete, which usually has a high w/c. On the other hand, autogenous shrinkage is significant for high strength concrete or high performance concrete as its w/c is low and cement content is high. It has been found that, besides w/c and cement content, various pozzolanic materials, such as silica fume, fly ash, slag, etc., also affect autogenous shrinkage behavior of mortar in different manner [3].

Autogenous Shrinkage of Concrete, edited by Ei-ichi Tazawa. Published in 1999 by E & FN Spon,
11 New Fetter Lane, London EC4P 4EE, UK. ISBN: 0 419 23890 5

Restrained crackings at early age of conventional concrete are mainly due to improper curing or plastic shrinkage and may be avoided by quality workmanship. However, the autogenous shrinkage in high strength concrete or high performance concrete is inevitable even if good curing process has been taken after placing [4]. The major reason for the inevitable restrained cracking is due to the autogenous shrinkage at the early age. As a result, it can only be resolved by means of adjusting mix proportion, including cement content, w/c, total aggregate volume, etc., to prevent excessive autogenous shrinkage from happening in the case plain concrete [5,6]. It may, however, also be possible solution to introduce fibers in the concrete to reduce autogenous shrinkage or chemical shrinkage so as to prevent early-age cracking [7,8].

The development of high-workability concrete, one of the species of high performance concrete (HPC), was originated to resolve technical difficulties in job-site casting of concrete structures and to ensure high-quality construction. The high workability of HPC is usually achieved by using low water/cement ratio, high cement content and appropriate amount of pozzolanic materials. It was found that the autogenous shrinkage in high-workability concrete or HPC is significant as compared with ordinary concrete. In this paper, the influences of paste/aggregate ratio and content of pozzolanic materials, including slag, fly ash, and their combination, on autogenous shrinkage of HPC were experimentally studied.

It is the purpose of this investigation to study the ways to control autogenous shrinkage of high performance concrete, apart from mix proportioning, by means of introducing fibers or low shrinkage cement. The mix proportions of concrete were designed to exhibit high flowability. Several trial batches were conducted for each mix proportion to ensure the workability and to fine-tune the superplasticizer dosage.

2 Experimental program

2.1 Mix proportions

Measurements of both autogenous shrinkage and drying shrinkage were conducted to study the effects of various pozzolan contents (slag, fly ash and their combinations) and cement paste-to-total aggregate weight ratio (P/A) on the shrinkage behavior of concrete. The concrete mix proportions were designed to exhibit high workability or self-compaction capability. The slump measurements of these mixes were controlled at the range of 23 to 27 cm. The water-to-total cementitious materials ratio was kept constant at 0.32 for all mixes. The design compressive strength at 28 days of these mixes is 60 MPa. The actual compressive strengths of different mix proportions, however, fluctuated more or less due to various content of pozzolans. The weight ratios of cement replaced by pozzolans are 0%, 20%, and 40%. For the content of pozzolan to replace part of cement, three combinations of slag and fly ash are selected, including 100% slag, 100% fly ash, and 43% slag/57% fly ash. The paste-to-aggregate ratios adopted are 0.44, 0.41, and 0.38 in this test program. The content of superplasticizer was finely tuned for each individual mix proportion such that the slump and slump flow were kept at 25±2 cm and 60±5 cm to ensure high flowability. The mix proportions are listed in Table 1.

2.2 Materials

The reference concrete mix was based on ordinary Portland cement (ASTM Type I). The slag used in this study was ground granular blast furnace slag. The specification of slag is given in Table 2. The fly ash used in this study was supplied by a major local power plant. The specification of fly ash is given in Table 3. The superplasticizer is a melamine-based high-ranged water reducer from a local supplier.

Table 1 -- Mix Proportions of Concrete

(in Kg/m³)

No.	Cement	Water	Slag	Fly Ash	Sand	Gravel	SP
1A-44	550	176	0	0	787	853	1.6%
2A-44	330	176	220	0	787	853	2.4%
2B-44	330	176	95	125	787	853	3.0%
2C-44	330	176	0	220	787	853	4.4%
3A-44	220	176	330	0	787	853	3.0%
3B-44	220	176	142	188	787	853	4.4%
3C-44	220	176	0	330	787	853	4.8%
1A-41	520	166	0	0	803	871	1.6%
2A-41	312	166	208	0	803	871	2.4%
2B-41	312	166	89	119	803	871	3.0%
2C-41	312	166	0	208	803	871	4.4%
3A-41	208	166	312	0	803	871	3.0%
3B-41	208	166	134	178	803	871	4.4%
3C-41	208	166	0	312	803	871	4.8%
1A-38	493	158	0	0	822	890	1.6%
2A-38	296	158	197	0	822	890	2.4%
2B-38	296	158	85	112	822	890	3.0%
2C-38	296	158	0	197	822	890	4.4%
3A-38	197	158	296	0	822	890	3.0%
3B-38	197	158	127	169	822	890	4.4%
3C-38	197	158	0	296	822	890	4.8%

*The first character in No. in column 1 denotes *cement-to-total cementitious materials ratio* of 1, 0.6, and 0.4 for 1, 2, and 3 respectively; the second character denotes various *pozzolan combinations* of 100% slag, 43% slag/57% fly ash, and 100% fly ash for A, B, C respectively; and the extension in No. denotes *paste-to-aggregate ratio* of 0.44, 0.41, and 0.38 for 44, 41, and 38 respectively.
**SP is expressed in weight ratio of cement

2.3 Measurement

For measurement of observed shrinkage of concrete, ϕ10cm ×30cm cylindrical specimens were used for concrete. In each single case, at least 3 specimens were cast. All specimens were demoulded after 24 hours. Specimens for the measurement of autogenous shrinkage were sealed with aluminum tape in order to prevent water

Table 2 Specification of Slag

Physical Properties				
Fineness cm^2/g	Seive#325 %	Activity Index %	Specific Weight	Loss on Ignition
4265	7	68	2.9	0.43

Chemical Analysis					
SiO_2	Al_2O_3	Fe_2O_3	CaO	MgO	SO_3
34	14	0.3	41	7.6	2.1

evaporation and the others were kept in fog room (23°C, R.H.100%) for further curing. At the age of 7 days, specimens for measurement of drying shrinkage were moved into drying room (23°C, R.H.50%). The longitudinal length change of the specimens was measured with metal plugs attached to opposite sides of the specimens using a 1/1000-mm dial gauge. The measurement of length change was started from 24 hours and 7 days for autogenous shrinkage and for drying shrinkage respectively.

Table 3 Specifications of Fly Ash

Physical Properties				
Fineness cm^2/g	Seive#325 %	Activity Index %	Specific Weight	Loss on Ignition
2070	15		2.12	5.02

Chemical Analysis					
SiO_2	Al_2O_3	Fe_2O_3	CaO	MgO	SO_3
47	28	5.2	6.3	1.6	0.8

4 Results and discussions

The main results in this experimental program are summarized in Table 4, including the average autogenous shrinkage at the age of 7 and 28 days and the drying shrinkage at the age of 28 days. The ratio of autogenous shrinkage of 7 days to that of 28 days and the ratio of autogenous shrinkage to drying shrinkage at 28 days are also given in the table for each individual mix proportion.

1. According to the test results, the autogenous shrinkage of HPC increases when part of OPC replaced by slag. As indicated in Figure 1, for each P/A ratio at the same age, the autogenous shrinkage of concrete with 40% of slag out of total cementitious materials is significantly higher than that without slag. When the slag content further increases, the autogenous shrinkage decreases slightly although it is still much higher than that without any slag, especially at the age of 28 days. The effect of slag in increasing the autogenous shrinkage is also observed in drying shrinkage according to Table 4.

2. When the pozzolan is substituted by fly ash only, the result becomes a different scenario. As shown in Figure 2, a summary of autogenous shrinkage at 7 and 28

days, the incorporation of fly ash in HPC leads to decrease in autogenous shrinkage. The higher the fly ash content, the lower the shrinkage. However, fly ash helps to increase the drying shrinkage the same way as slag does according to Table 4. This finding may be related to the difference in the mechanism between the two types of shrinkage. As part of OPC is replaced by fly ash, the amount of cementitious materials that undergoes hydration reaction was reduced and water migration in the pore structure is reduced substantially. Therefore, a great amount of autogenous shrinkage was prevented. However, drying shrinkage is driven by water migration due to moisture evaporation in pore structure. When concrete is exposed to low humidity, drying shrinkage occurs as evaporation of moisture takes place. This drying process occurs no matter the hydration reaction in the concrete is active or not.

Table 4 Measurements of autogenous shrinkage and drying shrinkage

$(\times 10^{-6})$

No.	Autogenous Shrinkage		Drying Shrinkage	a7/a28	a28/d28
	7 days	28 days	28 days	(1)	(2)
1A-44	96	250	482	0.38	0.52
2A-44	196	396	629	0.49	0.63
2B-44	171	362	605	0.47	0.60
2C-44	86	188	608	0.46	0.31
3A-44	154	359	554	0.43	0.65
3B-44	139	329	520	0.42	0.63
3C-44	59	98	543	0.60	0.18
1A-41	84	189	396	0.44	0.48
2A-41	165	343	548	0.48	0.63
2B-41	140	321	460	0.44	0.70
2C-41	76	130	507	0.58	0.26
3A-41	140	321	478	0.44	0.67
3B-41	127	297	420	0.43	0.71
3C-41	41	74	471	0.55	0.16
1A-38	66	127	330	0.52	0.38
2A-38	125	293	452	0.43	0.65
2B-38	120	272	380	0.44	0.72
2C-38	59	100	416	0.59	0.24
3A-38	127	307	420	0.41	0.73
3B-38	106	262	375	0.40	0.70
3C-38	36	62	400	0.58	0.16

3. When slag and fly ash are both incorporated, the effect of slag appears to dominate the autogenous shrinkage characteristic of concrete. The trend in Figure 3 is similar to that of Figure 1. It is interesting to note that the slag adopted in this study plays an important role in the autogenous shrinkage behavior of concrete not only due to the

active pozzolanic reaction because of its high fineness but also due to the possibly refined pore structure that aggravates the self-desiccation process.

4. Figure 4 indicates the effect of P/A ratio on the shrinkage of concrete. As shown in the figure, the higher the P/A ratio, the lower the value of a7/a28. It suggests that the concrete with higher P/A ratio may undergo more autogenous shrinkage as the age increases, while concrete with lower P/A ratio does not exhibit as much shrinkage due to the restrain from the relatively higher amount of aggregates. On the other hand, that the value of a28/d28 increases with P/A ratio indicates that the autogenous shrinkage is closer to the total possible amount of shrinkage in the concrete when concrete has more paste volume.

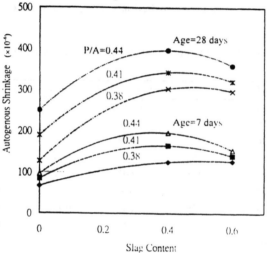

Figure 1 Effect of slag on autogenous shrinkage

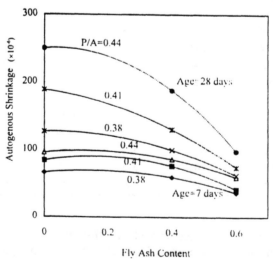

Figure 2 Effect of fly ash on autogenous shrinkage

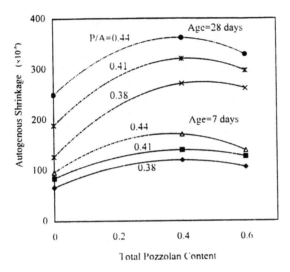

Figure 3 Effect of slag/fly ahs on autogenous shrinkage

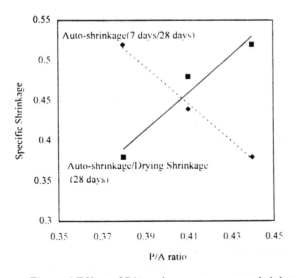

Figure 4 Effect of P/A ratio on autogenous shrinkage

5 Conclusions

The results of this investigation indicate clearly that the increase in P/A (paste-to-aggregate) ratio and the incorporation of slag lead to increase in autogenous shrinkage of HPC. In this regard, the effects of P/A ratio and slag on autogenous shrinkage are the same as on drying shrinkage. The incorporation of fly ash decreases the autogenous shrinkage although it increases the drying shrinkage just the same as slag does. However, it is important to note that the slag adopted has high fineness and is supposed to be active in pozzolanic reaction. On the other hand, the fly ash used in this study is Type F and has a low fineness modulus. Therefore, it is relatively inert in terms of hydration.

Acknowledgement

The authors are grateful to the National Science Council of Taiwan for the financial support to this research.

Reference

1. Powers, T. C. (1935). "Absorption of Water by Portland Cement Paste during Hardening Process," *Ind. and Eng. Chem.*, **27**(7), 790-94.
2. Wittmann, F. H. (1976). "On the Action of Capillary Pressure in Fresh Concrete," *Cem. and Conc. Res.*, **6**, 49-56.
3. Tazawa, E. and Miyazawa, S. (1992). "Autogenous Shrinkage Caused by Self Desiccation in Cementitious Materials," 9th *Int. Cong. On the Chemistry of Cement*, **IV**, 712-18.
4. Holland, T. C., Krysa, A., Luther, M. D., and Liu, T. C. (1986). "Use of Silica-fume Concrete to Repair Abrasion-erosion Damage in the Kinzua Dam Stilling Basin," 2nd *CANMET/ACI Int. Conf. on Fly Ash, Silica Fume, and Natural Pozzolans in Concrete*, **SP-91**, 841-63.
5. Tazawa, E. and Miyazawa, S. (1997). "Influence of Constituents and Composition on Autogenous Shrinkage of Cementitious Materials," *Mag. of Conc. Res.*, **49**(178), 15-22.
6. Dilger, W. H., Wang, C., and Niitani, K. (1996). "Experimental Study on Shrinkage and Creep of High-performance Concrete," 4th *Int. Symp. On Utilization of High-strength/High-performance Concrete*, 311-19.
7. Tazawa, E. and Miyazawa, S. (1996). "Influence of Autogenous Shrinkage on Cracking in High-strength Concrete," 4th *Int. Symp. On Utilization of High-strength/High-performance Concrete*, 321-30.
8. Chan, Y.W. and Lu, Yu-Shan (1998). "The Control of Early-age Cracking in High Strength Concrete," 6th *East Asia-Pacific Conference on Structural Engineering & Construction*, 1289-94.

15 EXPERIMENTAL EVALUATION OF AUTOGENOUS SHRINKAGE OF LIGHTWEIGHT AGGREGATE CONCRETE

K. Takada
Kajima Technical Research Institute, Tokyo, Japan
K. Van Breugel and E.A.B. Koenders
Delft University of Technology, Delft, The Netherlands
N. Kaptijn
Ministry of Transport, Public Works and Water Management, The Netherlands

Abstract
Autogenous length changes of concrete mixes in which the dense aggregates were replaced partly or wholly by lightweight aggregate particles are described. Replacement percentages were 10%, 17.5%, 25% and 100%. The aggregates of the full lightweight aggregate mixes were saturated completely or only up to 20 or 60%. The test specimens were cured under sealed conditions at 20 °C. Besides the length changes during hardening the strength development was determined. The tests confirmed that use of water containing lightweight aggregates affect the early volume changes drastically. The potentialities of modeling the phenomena which are expected to be responsible for the reduction of autogenous volume changes are dealt with.
Keywords: Autogenous shrinkage, lightweight aggregate, blended aggregate, degree of saturation, self desiccation, numerical modeling.

1 Introduction

The increasing use of low water/binder ratio concretes has forced both practical engineers and scientists to pay attention to the magnitude of volume changes in hardening concrete at early ages. This early age shrinkage may cause microcracking in the concrete and even macro crack in concrete structures. These cracks may jeopardize the short term performance and durability of the concrete. Many researchers have considered self desiccation of low water/binder ratio concretes to be the major reason of large shrinkage of these concretes during hardening. This self-desiccation is accompanied with a decrease of the relative humidity in the pore system and an associated increase of the surface tension in the absorption layer. Due to the increase of the surface tension the microstructure will contract, which contraction may cause the cracking of the concrete in case the shrinkage defor-

Autogenous Shrinkage of Concrete, edited by Ei-ichi Tazawa. Published in 1999 by E & FN Spon, 11 New Fetter Lane, London EC4P 4EE, UK. ISBN: 0 419 23890 5

mations are restrained. From the scientific point of view it is a challenge to evaluate and model the shrinkage mechanism triggered by self-desiccation. From the practical point of view the major question is how to get rid of this shrinkage phenomenon.

If self-desiccation of the hardening paste is a major cause of autogenous shrinkage, a logic way to mitigate or eliminate this type of shrinkage is to prevent the occurrence of self-desiccation. In order to achieve this it has been proposed to add saturated lightweight aggregates to the mix. The saturated aggregate particles will act as water reservoirs which release their water at the moment that the relative humidity in the concrete drops [1][2]. The use of lightweight aggregate instead of dense aggregate, however, is expected to result in a lower strength. The low water/binder ratio concrete, however, had just been adopted because of their high strength and high density. Here we arrive at an optimization problem, namely, whether there is an optimum percentage of the dense aggregate to be replaced by lightweight aggregate at which a substantial reduction of the autogenous shrinkage is obtained with only a minor or no reduction of the strength. This question will be considered in the next paragraphs. Moreover, the potential of numerical simulation programs to elucidate and quantify the mechanisms underlying autogenous shrinkage will be discussed.

2 Experimental determination of autogenous volume changes

2.1 General outline of experimental program
Although self desiccation is in essence an inherent feature of all hardening concrete mixes, substantial volume changes associated with self desiccation will generally only play a role in low water/binder ratio concretes. In the experimental program the focus of attention was, therefore, on mixes with water/binder ratios below 0.4. In this paper, mixes with a water/cement ratio of 0.37 will be considered.

Two test series were investigated. In the first series the effect on partial replacement of dense aggregate by lightweight aggregate was investigated. In the second series a full lightweight aggregate concrete is considered. In that series the effect of initial water content of the lightweight aggregate on autogenous volume changes was considered.

2.2 Mix compositions of the concrete

2.2.1 Mixes with partial replacement of dense aggregate by lightweight aggregate

The basis for the mix used in this series consists of a mix with a cement content of 475 kg/m³. This amount consisted for 50% of a blast furnace slag cement (CEM III/B52.5) and 50% Portland cement (CEM I/52.5 R). The slag cement was added in order to reduce the rate of reaction in the early stage of hydration. Water/cement ratio was 0.37. A silica-slurry, 50% water and 50% silica fume, was added in order to enhance the final strength. In the reference mix a siliceous aggregate was used, consisting of 45% sand (1-4 mm) and 55% coarse aggregate (4-16 mm). Two types of plasticizers were used, i.e. lignosulfonate and naphthalene sulfonate. The air content was 2.5%. The water/cement ratio amounted 0.37. With this mix it was aimed to realize a 85MPa compressive strength at 28 days.

This basic mix was modified by replacing a part of the dense aggregate by lightweight aggregate. The lightweight aggregate used was the Liapor F10, a product of Germany. The

Table 1 Mix composition of partial replacement mixes, series I

Mix component	unit	Replacement factor			
		0%	10%	17.5%	25%
Cement CEM III/B42.5 LH HS	kg	237.0	237.0	237.0	237.0
Cement CEM I / 52.5 R	kg	238.0	238.0	238.0	238.0
Water (incl. water in admixtures)	kg	175.75	175.75	175.75	175.75
Crushed gravel 4-16 mm	kg	944.18	849.76	778.95	708.14
Sand 0-4 mm	kg	772.51	772.51	772.51	772.51
Liapor F10, 4-8 mm (100% saturated)	kg	---	60.84	106.46	152.09
Lignosulfonate (dry substance 36%)	kg	0.95	0.95	0.95	0.95
Naphthalene sulfonate(dry substance 40%)	kg	7.60	7.60	7.60	7.60
Silicafume-slurry (50% water)	kg	50.0	50.0	50.0	50.0

replacement percentages were 10%, 17.5% and 25%. The aggregate was saturated with water at the moment of mixing. Besides the volume changes the compressive strength was measured in order to quantify the expected strength reduction caused by replacement of dense aggregate by lightweight aggregate. Details of the mix composition of the various mixes are summarized in Table 1.

2.2.2 Full lightweight aggregate mixes

Lightweight aggregate mixes with water/cement ratio 0.37 were designed. The lightweight aggregate considered was Liapor F10. The cement content and the amount of silica fume were kept the same as in the series presented in section 2.2.1. The aggregates were used with different degrees of saturation. The target values were 100%, 60% and 20%. The exact degrees of saturation were determined on aggregate samples taken from the aggregate just prior to mixing. Different degrees of saturation were used in order to obtain a broader spectrum of data for validating numerical simulations of the effect of moisture exchanges from the aggregate to the paste with continuing hydration. Details of the mix are presented in Table 2.

2.3 Description of experimental set-up
The volume changes in the early stage of hardening were determined on concrete prisms, 1000 x 150 x 150 mm. The concrete was cast in a mould, which was provided with an

Table 2 Mix composition of full lightweight aggregate mixes: series II

Mix component	unit	Designed degree of saturation of Liapor		
		100%	60%	20%
Cement CEM III/B42.5 LH HS	kg	237.0	237.0	237.0
Cement CEM I / 52.5 R	kg	238.0	238.0	238.0
Water (incl. water in admixtures)	kg	175.75	175.75	175.75
Sand 0-4 mm	kg	772.51	772.5	772.51
Liapor F10, 4-8 mm	kg	613.72	583.57	553.42
(absorption capacity is 13%)				
Lignosulfonate (dry substance 36%)	kg	0.95	0.95	0.95
Naphthalene sulfonate(dry substance 40%)	kg	7.125	9.50	11.875
Silicafume-slurry (50% water)	kg	50.0	50.0	50.0

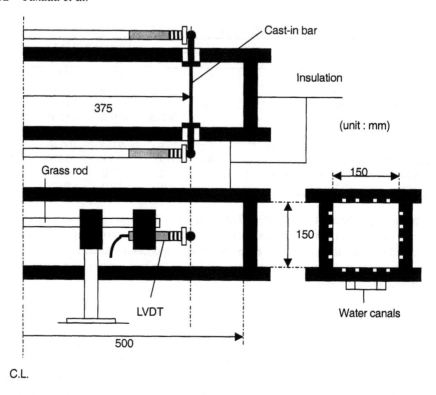

Fig. 1 Sketch of the mould for concrete specimens used for measuring autogenous deformation.
(the scale of the measuring devises is exaggerated to the dimension of the mould)

external insulation material (see Fig. 1). At the interior surface the mould consisted of thin steel plates. The steel plates could be cooled down or heated by a system of tubes located between the plates and the insulating material, through which tubes water of any desired temperature could be pumped. With this temperature control systems it is possible to impose an arbitrary temperature regime onto the concrete. Some minor temperature gradients within the concrete cross section are unavoidable with such a system. It was found, however, that even if a high strength concrete was used the temperature differences within the cross section could be kept less than 1.0 °C to 1.5 °C. The interior of the mould is covered by a plastic foil, and a piece of felt on the bottom, in order to avoid an external moisture exchange and to reduce friction between the concrete and the mould.

The length changes of the hardening concrete were measured over a length of 750 mm with a glass rod to which two LVDT's were connected. The glass rod was connected to two steel bars which were cast in the concrete at the required distance of 750 mm. The rod could be fixed to the cast-in bars after the concrete had gained sufficient strength. The volume changes in the very early stage of hardening, i.e. before setting, could not be measured in the way. In this research project this was not considered a problem, since these very early volume changes were considered not to generate significant forces in the structure which could give rise to cracking of the concrete. (Note: plastic shrinkage not considered)

After casting the top surface of the concrete was covered with a plastic foil in order to

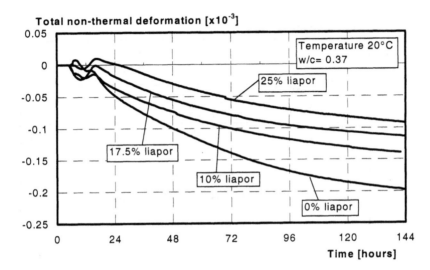

Fig. 2   Autogenous (non-thermal) deformation of different concrete mixes during hydration at early ages. Replacement percentages of 10%, 17.5% and 25%.

avoid moisture loss to the environment. The concrete was cured isothermally at 20 °C. Isothermal curing was adopted in order to avoid temperature effects which would complicate the interpretation of the results.

3   Test results

3.1   Series I: Partial replacement of dense aggregate by lightweight aggregate
In Fig. 2 the total deformation of four different concrete mixes is presented as a function of time. For the basic mix, which did not contain any lightweight material, shrinkage could be

Table 3   Compressive strength of mixes with partial and full replacement of dense aggregate by lightweight aggregate (Liapor F10) after 1, 2, 3, 7 and 28 days. w/c ratio = 0.37.

Replacement percentage [%]	Degree of saturation [%]	Cube compressive strength [MPa]				
		1d	2d	3d	7d	28d
0		41.9	52.0	57.5	70.4	82.5
10	100	38.8	51.0	57.7	70.0	89.6
17.5	100	43.4	52.5	58.9	71.7	94.3
25	100	37.6	47.4	53.1	67.5	87.8
100	100(97.7*)	24.4	-	46.5	58.2**	79.0
100	60(69.3*)	37.5	-	59.2	69.9**	85.2
100	20(29.4*)	32.1	-	55.2	61.3**	76.0

*measured   **6day

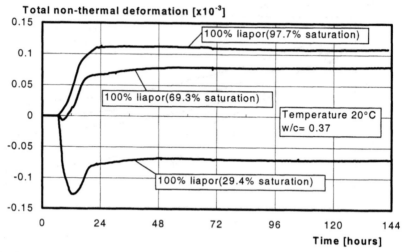

Fig. 3   Autogenous (non-thermal) deformation of full lightweight aggregate concrete mixes during hydration at early ages. Parameter: initial water content in the aggregate.

measured after about 6 hours after placing of the concrete. After 12 hours a slight expansion was measured, whereas after 15 hours continuous shrinkage is observed. For this particular mix the shrinkage after 144 hours reached 0.2×10^{-3}. Replacement of the dense aggregate by 10%, 17.5% and 25% Liapor F10 resulted in a substantial reduction of the shrinkage strains. With a replacement percentage of 25% the shrinkage after 144 hours was less than half the shrinkage of the basic mix.

In Table 3 the strength of mixes with and without replacement of dense aggregate by lightweight aggregate is presented. The strength data reveal that partial replacement of dense aggregate by lightweight aggregate, i.e. Liapor F10, by up to 25% by volume, had hardly noticeable effect on the compressive strength.

3.2   Series II: Full lightweight aggregate concrete

For full lightweight aggregate mixes the isothermal deformation curves are shown in Fig. 3. The mix made with saturated aggregate shows expansion right from the start of the measurements. The expansion reached its maximum of about 0.115×10^{-3} after 24 hours and then remained almost constant for more than 5 days. The mix with aggregate with a degree of saturation of 60% (69.3% measured) showed minor shrinkage in the very early stage of hardening. After 10 hours expansion was measured. The maximum was reached after 72 hours and amounted to about 0.08×10^{-3}. The mix made with dried aggregate - degree of saturation 20% (29.4% measured) - started with intensive shrinkage until about 13 hours. This shrinkage period was followed by expansion between 13 and 18 hours and moderate expansion in the next 5 days. In this case the net result of deformation is still shrinkage.

The compressive strength of the full lightweight aggregate concretes tends to be lower than no- and partial-lightweight aggregate concretes but the reduction is not large. The concrete which has full lightweight aggregate and medium degree of saturation showed comparable strength with the no- and partial-lightweight aggregate concretes.

4 Evaluation of experimental results

Among the possible mechanisms which have been considered as possible causes of autogenous deformations, the reduction of the relative humidity in the pore system with progress of the hydration process and the associated increase in the surface tension in the capillary water seems the dominant cause [3][4]. The increasing surface tension puts the microstructure under stress which is accompanied by a volume reduction of the system. The mechanism has been studied by several researchers. In case water saturated aggregate is used, the decrease of the relative humidity generates a moisture flow from water saturated aggregate particles towards the drying paste. This moisture flow causes a reduction of the decrease of the relative humidity in the paste compared to mixes in which only dense aggregates are used. The tests from the first series confirm the existence of the assumed mechanism. Partial replacement of the dense aggregate by lightweight aggregate up to 25% causes a reduction of the autogenous shrinkage by almost 50% in case Liapor F10 is used. This mechanism was explained and substantiated with a number of relevant formulae by Weber [5]. In her report the microstructure and changes thereof were, although considered, not quantified explicitly.

A noticeable effect of using saturated lightweight aggregate could be achieved already with partial replacement of the dense aggregate by lightweight aggregate. The advantage of only partial replacement of the dense aggregate is, that still high strength values can be reached. With replacement percentages up to 25% the strength reduction could be ignored. Mixes with replacement percentages of 10% and 17.5% even exhibited a small increase of the strength. This might be caused by some additional delayed hydration which is made possible by suction of the water from the aggregated towards the hydrating paste [2].

The efficiency of lightweight aggregates in terms of reducing the autogenous shrinkage strongly depends on the initial water content of the aggregate. In case water saturated aggregates are used, no autogenous shrinkage was observed at all. Even expansion occurred. Reducing the initial water content from 100% to 60% (69.3% measured) caused a reduction of the expansion, but still no autogenous shrinkage was observed. In mixes with an initial water content down to 20% (29.4% measured), the aggregate probably takes up some water from the paste. After a rapid shrinkage in the first stage of the hydration process, some swelling was observed as well as the other cases with higher initial water content. After 5 days the overall result of deformation was a shrinkage of the mix.

5 Towards numerical simulation of internal curing mechanisms

In a recent study Koenders [6][7] had investigated the possibilities to simulate the microstructural development and the related changes of the relative humidity and volume changes numerically. For his numerical analysis he used a modified version of the simulation program HYMOSTRUC, developed at the Delft University of Technology [8]. With this simulation program the hydration process and microstructural development in hardening pastes are considered as intimately connected processes. Changes of the state of water in the gradually changing pore system are determined explicitly with the model. One of the parameter studies described in Koender's thesis is of special interest here, viz. the numerical evaluation of ribbon paste hydration. In that parameter study the effect of a variable water content over the thickness of a layer of paste between two aggregate particles on the rate of hydration was investigated. The variable water content was considered to be caused by the stereological "wall effect". Due to this wall effect the water content of the paste just adjacent to the surface of the aggregate particles is higher than the water content at greater distance from the aggregate surface. Diagrammatically this is shown in Fig. 4. In this figure the local water/cement ratio is presented as a function of the distance from the

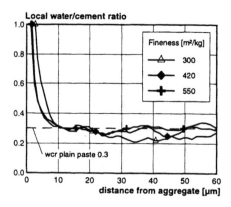

Fig. 4   Change of local water/cement ratio with distance from aggregate surface. Different fineness. W/C ratio = 0.3. (Koenders, [6]).

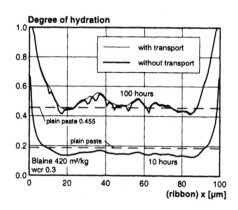

Fig. 5   Calculated effect of moisture transport on degree of hydration over the thickness of ribbon paste. Blaine=420 m²/kg. W/C ratio = 0.3. (Koenders, [6]).

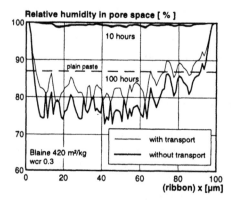

Fig. 6   Calculated effect of moisture transport on relative humidity in the paste. (Koenders, [6]).

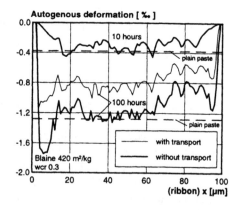

Fig. 7   Calculated autogenous shrinkage in ribbon paste with and without making allowance for the effect of moisture transport within the paste (Koenders, [6]).

aggregate surface for pastes made with cements with different fineness, i.e. 300, 420 and 550 m²/kg. The mean nominal water/cement ratio was 0.3. Due the less dense packing of the cement particles the local water/cement ratio at the surface substantially exceeds the nominal mean value, whereas at some distance from the aggregate surface the local water/cement ratio is lower than the nominal mean value. With the numerical simulation program HYMOSTRUC the progress of the hydration process over the ribbon thickness was investigated. Due to the high amount of water in the paste-aggregate interfacial zone the amount of cement that was converted into reaction product was higher than at some distance from the aggregate surface. For a paste with a water/cement ratio 0.3 the degree of hydration over the thickness of a 100 μm thick ribbon is shown in Fig. 5 for hydration times of 10 and 100 hours. The calculated degree of hydration in the plain paste is indicated as well.

Inherent to the variable degree of hydration over the thickness of the ribbon also the relative humidity in the pore space will differ. A variable degree of hydration will generate a moisture transport from the water-rich interfacial zone to the relatively dry bulk paste. The transport of water through the hydrating paste will affect the hydration process and the associated changes of the relative humidity in the paste. The effects of moisture transport on the development of the degree of hydration and the relative humidity of the ribbon paste are shown in the Figs. 5, 6 and 7. In Fig. 5 the degree of hydration after 100 hours is indicated with a bold and a thin line. The thin line represents the calculated degree of hydration in which the effect of moisture transport from the water rich zone to the drying bulk paste is allowed for. A little increase of the degree of hydration due to the moisture transport is observed that means a little part of the transported moisture is consumed by the hydration of the bulk paste. Fig. 6 shows an increase of the relative humidity caused by the supply of water from the water rich interfacial zone compared to the result calculated without the consideration of the moisture transport. The drop of the relative humidity in the bulk paste after 100 hours is lower than in the case no moisture transport is concerned. Less reduction of the relative humidity is accompanied by less autogenous shrinkage. This is shown in Fig. 7. This figure clearly shows that in this particular case the autogenous shrinkage of the bulk paste after 100 hours is reduced by about thirty percent in comparison with the situation with no supply of water from the water rich interfacial zone. The phenomenon that is observed here is, to a certain extent, similar to what is expected to happen in case a water saturated lightweight aggregate exists in concrete. In that case the water is not (only) supplied from the water rich interfacial zone, but primarily from the saturated aggregate particles.

6 Concluding remarks

Autogenous volume changes caused by self desiccation in low water/cement ratio concrete can reach values high enough to cause cracking of the concrete. A possible manner to reduce the effect of self desiccation exists in the use of water saturated lightweight aggregates instead of dense aggregate. In an experimental test program the efficiency of the use of saturated lightweight aggregates in terms of reducing autogenous shrinkage was verified. It was found that even partial replacement of the dense aggregate by lightweight aggregate resulted in a substantial reduction of the autogenous shrinkage. Full lightweight aggregate mixes, made with Liapor F10 and a water/cement ratio of 0.37, exhibited swelling and no externally measurable shrinkage at all. In the tests described in this paper partial replacement percentages up to 25% did not give rise to a noticeable reduction of the strength.

Besides showing experimental evidence of the effectiveness of saturated lightweight aggregate for reducing autogenous shrinkage, the potentialities were discussed of numerical models to simulate the mechanisms responsible for the shrinkage reducing effect. In a study on ribbon paste hydration, the effect of moisture transport within the paste volume on autogenous shrinkage was quantified numerically. It was explained how in isothermal tests moisture transport from water rich zones to dry zones in a hydrating paste could allow for a substantial reduction of the autogenous shrinkage. A similar phenomenon as evaluated numerically by Koenders [6] is expected to occur in mixes made with water saturated lightweight aggregates.

It is noticed that the test results described in this paper all refer to Liapor F10. Other lightweight aggregate may behave differently.

References

1. Weber, S., Reinhardt, H.W. (1995) "A blend of aggregates to support curing of concrete". Proc. Structural lightweight aggregate concrete. Int. Symposium Sandefjord, Norway.
2. Vaysburd, A.M. (1996). "Durability of lightweight concrete bridges in severe environments". Concrete International, Volume 18, No. 7, pp. 33-38.
3. Hua, C., Acker, P., Ehrlacher, A. (1995). "Analysis and models of the autogenous shrinkage of hardening cement paste Part I: Modelling at macroscopic scale". Cement and Concrete Research, Vol. 25, No. 7, pp. 1457-1468
4. Baroghel Bouny, V. (1996). "Texture and Moisture Properties of Ordinary and High Performance Cementitious Materials". Proc. RILEM-seminar Beton: du Materieau a la structure. Arles (Fr).
5. Weber, S. (1996) "Nachbehandlungsunemphindlicher Hochleistungsbeton". Institute fur Werkstoffe im Bauwesen, PhD, Stuttgart, 211 p.
6. Koenders, E.A.B. (1997). "Simulation of volume changes in hardening cement-based materials". PhD, Delft, 171 p.
7. Koenders, E.A.B., van Breugel, K. (1998) "Numerical modelling of dimensional changes in low water/cement ratio pastes and concrete". This symposium.
8. Breugel, K. van (1991). "Simulation of hydration and formation of structure in hardening cement-based materials". PhD, Delft, 295 p.

Discussions

Effect of chemical composition and particle size of fly ash on autogenous shrinkage of paste
S. Tangtermsirikul, Thammasat University, Thailand

Ø.Bjøntegaard	What isothermal temperature were the tests performed at?
S.Tangtermsirikul	Thailand is a tropical country. So, the temperature was $25\pm 2°C$.
Ø.Bjøntegaard	What was the temperature increase during the test?
S.Tangtermsirikul	The specimen size was too small. So, there was no problem with the temperature increase.
K. Torii	I totally agree on your opinion. Sulphate content is very important because the sulphates, Na_2SO_4, K_2SO_4 and $CaSO_4$, are attached on the surface of fly ash particles. On mixing, they dissolve very easily and react to produce ettringite much more. The sulphate content of fly ash is about 1 %, however it is very important.
M.S. Akman	I wonder the alkali content has influence on autogenous shrinkage. How did you think of that?
S.Tangtermsirikul	We have not investigated that aspect yet.

Influence of expansive additives on autogenous shrinkage
A. Hori, and M. Morioka, Denkikagaku Kogyo Co.,Ltd., Japan
E. Sakai and M. Daimon, Tokyo Institute of Technology, Japan

No question.

Factors affecting the autogenous shrinkage of silica fume high-strength concrete
J. J. Brooks, J.G. Cabrera and M.A. Megat Johari, University of Leeds, UK

No question.

Measurements of autogenous length changes by laser sensors equipped with digital computer system
  M. Morioka, A. Hori and H. Hagiwara, Denkikagaku Kogyo Co.,Ltd., Japan
  E. Sakai and M. Daimon, Tokyo Institute of Technology, Japan

Ø.Bjøntegaard	I assume that there are end-plates that take care of the hydrostatic pressure during the fresh state of the concrete. I wonder if these end-plates are removed when the hydration heat develops in order to allow free expansion?
M.Morioka	The end-plates were removed at initial setting time. And sponges were installed not to develop the friction resistance at both sides of specimen.

Chemical shrinkage of cement paste, mortar and concrete
  H. Justnes and T. Hammer, SINTEF, Norway
  B. Ardoullie, E. Hendrix, D. Van Gemert, and K. Overmeer, Katholieke Universiteit Leuven, Belgium
  E.J. Sellevold, The Norwegian University of Science and Technology, Norway

No question.

Effects of slag and fly ash on the autogenous shrinkage of high performance concrete
  Y.W. Chan, C.Y.Liu and Y.S. Lu , Taiwan University, Taiwan

M. S. Akman	You had many results on concrete specially, and you did test the quality and quantity of mortar and paste phase of your concrete. And you obtained the relationship between the shrinkage of the concrete and that of the paste with a formula like that of Picket Formula which is very useful to estimate concrete shrinkage by only making the shrinkage test of paste phase.
Y.W.Chan	In fact we have conducted to test only paste phase and try to construct the relationship between the shrinkage of concrete and that of paste phase, and we are hoping more data coming up.
Ø.Bjøntegaard	In the concrete mix proportion table it seems that a decrease in water-to-binder ratio is accompanied by a decrease in the total cement paste volume. Is that so?
Y.W.Chan	We tried to keep water-cement ratio, however, we did vary the paste aggregate ratio, so the total cementitious materials is content for different paste-aggregate ratio.

E.Tazawa	You used a melamine-based high-range water reducer. It seems the superplasticizer content is a little larger than the normal content which we use. Do you have any idea for the actual concentration of melamine?
Y.W.Chan	The solid content of the particular superplasticizer is 43%, so in fact the presented data, some water has been counted in water. I think that the use of fly ash requests more superplasticizer. We tried to keep workability and slump-flow.
E.Tazawa	We do not have enough data. But I think that the setting time have delayed to some extent in case of using naphthalene type. Do you have any data of the setting time of concrete?
Y.W.Chan	I did not measure the setting time of concrete. I agree that the setting time is different when we use different dosage of chemical admixture. I have kept the conditions of fresh concrete as possible so unfortunately I did not measure the setting time.
P.-C.Aïtcin	If you are modifying the rate of hydration, you will also modify rate of autogenous shrinkage at early ages?
Y.W.Chan	I have measured the shrinkage after 24 hours.
S.Hanehara	How do you think about the mechanism and effect of slag that increases autogenous shrinkage?
Y.W.Chan	I think there are two effects. One is that the high fineness of slag increases the hydration greatly at the very beginning and the second is that the particle size of slag is smaller than cement. When we use the slag replaced cement, the microstructure is made denser than that of without slag, and so this effect tends to increase the autogenous shrinkage.

Experimental evaluation of autogenous shrinkage of lightweight aggregate concrete
K. Takada, Kajima Technical Research Institute, Japan
K. van Breugel and E.A.B. Koenders, Delft University of Technology, The Netherlands
N. Kaptijn, Ministry of Tansport, The Netherlands

R. Sato	How do you evaluate the change of mortar properties in concrete containing water-saturated aggregate other than autogenous shrinkage? For example, density, permeability, strength and so on?

K.van Breugel   Nice question. So far we have not evaluated these points. They are being investigated within the frame work of a large international research project. Within this research project the properties you mention are being looked at in detail, from the engineering point of view.

P.-C.Aïtcin   For the first time, we have seen some laser device used to measure autogenous shrinkage successfully, but hopefully autogenous shrinkage is not an unavoidable phenomenon, it can be controlled.

PART FIVE

SEPARATION OF VOLUME CHANGE COMPONENTS AND PREDICTION OF SHRINKAGE

16   THERMAL DILATION–AUTOGENOUS SHRINKAGE: HOW TO SEPARATE?

Ø. Bjøntegaard and E.J. Sellevold
The Norwegian University of Science and Technology (NTNU), Trondheim, Norway

Abstract
Thermal dilation (TD) and autogenous shrinkage (AS) generate stresses in concrete members hardening under restraint. The sum of TD and AS may readily be determined in the laboratory for realistic temperature histories, and may serve as an accurate basis for stress calculations in structures. However, for general calculation programs it is desirable to have individual material models for TD and AS. An experimental strategy is proposed to obtain a basis for such models. Preliminary results are given, which demonstrate that AS depends strongly on temperature history, and cannot be realistically estimated from isothermal tests.
Keywords: Autogenous shrinkage, high performance concrete, temperature dependence, thermal dilation

1   Introduction and background

Hardening concrete will generate stresses if the movements caused by hydration reactions are restrained. There are two active mechanisms producing movements: Thermal Dilation (TD) and Autogenous Shrinkage (AS). Traditionally, only thermally induced stresses have been considered when estimating cracking risk in young concrete. However, with increased use of high performance concrete (water-to-binder ratio below 0.45) it has become clear that AS may contribute significantly to stress generation, and any serious approach to estimate cracking risk by calculations must take both TD and AS into account as "driving forces" to stress generation.

In a realistic case, where the concrete goes through a natural heating-cooling cycle the first days, of course TD and AS occur simultaneously, and the sum of the two deformations may easily and accurately be measured in the laboratory for the given

Autogenous Shrinkage of Concrete, edited by Ei-ichi Tazawa. Published in 1999 by E & FN Spon, 11 New Fetter Lane, London EC4P 4EE, UK. ISBN: 0 419 23890 5

temperature history. This total deformation is the driving force to stress generation, and from a calculation point of view, there is no need to separate TD and AS. However, a more general stress calculation procedure applicable to any temperature development requires a mathematical model for each mechanism. To achieve this goal it is necessary to design and perform experiments that makes a separation of TD and AS possible. Our effort to do this is the topic of the present paper.

To our knowledge there is, surprisingly, no systematic attempt to formulate general models for TD and AS reported in the literature. Information is available on the Thermal Dilation Coefficient (TDC), and a large number of papers have been published in recent years on AS, but the vast majority on isothermal tests at 20°C.

AS is the <u>external</u> consequence of the chemical shrinkage associated with the hydration reactions, which amounts to about 25% by volume of the chemically bound water. The first few hours the concrete behaves as a liquid, and as chemical shrinkage takes place it is directly reflected in the AS measured externally. During setting a solid skeleton is formed, which allow empty pores to form (self-desiccation) and, as a consequence, AS becomes much smaller than the chemical shrinkage. Thus, initially AS (measured volumetrically) is equal to the chemical shrinkage, while AS is reduced strongly through setting. To measure AS accurately (both linearly and volumetrically) at early times (before and during setting) is very difficult. It is of practical importance, however, since Norwegian experience shows that high performance concrete is sensitive to cracking during this period [1]. The problems associated with AS and very early cracking is a topic discussed separately [2] at this conference.

The present paper only considers AS in the period after setting, i.e. when the elastic modulus is sufficiently developed so that stresses are produced in the Stress-Rig (TSTM). Even in this period there are experimental pitfalls, mainly associated with water supply. If bleeding has taken place before setting, then the bleed water will be reabsorbed by the concrete as self-desiccation occurs, resulting in reduced shrinkage or even expansion. Similarly, water available in the aggregates may play a role, since very little water is required to "refill" the self-desiccation pores and thereby eliminate AS. For instance, in a concrete with 400 kg cement, a 50% degree of hydration corresponds to a volume of self-desiccation pores of: $400 \cdot 0.24 \cdot 0.25 \cdot 0.5 = 12 \ l/m^3$. This amount of water can easily be available in the aggregates. We have demonstrated that AS is eliminated in concretes with lightweight aggregate [6]. We believe that variations both in amounts and availability of water in aggregates may explain the large differences in AS found in published reports.

The information on TD in the literature is also very limited, but provides certain expectations to the TDC development during the hardening phase. In the fresh (liquid) state the water phase dominates and TDC is very high compared to solids (possibly $60 \cdot 10^{-6}/°C$). When a skeleton is formed we expect solid behavior with much lower TDC. Wittman [3] found that for cement paste (w/c=0.40, saturated) that TDC increased from about $10 \cdot 10^{-6}/°C$ at 3 days hydration fairly linearly up to about $18 \cdot 10^{-6}/°C$ at 60 days. The moisture content is also important. For mature cement paste Meyers [4] found that TDC depend strongly on ambient Relative Humidity (RH), in that TDC increased from $12 \cdot 10^{-6}/°C$ at 100% RH, to about $18 \cdot 10^{-6}/°C$ at 70% RH, decreasing to $12 \cdot 10^{-6}/°C$ again at 40% RH. It should be added that both publications do present contradictions and inconsistencies, but we conclude that a minimum TDC is expected after setting, followed by a gradual increase over time both due to increased hydration and reduced RH because of self-desiccation. The determination of TDC also depends strongly on experimental conditions and rate of temperature change. Cement paste contains pore water in different states: interlayer-, gel-, adsorbed-, capillary- and "free" water in the largest pores. Temperature changes will induce redistribution of water between the different phases, for example, heating will drive water from small pores (gel-adsorbed water) to larger pores (capillary water) since the entropy of water in small pores is lower than in large pores. Thus, depending on the time scale, the observed thermal dilation will consist of a "pure" temperature effect (fast), a moisture redistribution effect (slower) and finally both superimposed on a basic AS as long as the hydration reaction is going on.

Based on the preceding considerations the following approach was taken to obtain generalized models for TDC and AS during concrete hardening under realistic temperature conditions. The steps outlined below developed as consequence of the results obtained "on the way". The work has by no means reached the final stage yet. Hence, this paper is a report on work in progress, for the purpose of clarifying the present situation for ourselves as well as for colleagues involved in the same process. Note that all the experiments are carried out using one particular high performance concrete mix.

1. AS at a series of isothermal temperatures from mixing. Purpose: to establish if a simple maturity concept can be used to describe AS. This proved to be impossible.
2. AS at a series of isothermal temperatures, but with fixed starting temperatures (13°C and 20°C) from mixing until 8 hours. At 8 hours the temperature was increased in steps of 7-10°C until the desired isothermal temperatures were reached. This series is referred to as "Poly-isothermal".

The results of steps 1 and 2 demonstrated that a simple generalized description of AS in terms of maturity is not possible. The next step was therefore to determine TDC as directly as possible during isothermal and realistic temperature histories.

3. TDC measured "directly". The principle is to superimpose steps of ±3°C for some hours duration on either isothermal or realistic temperature developments. This scheme allows the calculation of TDC at each step, and then to calculate AS by removing the TD-contribution to the total measured movement. This series is referred to as "Saw-toothed".

Work with this latter approach is presently going on, and the resulting TDC and AS are being analyzed with a view to obtaining generalized descriptions. The work is discussed below.

2   Experimental

2.1   Test method

The experimental work is based on the use of a concrete dilatometer which measures the free length change of a prismatic specimen with dimension 100 mm • 100 mm •

500 mm (length), see Fig. 1. At each end of the specimen an inductive displacement transducer measures the length change. The signals are recorded separately and added to obtain the total length change. The transducers are connected to the specimen by thin invar rods reaching 10-15 mm into the specimen and, hence, the "active" length of the specimen is 470-480 mm. At the end of each invar rod a thin steel disc is fixed as "anchorage". A thermo-couple measures the temperature in the center of the specimen. Registrations of both length change and temperature start within 1 hour after concrete mixing. The temperature control of the specimen is provided by water that is circulated in copper tubes fixed outside 5 mm thick copper plates forming the inner walls of the mould. The plates are heavily insulated on the outside. The water temperature is controlled by a computer.

When the fresh concrete is placed in the mould it is covered with a Al/plastic foil impermeable to moisture. The 5 mm copper plate cover is placed on top and weighted down to provide a seal against moisture loss. The ability of the equipment to prevent moisture loss during the experiments is presumably high, however, not yet determined by measurements. This important question will be resolved by experiments in the near future.

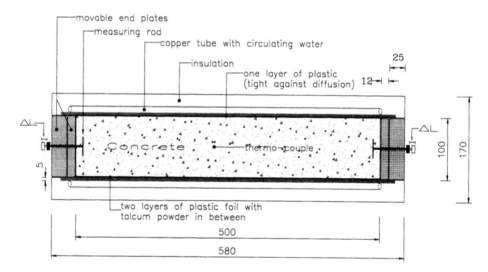

Fig. 1 The concrete dilatometer

2.2 Temperature histories

All the experiments reported here were carried out using one particular high performance concrete mix. Up till now a total of 19 tests have been performed, of which 16 with different temperature histories. Table 1 shows the different temperature histories employed. The initial temperature at each test is given in the left column of the table and the subsequent temperature "treatment" is indicated in the other columns. As can bee seen from the table, most of the tests are performed with initial temperatures of 20°C. Fig. 2 shows measured temperatures during some of the tests with initial temperature of 20°C.

Table 1. Survey of temperature histories (°C)

Initial temperature	Isothermal	Isothermal (saw-toothed) ±3	Poly-isothermal Final temperature				Realistic (smooth) Max.temperature				Realistic (saw-toothed) Max.temperature	
			20	27	35	45	29	32	47	62	50	60
5	XX											
13	X		X		X							
20	XXX	X		X	X	X	X		X	X	X	X
26								X				
45	X											

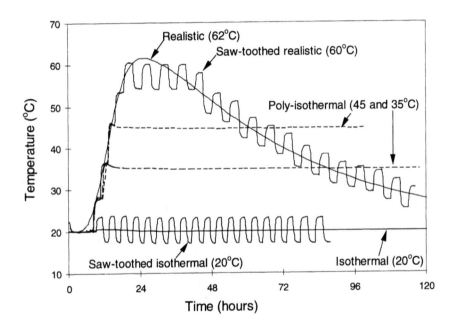

Fig. 2 Examples of different temperature histories (measured)

3   Mix proportions

The same mix is used in all tests. The water-to-binder ratio is 0.40 and the silica fume content is 5% of cement weight. The concrete represents a typical Norwegian concrete used for bridge constructions. Mix proportions are given in Table 2. The slump had a mean value of 18 cm with a coefficient of variation of 15% for the 12 batches at 20°C. The air content varied from 2% to 3.1% for the 19 concrete batches.

Table 2 Mix proportions (kg)

Cement (CEM I-52.5 LA)	368
Water	155
Silica fume	18.4
Sand 0-2 mm	150
Sand 0-8 mm	767
Stone 8-11 mm	636
Stone 11-16 mm	318
Plasticizer (Scancem P)	1.93
Superplasticizer (Mighty 150)	3.09

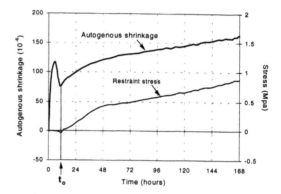

Fig. 3 Autogenous shrinkage and generated stresses (100% restraint) during 20°C isothermal tests. t_o indicates where AS is zeroed.

4   Results

4.1   General behavior of autogenous shrinkage

Fig. 3 shows typical 20°C isothermal results both for AS and the corresponding stress development under full restraint in the TSTM. AS is recorded from time zero, i.e. about ½-1 hour after mixing. The TSTM only provides full restraint after the feedback system is activated at 6 hours after mixing (the feedback system requires a certain concrete stiffness before it is activated). The AS curve shown in Fig. 3 is quite typical: The initial strong contraction (whose physical reality is questionable) is followed by expansion. The expansion is real as evidenced by the (small) compressive restraint stress, and is possibly caused by reabsorption of bleed water. We have chosen to zero the AS curves at the end of the expansion (10 hours age at 20°C). This is indicated as t_o in Fig. 3, and, as can bee seen, AS developing beyond t_o generates significant stresses in the TSTM. The behavior before t_o is discussed in detail at this conference [2].

4.2   Reproducability

As shown in Table 1 a number of repeat tests have been made. Fig. 4 shows such data for isothermal tests at 5°C and 20°C. There is deviation between the curves, but the main features of the behavior are reproduced. Note that within the IPACS-program [5] there is a Round-Robin test under way both concerning AS and stressgeneration under full restraint.

4.3   Isothermal tests

The concretes were mixed at temperatures as close to the test temperatures as possible. Fig. 5 shows the AS results at 5, 13, 20 and 45°C. The desired temperatures were in all cases reached within one hour after placing the concrete in the temperature controlled

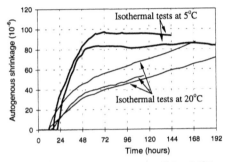

Fig. 4 Repeated isothermal tests at 20°C and 5°C

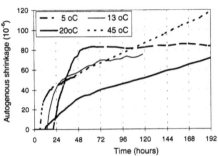

Fig. 5 Autogenous shrinkage during isothermal tests at different temperatures

dilatometer i.e. long before the zero-time for the AS-plots. The figure shows very unsystematic behavior; particularly the first 48 hours. Clearly, a simple time shift using the maturity principle will not be able to characterize the results coherently. From a practical point of view, fully isothermal tests are not very relevant, hence we abandoned this approach and proceeded to the more realistic "poly-isothermal", see below.

4.4 Poly-isothermal tests

This series was intended to be more realistic in terms of practical concreting, using fixed temperatures of 13°C and 20°C the first 8 hours, and then increasing the temperature in steps to the desired level, see Fig. 2. Fig. 6 gives the results vs real time. In Fig. 7 a "normal" activation energy (ΔE) for strength development is used to transform real time to maturity. Note that the expression ΔE=A+B(20-T) is used. In Fig. 8 a much higher ΔE value is used. The thermal dilations due to the temperature steps are not shown in the figures, they are eliminated by a curve fitting procedure. Note also that the temperatures are not "realistic", in the sense that a given final temperature is maintained till the end of the test, and not reduced after the heating phase as would be the case in a structure.

The raw data given in Fig. 6 is fairly systematic for each initial temperature, increased temperature leads to increased magnitude and rates of AS. The time shift provided by the normal activation energy is not enough to make them coincide (Fig. 7). The very high ΔE-value used in Fig. 8 makes the curves coincide in the period up to about 1 week maturity time, but with significant deviation after that. Furthermore, the two different initial temperatures (13 and 20°C) give quite different results both for the rate of AS development and total AS after a certain time. In practical terms, a generalized description of AS using maturity can be given, but only for a given initial temperature and up to about 1 week maturity time. Note that with this high ΔE-value 1 hour at 45°C correspond to 19 hours at 20°C, hence agreement up to 1 maturity week means only about 9 hours at 45°C, something which is much to short time to be practically useful. This is not a very general, nor a very practical result, and a different approach was chosen for the next stage.

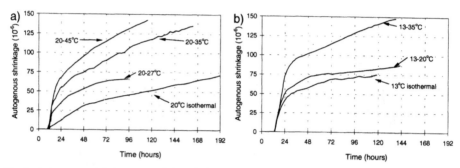

Fig. 6 Autogenous shrinkage vs real time during poly-isothermal tests with initial temperature of (a) 20°C and (b) 13°C.

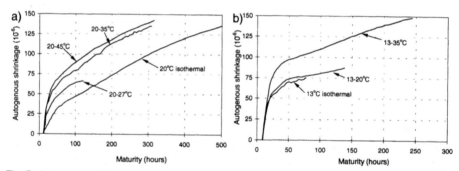

Fig. 7 Autogenous shrinkage vs maturity time during poly-isothermal tests with initial temperature of (a) 20°C and (b) 13°C. Maturity time is the equivalent time at 20°C isothermal conditions (A=33500 J/mole and B=1470 J/mole·°C)

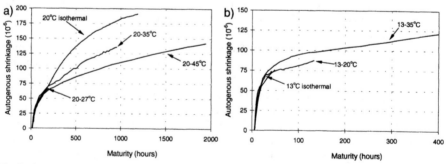

Fig. 8 Autogenous shrinkage vs maturity time during poly-isothermal tests with initial temperature of (a) 20°C and (b) 13°C. Maturity time is the equivalent time at 20°C isothermal conditions (A=91500 J/mole and B=0)

AS is clearly a very complicated concrete "property", in that what is measured in the laboratory is the <u>overall</u> result of a large number of physical processes: a) the various hydration reactions and their dependence on temperature, time and the state of stress existing in the various phases; b) the state of internal stress generated by self-desiccation through a combination of capillary and Gibbs-Bangham mechanisms; and 3) the viscoelastic response of the skeleton matrix which changes very strongly both with degree of hydration, temperature level and the rate of temperature change.

Thus there is no reason to expect simple behavior of AS even under the relatively simple isothermal and poly-isothermal conditions discussed so far. In a realistic case where the temperature development is analogous to the one shown in Fig. 2, the situation is even more complicated and to base a model on the isothermal and poly-isothermal results is clearly not reasonable. The next step was therefore to determine AS as directly as possible under realistic conditions. This was done by determining the TDC during an experiment using the "saw-tooth" principle illustrated in Fig. 2, see below.

4.5   Realistic- and saw-tooth temperature tests

The detailed temperature program for the saw-tooth tests are given in Fig. 9, together with the set of smooth realistic temperature curves which have also been applied in the AS and TSTM equipment. Fig. 10 gives the saw-tooth results in terms of calculated TDC vs real time. Fig. 11 shows an example of the relationship between total measured deformation and the two contributing parts TD and AS, as calculated. Fig. 12 shows three sets of AS curves, each set comparing AS calculated from smooth and saw-tooth temperature developments shown in Fig. 9.

The isothermal results form a consistent picture; Fig. 12 shows that the AS for smooth and saw-tooth correspond well, demonstrating that the ±3°C steps did <u>not</u> change the overall AS. The TDC (Fig. 10) behave as expected; strong initial decrease followed by slow increase as the skeleton matrix dominates and RH is decreased by self-desiccation.

It is worth noting the following consistent observation in the isothermal saw-tooth test: In each step the temperature in the concrete reached thermal equilibrium after about 1 hour and the temperature was maintained, at least 1 more hour before the next step. In a COOL-step (23-17°C) the second hour produced a AS slope similar to the 20°C isothermal "background"; while in a HEAT-step the initial expansion was followed by a <u>very high </u>rate of contraction each time, several time as high as the "background". Thus, if moisture redistribution is the main mechanism for the time dependence it is slower during HEAT (gel→capillary) than the opposite way during COOL. The fact that the mean saw-tooth AS curve corresponds to the smooth isothermal demonstrates that no extra irreversible structural changes are introduced by the temperature steps. These results also demonstrate that there is no "unique" TDC for concrete; its magnitude depends on the time allowed for the time dependent part. Much more work is clearly required to reach a better understanding of TD.

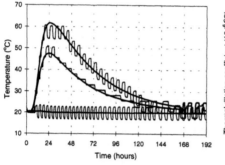

Fig. 9 Measured concrete temperatures: realistic (smooth and saw-toothed) and isothermal (smooth and saw-toothed)

Fig.10 Development of the TDC at 3 different saw-toothed temperature histories, see Fig. 9

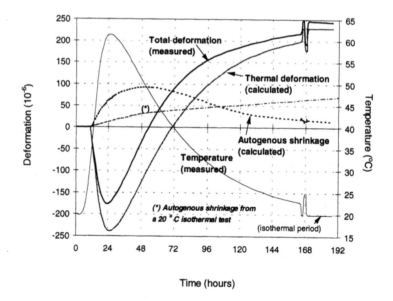

Fig. 11 Example of the relationship between total measured deformation and the two contributing parts TD and AS (smooth temperature, 62°C maximum)

The results in Fig. 10 for the two saw-toothed realistic temperature histories given in Fig. 9 are less consistent. The TDCs for the 50°C maximum curve correspond very directly to the isothermal, but the 60°C maximum curve is different. It is, however, clear from Fig. 9 that the saw-tooth pattern also is different in the two tests. These are the only saw-toot tests run so far; the need for more work is obvious. In Fig. 12 the AS are calculated for the two sets of realistic tests using the TDC given in Fig. 10. The principle of calculation is illustrated in Fig. 11 for the smooth realistic case with 62°C maximum. Fig. 12 shows fairly bad agreement for the two sets of smooth and saw-tooth curves. Relatively small changes in the TDCs have been shown to be able to produce much better correspondence. The main point, however, is that each set of

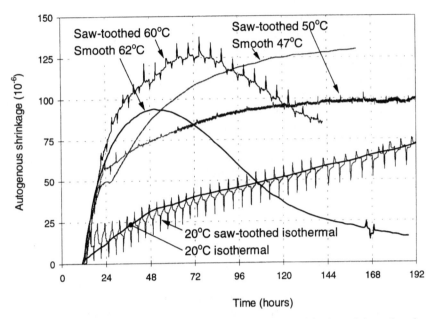

Fig. 12 Development of AS at realistic- (smooth and saw-tooted) and isothermal (smooth and saw-toothed) temperatures.

curves display the same major features: The 50°C maximum curves show that AS flatten out after about 4 days. The 60°C maximum curves show that AS becomes negative (i.e. expansion!) after 2 to 3 days.

The latter observation is surprising, but possible real as indicated in Fig. 11 after 168 hours: Here the temperature is constant and the total measured deformation is expansion. This has been shown in two experiments, but not yet by the stress development in the TSTM, since the specimens break very early in the cooling phase with 100% restraint. A system is now ready to provide less than 100% restraint to enable stresses to be measured over long times. The expansion after 2-3 days is of practical importance, since the critical time for cracking in many cases occurs later than this.

5   Conclusions

These preliminary results demonstrate very clearly a few main points:

- At our present state of knowledge the only certain way to estimate deformations in order to predict stresses in a structure by calculations, is to measure thermal dilation (TD) + autogenous shrinkage (AS) in the laboratory at the relevant temperature development. Any generalized TDC and AS models are not trustworthy today.

- The method (saw-tooth pattern etc) to determine TDC experimentally must be refined through further experiments.
- The determination of AS as the difference between the total measured deformation and the calculated TD (based on TDC from saw-tooth tests) appears the most promising at the moment. Efforts should be concentrated on this area first, using realistic temperature histories. The primary questions to be answered are:
  - Does increased maximum values in realistic temperature histories lead to decreased AS after a few days culminating at around 60°C in AS-expansion?
  - Is this a general phenomenon or dependent on binder components?
- A basic isothermal AS-test at 20°C is considered useful as reference point for a given mix. However, the primary questions should, in our opinion, be answered in as broad a way as possible (internationally), before realistic material models can be formulated.

6 Acknowledgement

The present work is part of the Norwegian project NOR-IPACS supported by NFR (Norwegian Research Council) with the following partners: Selmer ASA (project leader), Elkem ASA Materials, Norcem A/S, Fesil ASA and NTNU. The national project is from mid 1997 integrated with a new European Brite-Euram project (IPACS) with Sweden (Scancem) as project leader and partners from Norway, Germany, Italy and The Netherlands.

7 References

1. Kompen, R., (1994) High performance concrete: Field observations of cracking tendency at early age. *Proceedings of the International Rilem Symposium: Thermal Cracking in Concrete at Eearly Ages,* Munich, Rilem Proceedings 25, Oct. 1994
2. Hammer. T.A. (1998) Test method for linear measurement of autogenous shrinkage before setting. *Paper submitted to this conference*
3. Wittman, F. and Lukas, J. (1974) Experimental study of thermal expansion of hardened cement paste. *Materals and Structures*, no. 40, pp. 247-52
4. Meyers S.L. (1951) How temperature and moisture changes may effect the durability of concrete. *Rock Products*, pp. 153-7, Chicago, Aug. 1951
5. Bjøntegaard Ø., Sellevold E.J. and Hammer T.A. (1997) High performance concrete at early ages: Selfgenerated stresses due to autogenous shrinkage and temperature. *Third CANMET/ACI International Symposium on Advances in Concrete Technology*, Auckland, New Zeeland, Aug. 25-27
6. Hammer T.A., Bjøntegaard Ø. and Sellevold E.J. (1998) Cracking tendency of high strength lightweight aggregate concrete at early ages. *CANMET/ACI/JCI Fourth International Conference on Recent Advances in Concrete Technology*, Tokushima, Japan, June 7-11

17 AUTOGENOUS SHRINKAGE MEASUREMENT

P.-C. Aïtcin
Department of Civil Engineering, University of Sherbrooke, Sherbrooke (Québec) Canada

Abstract
The measurement of drying shrinkage of ordinary concrete is performed under well established standard methods, which give reliable and reproducible results. However, these methods are inappropriate when used to measure the shrinkage of a low water/binder concrete specimen (high performance concrete) because, during the first 24 hours when the specimen is kept in the mould, there takes place, to a varying extent, autogenous shrinkage due to the development of self-desiccation.

Self-desiccation develops in any concrete that is not water cured because it is a direct consequence of hydration reactions, but the magnitude of autogenous shrinkage varies. In fact, autogenous shrinkage is a consequence of the development of menisci in the capillary network in the hydrating cement paste. The smaller the menisci the larger the autogenous shrinkage. This translates, in terms of the water/binder ratio: the smaller the water/binder ratio the greater autogenous shrinkage.

The essential differences and similarities between drying shrinkage and autogenous shrinkage are discussed, and the main factors influencing autogenous shrinkage are presented. Based on the experience of the author and on theoretical considerations, the principle of a new procedure for the evaluation of the effects of different types of shrinkage on the length change of concrete is presented.
Keywords: Autogenous shrinkage, drying shrinkage, high-performance concrete, self desiccation, measurement of shrinkage, hydration reaction, heat of hydration, temperature rise

1 Introduction

The measurement of drying shrinkage of ordinary concrete is easy and does not present

Autogenous Shrinkage of Concrete, edited by Ei-ichi Tazawa. Published in 1999 by E & FN Spon, 11 New Fetter Lane, London EC4P 4EE, UK. ISBN: 0 419 23890 5

any problems: it is only necessary to be patient and methodical. Standard methods, like ASTM C 157 "Standard Test Method for Length Change of Hardened Hydraulic Cement, Mortar and Concrete", are well established to achieve a reliable and reproducible measurement of length changes of specimens of ordinary concrete subjected to drying. This fortunate situation arises from the fact that, from a practical point of view, drying shrinkage is the only type of shrinkage which develops significantly in ordinary concrete. However, when such a standard method is used to measure the shrinkage of high performance concrete with a low water/binder ratio, it will be shown that this is totally inappropriate. The reason for this is that the standardized methods of shrinkage measurement ignore the autogenous shrinkage that develops in a high performance concrete specimen during the first 24 hours after casting.

The autogenous shrinkage developed during the first 24 hours in a specimen of ordinary concrete is negligible compared with the drying shrinkage that will develop when the concrete dries. Hence, the initial length measured when the specimens are demoulded 24 hours after casting can be used as a reference length. However, this is not so in the case of high performance concrete that has already undergone some contraction due to the development of autogenous shrinkage.

In order to understand better the difference in the shrinkage behavior of ordinary concrete and high performance concrete, it is necessary to go back and look at the essential differences between drying shrinkage and autogenous shrinkage. It is only on that basis that it is possible to define the principle of an appropriate testing method to measure properly the different types of shrinkage that can develop in concrete.

The differences and similarities between drying shrinkage and autogenous shrinkage will be appreciated from the description of these two types of shrinkage.

2   Drying shrinkage

As a result of a disequilibrium in the relative humidity between concrete and its environment, any concrete loses water when exposed to an environment whose relative humidity is lower than that existing in the capillary network in the concrete. As a result of this water loss, concrete shrinks. In common usage, the term shrinkage is a shorthand expression for drying shrinkage of hardened concrete exposed to air with a low relative humidity. The magnitude of drying shrinkage depends on many factors, including the properties of the materials, temperature, relative humidity of the environment, the age at which concrete is first exposed to a drying environment, and the size of the concrete element that is drying [1][2].

The driving force for drying shrinkage is the evaporation of water from the capillary network in the concrete at the menisci which are exposed to air with a relative humidity lower than that within the capillary pores. As the free water contained in the capillary pores is held by forces which are inversely proportional to the diameter of the pores, the loss of water is progressive and proceeds at a decreasing rate. Researchers still disagree about the fundamental mechanism responsible for drying shrinkage. These mechanisms include the capillary tension, the surface adsorption, and the interlayer water removal [3][4][5][6][7].

Drying shrinkage occurs first at the surface of concrete, and indeed only at a surface exposed to dry air; this could be a single surface or all external surfaces of a concrete element. Drying shrinkage then progresses toward the core of the element so that the loss of water expressed as a percentage of the apparent volume of concrete is smaller for concrete elements with a lower surface/volume ratio. Factors influencing the magnitude of the loss of water are the porosity of the concrete and the characteristics of the capillary pore network such as the size and shape of the pores and their continuity.

3   Autogenous shrinkage

The expression autogenous shrinkage was first mentioned in the literature in 1934 [8] and positively defined by Davis [9] as a macroscopic volume change caused by the hydration of cement. The expression chemical shrinkage is also found in the literature because autogenous shrinkage is a consequence of the decrease in the absolute volume of the hydrated cement paste during its hydration [10]. Jensen and Hansen [11] found in the literature eight different expressions to describe autogenous shrinkage. In this paper, we shall use the most widely used expression: autogenous shrinkage [12].

It may be worth noting that, because the products of hydration can form solely in water-filled space, only a part of the water in the capillary system can be used in hydration. Thus, for hydration to take place, there must be enough water present both for the chemical reactions and for filling of gel pores. These gel pores are formed by the reactions of hydration of cement. They are much finer than capillary pores so that they drain water from the coarsest capillary pores. As a consequence in the absence of any external water supply, the coarsest capillary pores start to dry in a manner similar to drying by evaporation. This is why this phenomenon is called self-desiccation.

The causes of autogenous shrinkage are the same as the causes of drying shrinkage because it is the same physical phenomenon that is developing within concrete: the creation of menisci within the capillary system and the resulting tensile stresses. Nevertheless, there are some major differences between autogenous shrinkage and drying shrinkage:

- autogenous shrinkage develops without any mass loss, unlike drying shrinkage;
- autogenous shrinkage develops isotropically within the concrete while drying shrinkage progresses from the drying surface to the core of the element;
- autogenous shrinkage does not develop any relative humidity gradients, drying shrinkage yes;

The development of autogenous shrinkage, which is linked directly to cement hydration, starts to develop uniformly and isotropically in a matter of hours after the casting of concrete (almost always before 24 hours), whereas drying shrinkage starts to develop more or less slowly at the surface when hardened concrete is exposed to a dry environment (which is usually a matter of days rather than hours).

Autogenous shrinkage is not an unavoidable phenomenon if the coarse capillaries are refilled by an external supply of water. As long as the gel pore system and capillary pore systems are connected to this external supply of water, no menisci are formed within the capillary system, so that there are no induced tensile stresses and no autogenous shrinkage.

The ingress of external water favors further hydration of the still unhydrated parts of cement particles but, at this stage, hydration results in an increase in the absolute volume because the water ingresses from outside the concrete. In consequence, the gel pore system and the capillary pore system may become disconnected, depending on the local pore size distribution in the hardening concrete. As soon as the water phase has ceased to be continuous, water ingress from outside stops and conditions for the development of autogenous shrinkage are created. This explains why specimens of the same cement paste (with a water/cement ratio of 0.30) but of different size, deform in different ways: if thin, they can expand, and they can contract if they are larger (Fig. 1) [13][14]. It has also been observed that, in some cases, very small concrete specimens with a very low water/cement ratio can crack when immersed in water immediately after casting [15]. Here, only the near the surface pore system remains connected to the external supply of water, so that only a thin layer of concrete is free from autogenous shrinkage, while in the core of the specimen autogenous shrinkage develops [15]. Such a situation is far from ideal with respect to cracking as it will be seen in the next section.

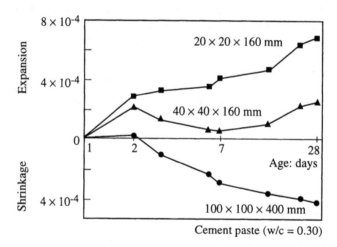

Fig. 1. Influence of the size of the specimen on its volumetric change (after Tazawa and Miyazawa, 1997).

4 Influence of the temperature of concrete on volumetric changes during hydration

Let us now look at the influence of the evolution of the temperature of hardening concrete on autogenous shrinkage (if it occurs). Hydration reactions are accompanied by a heat release which, depending on environmental conditions and geometrical factors, results usually in an increase of temperature of the concrete, more rarely in a decrease of temperature, or even in isothermal curing conditions.

Let us suppose first that the concrete temperature rises. The rise is accompanied by an increase in the absolute volume concurrent with the autogenous shrinkage of the hydrated cement paste. It is, therefore, possible that, in certain circumstances, the volume increase due to thermal expansion is equal to the autogenous shrinkage. In that situation, cement hydration takes place without any change in absolute volume. It is often observed that, during the very first hours of hardening, concretes with a very low water/binder ratio swell as long as the thermal expansion is larger than autogenous shrinkage. However, usually, autogenous shrinkage overtakes the thermal expansion quite rapidly, so that low water/cement ratio concretes shrink after the initial swelling phase.

If the concrete temperature decreases while hydration is proceeding, which can be the case in unprotected thin slabs cast in winter [16][17], then the thermal contraction of concrete has to be added to the autogenous shrinkage. This is the worst situation as far as shrinkage cracking is concerned.

Isothermal hydration is an intermediate, well-defined curing condition, during which autogenous shrinkage develops in the absence of any thermal volume change. It follows that isothermal hydration of concrete not exposed to drying can be used to determine the maximum autogenous shrinkage that can develop in a given cement paste or concrete. It should be emphasized that, when concrete specimens used to measure shrinkage are cast in $100 \times 100 \times 375$ mm steel moulds and cured for 24 hours at room temperature, hydration proceeds under quasi-isothermal conditions.

Measuring volumetric changes under isothermal conditions, in the absence of an external supply of water, represents a well-defined curing condition that can be used to measure the maximum autogenous shrinkage that can develop in a given cement paste or concrete.

It is very important to look further at the influence of the temperature of concrete during the development of autogenous shrinkage with respect to its major consequence: shrinkage cracking. Let us look at the results of a very simple experiment in which concrete with a water/binder ratio of 0.30 was cast in a $300 \times 300 \times 900$ mm plywood form, 20 mm thick. Two vibrating wires (Geokon type VCE 4202 and 4204), one 100 mm and the other 200 mm long, were placed on the longitudinal axis specimen, as shown in Fig. 2. The temperature of the concrete was monitored by thermocouples near the vibrating wires (and also through the readings of the vibrating wires) since casting up to the age of 35 days. There were no discrepancies between the values of temperature given by the thermocouples and the vibrating wires.

The top part of the figure is a plot of the temperature of the concrete near the vibrating wires. Part AB represents the temperature rise of concrete cured in the quasi-adiabatic condition of the early hydration. This condition is followed by a curing condition in which the heat losses reduce the heat generated by hydration until the temperature of the concrete reaches its peak (Part BD). Finally, concrete cools down when heat losses are greater than the heat generated by the hydration of cement particles (Part DC).

The actual strain given by the vibrating wire, without any temperature correction, is shown in the lower part of Fig. 2 by Curve I. It can be seen that during the dormant period no strains were recorded. Thereafter, the concrete experienced some expansion. If we look at the temperature/time curve, we find that this expansion occurs at the early stage of hydration when the concrete is heating up and the autogenous shrinkage is starting to develop. However, this expansion period did not last very long, (from 6 h to 10 h) because autogenous shrinkage then overtook the thermal expansion so that the initial expansion gave way to contraction. It is interesting to note that it is only after about 15 hours that the concrete specimen returned to its initial size. After the peak temperature, autogenous shrinkage as well thermal shrinkage continue to take place. Both effects are cumulative at this stage until concrete returns to ambient temperature. After 2 days, thermal contraction stopped, but autogenous shrinkage did not.

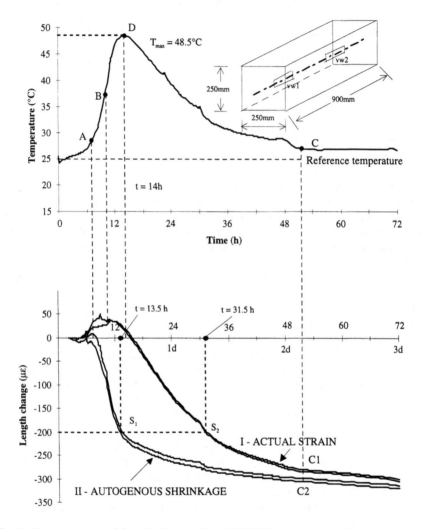

Fig. 2. Temperature and length changes in a 0.30 W/B concrete specimen.

If the vibrating wire reading is corrected in order to eliminate the effect of the temperature (Curve II) in Fig. 2, then the corrected strain is the same (as a first approximation) as the strain in a concrete specimen cured under quasi-isothermal conditions. (The departure from true isothermal conditions is due to the self-accelerating nature of the hydration reactions).

Thus it can be said that, in the experiment being discussed, as a first approximation, the actual strain/time curve corresponds to volumetric change of concrete under quasi-adiabatic conditions. The strain/time curve corrected for the temperature effect corresponds to the volumetric change of concrete under quasi-isothermal conditions. In order to apply the temperature correction to the reading of the strain gage, a reference temperature has to be selected; in our case, it was 24° C. Moreover, an average thermal expansion coefficient has to be used; like many other researchers we selected 10×10^{-6} per °C. When autogenous shrinkage is monitored in small $100 \times 100 \times 375$ mm specimens, the conditions are truly isothermal because the hydration reactions are not accelerated by the heat of hydration.

Let us now use the actual and corrected strain curves to find the time when a strain of 200×10^{-6} has developed. For the corrected curve (isothermal conditions), this strain is reached 13 hours after placing the concrete, whereas for the uncorrected curve it occurs after 31.5 hours. From a consideration of cracking this is a major difference. Specifically, under isothermal conditions, a strain of 200×10^{-6} occurs when the concrete has just left the plastic state but has not yet acquired very strong cohesion; such concrete is very weak in tension and therefore prone to cracking. On the other hand, under quasi-adiabatic conditions, a strain of 200×10^{-6} is developed in very strong concrete. In the case under study, the 31.5 hour compressive strength could be estimated (from an interpolation between the 1-day and 7-day values) to be 45 MPa. The corresponding tensile strength is estimated to be 6.3 MPa.

It follows that, in high performance concrete (in which autogenous shrinkage is significant when early water curing is not possible) the increase in the temperature of concrete is very helpful as it delays the development of effective strains. When water curing is used to limit the development of autogenous shrinkage, it is important that only a small amount of water is used in order to avoid transforming quasi-adiabatic curing conditions into quasi-isothermal curing conditions, which could be worse than no curing at all in so far as cracking is concerned. During this water curing, it is necessary to supply only a very small amount of water over the amount just needed to limit the development of autogenous shrinkage.

In the case of a high water/cement ratio concrete, where autogenous shrinkage is negligible, the heat of hydration results in an expansion of the concrete specimen as soon as the concrete temperature increases, and in thermal contraction after the peak temperature.

In all concretes, regardless of the water/cement ratio, it is important to limit as much as possible the peak temperature in order to limit the final thermal contraction. However, it must be emphasized that thermal contraction occurs when concrete is already quite strong in tension so that it can resist the stresses induced by thermal contraction.

Let us now look at what happens to concrete specimens used to measure shrinkage when they are demoulded after 24 hours and cured under different conditions.

5 Volume variation of hydrated cement paste or concrete cured under water

It is common knowledge that, when hydrated cement paste or concrete with a high water/cement ratio is cured under water, slight swelling occurs. This swelling has been explained by the introduction of water molecules within the layered structure of the C-S-H. In the case of concrete with a low water/binder ratio, the situation is more complex. Depending on the degree of connectivity between the gel pores and the capillary network, autogenous shrinkage may or may not develop. The determining factor is the water/cement ratio. The size of the specimen also plays an important role because the ingress of water into concrete is a direct function of its porosity, which largely depends on the value of the water/cement ratio. Therefore, in thin specimens made of low water/cement ratio concrete, autogenous shrinkage can be negligible whereas in large concrete elements, a more or less thick surface layer of concrete is free from autogenous shrinkage; this is not the case in the interior of the elements. These two different shrinkage behaviors of concrete within an element result in a shrinkage gradient [11][19][20][21].

If a reference specimen is to be permanently immersed after demoulding, it is very important to immerse it in water before hydration begins. The concrete setting time, as measured by ASTM C403, can be used to determine when immersion of the mould containing the concrete should take place.

6 Drying shrinkage time dependence

Any concrete shrinks when subjected to drying, what can vary is the rate and the ultimate value of the shrinkage. The ultimate value of shrinkage depends primarily on the size of the capillaries that have emptied (as a function of the relative humidity of the ambient air), the water/cement ratio (which influences the tensile strength of concrete), and on the aggregate skeleton (which resists drying shrinkage).

It is very often said that low water/cement ratio concretes develop lower drying shrinkage than high water/cement ratio concretes. This can be true at the age of 28-days, or even at 1 year, but such a statement reflects an ignorance of the fundamental origin of drying shrinkage. Admittedly, a low water/cement ratio concrete exhibits a lower drying shrinkage rate than a high water/cement ratio concrete because that self-desic-cation may have already dried a significant volume of the capillary network. Also, a low water/cement ratio concrete offers a higher tensile strength which can resist the tensile stresses developed by the menisci. However, the total shrinkage at an infinite time should not be too different and is affected only by the tensile strength of concrete.

7 Volumetric changes and standard testing methods

Volumetric contraction accompanying the hydration reactions may or may not result in autogenous shrinkage. The governing factors are the presence or absence of an external source of water and also whether the gel pores system remains or not connected to this external source of water through the capillary pore system.

When self-desiccation occurs, it may or may not result in overall contraction because autogenous shrinkage can be counterbalanced by thermal expansion. If autogenous shrinkage is not fully counterbalanced by thermal expansion, water curing at a later stage results in an expansion of the hydrated paste which can counterbalance the residual autogenous shrinkage and thermal contraction as long as water can ingress into the concrete.

It is evident from the above that the measurement of shrinkage of low water/cement ratio concrete is not an easy task: the length change is a combination of the development of autogenous shrinkage, of thermal volumetric changes and of drying shrinkage, all of which occur sequentially or simultaneously within the given concrete specimen. Thus, the total shrinkage depends on:

- the pore size distribution in the concrete which is affected by the water/cementitious material ratio as well as the water/cement ratio,
- the connectivity of the gel pore and capillary pore systems,
- the size of the specimen,
- the early curing conditions after the start of the hydration reactions,
- the heat developed during the cement hydration and the heat loss at the surfaces of the sample,
- the type of curing applied when concrete is cooled back to the ambient temperature.

In the case of concrete with a very high water/binder ratio (0.60 to 0.80) cured in the absence of an external source of water, there is practically no autogenous shrinkage because, following the volumetric contraction of the hydrated cement paste, the high porosity drains the water from the numerous large capillary pores. Moreover, the number of cement particles hydrating per unit volume of concrete is small. Consequently, the menisci originating from self-desiccation have large diameters and result in very weak tensile stresses. There is, therefore, no problem in measuring the reference length of the concrete sample at 24 hours when the specimen is demoulded according to ASTM C157 and exposed later on to dry air in order to measure shrinkage. The ASTM C157 method is, therefore, perfectly valid to measure the drying shrinkage of high water/binder ratio concrete.

But problems start when this testing method is blindly extended to measure the shrinkage of a low water/binder ratio concrete. If we have, for example, a 0.30 water/binder ratio concrete specimen, cast and cured during the first 24 hours according to ASTM C157, as soon as hydration begins, self-desiccation starts to develop under almost isothermal conditions, without a source of external water until the demoulding, usually at 24 hours. As previously said, as the specimen is small, its temperature does not rise much because the heat loss is relatively large. Moreover, when using the ASTM C157 method, concrete specimens are protected from drying during the first 24 hours only by being placed in a moist curing room at a minimum relative humidity of 95 percent. Consequently, during the first 24 hours, there is no external water supply that can compensate for self-desiccation.

The magnitude of autogenous shrinkage developed during the first 24 hours is a function of the water/binder ratio, but it depends also on the time when the hydration reactions begin, on the type of cement, and on the type of admixture used. For example, with an accelerator, hydration starts earlier and results, other things being equal, in a larger autogenous shrinkage at 24 hours because more cement is hydrated when demoulding occurs. On the contrary, if a retarder is used, the hydration reactions are delayed and, when the specimen is demoulded at 24 hours, smaller autogenous shrinkage has developed, in spite of the fact that the water/binder ratio of the two concretes was the same.

Consequently, if the magnitude of autogenous shrinkage at 24 hours has to be determined precisely, it is necessary to embed a length gage within the specimen at the time of casting or alternatively to develop a new method of measurement of autogenous shrinkage as proposed by Tazawa and Miyazawa [18], Persson [22] or Jensen and Hansen [23]. It is also useful to install a thermocouple in the specimen in order to monitor the variation in temperature during the first 24 hours in order to have a better knowledge of the development of the reactions of hydration during this critical period.

From the above, it follows that the published data on autogenous shrinkage or drying shrinkage of high performance concrete obtained using ASTM C157 standard procedure are of no value because they all missed the initial autogenous shrinkage during the first 24 hours, as shown in Fig. 3.

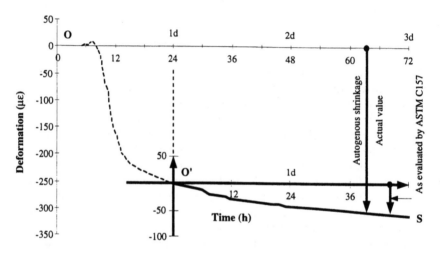

Fig. 3. Why ASTM C157 test method cannot be used to measure autogenous shrinkage of low water/binder ratio concrete.

8   Proposed specimen preparation for the measurement of autogenous shrinkage in low water/binder concrete

As isothermal curing in the absence of an external source of water represents a well defined curing condition, easily reproducible in the laboratory, such curing can be used to measure the **maximum potential autogenous shrinkage** that can develop in a low water/binder ratio concrete specimen without moist curing. It is necessary to measure the autogenous shrinkage developed during the first 24 hours [7][13][24]. In order to continue the measurement of autogenous shrinkage after demoulding at the age of 24 hours, the specimens have to be sealed at this stage. This is necessary to prevent the development of drying shrinkage. Sealing can be done, for example, by the application of two layers of adhesive aluminum foil. Once the autogenous shrinkage has stabilized, the foils can be removed. Henceforth, drying shrinkage can be measured.

9 Summary

Whereas it is easy to measure drying shrinkage of high water/binder ratio concrete, this is not the case for low water/binder ratio concrete because of the very early development of autogenous shrinkage during the early stages of hydration. It is, therefore, inappropriate to establish the reference length of a concrete specimen for the determination of shrinkage as late as 24 hours after casting because, in some cases, a large part of autogenous shrinkage would have occured by then.

Autogenous shrinkage is influenced by the water/binder ratio, the type of cement and the various cementitious materials, the type of admixtures, and the curing conditions.

The influence of the temperature on length change is such that in the core of a concrete element shrinkage usually develops under quasi-adiabatic conditions. In a small specimen, autogenous shrinkage occurs under quasi-isothermal conditions.

A simple procedure of the treatment of shrinkage test specimens is suggested in order to measure the **maximum autogenous shrinkage** that can develop in any concrete. It is important that a well defined and accepted procedure be used by researchers and engineers, if we want to progress in mastering concrete shrinkage. Mastering concrete shrinkage means controlling concrete cracking, and controlling concrete cracking means improving the durability of concrete structures.

References

1. Neville, A.M. (1995) *Properties of Concrete*, Longman Group Limited, Harlow Essex England and John Wiley & Sons, New York, pp. 426-448.
2. Wiegrink, K., Marikunte, S., Shah, S.P. (1996) Shrinkage Cracking of High-strength Concrete. *ACI Materials Journal*, Vol. 93, No. 5, pp. 409-415.
3. Wittmann, F. (1968) Surface Tension Shrinkage and Strength of Hardened Cement Paste. *Matériaux et Constructions*, Vol. 1, No. 6, pp. 547-552.
4. Wittmann, F.H. (1976) On the Action of Capillary Pressure in Fresh Concrete. *Cement and Concrete Research*, Vol. 6, No. 1, pp. 49-56.
5. Buil, M. (1979) Contribution à l'étude du retrait de la pâte de ciment durcissante, *Rapport de Recherche LPC No. 92*, December, 72 p.
6. Baron, J. (1982) Les retraits de la pâte de ciment, chap. 27; *Le béton hydraulique, Connaissance et Pratique*, (ed. J. Baron and R. Sautery) Les Presses de l'École Nationale des Ponts et Chaussées, Paris, pp. 485-501.
7. Nagataki, S., Yonekura, A. (1983) The Mechanisms of Drying Shrinkage and Creep of Concrete. *Transactions of Japan Concrete Institute*, Vol. 5, pp. 127-140.
8. Lyman, C.G. (1934) *Growth and Movement in Portland Cement Concrete*, Oxford University Press, London, U.K., pp. 1-139.
9. Davis, H.E. (1940) Autogenous Volume Change of Concrete, *Proceedings of the 43rd Annual American Society for Testing Materials*, Atlantic City, N.J., June, pp. 1103-1113.
10. Le Chatelier, M.H. (1900) Sur les changements de volume qui accompagnent le durcissement des ciments. *Bulletin de la Société de l'Encouragement pour l'Industrie Nationale*, 5e série, Tome 5, Jan., pp. 54-57.

11. Jensen, O.M., Hansen, P.F. (1996) Autogenous Deformation and Change of the Relative Humidity in Silica Fume-Modified Cement Paste. *ACI Materials Journal,* Vol. 93, No. 6, pp. 539-543.

12. Aïtcin, P.-C., Neville, A., Acker, P. (1997) Integrated View of Shrinkage Deformation. *Concrete International*, Vol.19, No.9, pp. 35-41.

13. Tazawa, E. I., Miyazawa, S. (1995) Experimental Study on Mechanism of Autogenous Shrinkage of Concrete. *Cement and Concrete Research*, Vol. 25, No. 8, pp. 1633-1638.

14. Tazawa, E.I., Miyazawa, S. (1997) Influence of Constituents and Composition on Autogenous Shrinkage of Cementitious Materials. *Magazine of Concrete Research*, Vol. 49, No. 178, pp. 15-22.

15. Tazawa, E.I., Miyazawa, S. (1996) Influence of Curing Conditions on Autogenous Shrinkage of Concrete, *International Conference on Engineering Materials, Ottawa, Canada,* (ed. A. Al-Manaser, S. Nagataki, R. Joshi) CSCE/JSCE, Ottawa, Tokyo 1997, pp.373-384.

16. Lessard, M., Dallaire, E., Blouin, D., Aïtcin, P.-C., (1994) High-Performance Concrete Speeds Reconstruction for McDonald's. *Concrete International*, Vol. 16, No. 9, pp. 47-50.

17. Lachemi, M., Bois, A.-P., Miao, B., Lessard, M., Aïtcin, P.-C. (1996) First Year Monitoring of the First Air-Entrained High-Performance Bridge in North America. *ACI Structural Journal*, Vol. 93, No. 4, pp. 379-386.

18. Tazawa, E.I., Miyazawa, S. (1993) Autogenous Shrinkage of Concrete and Its Importance in Concrete Technology, in *Creep and Shrinkage,* (ed. Z.P. Bazant and I. Carol) E & FN SPON, London, pp. 159-168.

19. Attolou, A. (1985) *Étude du séchage du béton par gammadensimétrie*, compte rendu LCPC-AER 130114, June 1985, 22 p.

20. Alvaredo, A.M., Wittmann, F.H. (1995) Shrinkage and Cracking of Normal and High Performance Concrete, in *High Performance Concrete: Materials, Properties and Design,* (ed. F.H. Wittmann and P. Schewesinger) Aedificatio Verlag, Freiburg, pp. 91-110.

21. Baroghel-Bouny, V., Godin, J., Gawsewitch, J. (1996) Microstructure and Moisture Properties of High-performance Concrete, *4th International Symposium on Utilization of High-strength/High-performance Concrete,* (ed. F. de Larrard and R. Lacroix) Les Presses de l'Ecole Nationale des Ponts et Chaussées, Vol. 2, pp. 451-461.

22. Persson, B. (1997) Self-desiccation and its Importance in Concrete Technology. *Materials and Structures*, Vol. 30, No. 199, pp. 293-305.

23. Jensen, O.M., Hansen, P.F. (1995) A Dilatometer for Measuring Autogenous Deformation in Hardening Portland Cement Paste. *Materials and Structures*, Vol. 28, No. 181, pp. 406-409.

24. Lepage, S., Baalbaki, M., Dallaire, E., Aïtcin, P.-C., Early Shrinkage Development in a High Performance Concrete, submitted to *Cement, Concrete and Aggregate.*

18 EFFECT OF CONSTITUENTS AND CURING CONDITION ON AUTOGENOUS SHRINKAGE OF CONCRETE

E. Tazawa
Department of Civil Engineering, Hiroshima University, Hiroshima, Japan
S. Miyazawa
Department of Civil Engineering, Ashikaga Institute of Technology, Ashikaga, Japan

Abstract
Various factors influencing autogenous shrinkage of concrete, such as the type of cement, water-cement ratio, volume concentration of aggregate are experimentally investigated. The prediction model for autogenous shrinkage of concrete is proposed based on experimental data. The application of maturity concept to estimate the effects of specimen size on autogenous shrinkage is studied. Shrinkage of concrete specimens exposed to the atmosphere with different relative humidity is also discussed.
Keywords: curing, prediction, relative humidity, type of cement, water-cement ratio

1 Introduction

It has been proved from recent studies that autogenous shrinkage of concrete due to cement hydration is considerably large for high-strength concrete[1][2]. It has also been demonstrated that autogenous shrinkage should be taken into account for crack control and design of high-strength concrete structures[1]. For example, autogenous shrinkage stress due to restraint by embedded reinforcing bars can be very large[3]. When high-strength concrete is used for structures with large members, not only thermal stress due to cement hydration but also autogenous shrinkage stress should be considered[4].

A prediction equation for estimating shrinkage strain of normal strength concrete has been proposed by Japan Society of Civil Engineers(JSCE)[5]. The shrinkage strain calculated with the JSCE equation, which includes autogenous shrinkage and drying shrinkage, is a function of water content of the concrete and the ratio of the surface area to the volume of the member, because it has been derived under conditions in which drying shrinkage is dominant. However, autogenous shrinkage occurs even

Autogenous Shrinkage of Concrete, edited by Ei-ichi Tazawa. Published in 1999 by E & FN Spon, 11 New Fetter Lane, London EC4P 4EE, UK. ISBN: 0 419 23890 5

in concrete without evaporation, and is significantly influenced by water-cement ratio rather than water content. And cement type has great influence on autogenous shrinkage[6], as is not the case in drying shrinkage.

For massive concrete structures, autogenous shrinkage occurs simultaneously with thermal strain during a heating and a cooling periods. It is convenient that autogenous shrinkage strain and thermal strain can be determined separately in order to analyze stress generation for crack control.

Most part of shrinkage has been consider as drying shrinkage for ordinary concrete. For high-strength concrete, however, both shrinkage during the curing period and shrinkage after curing should be considered. Influence of ambient relative humidity on shrinkage of high-strength concrete has not been clear.

In this study, effect of aggregate content on autogenous shrinkage is experimentally studied and the estimation by an existing composite model is proposed. A prediction equation for autogenous shrinkage of concrete with normal content of aggregate is proposed on the basis of experimental data. Influence of specimen size and ambient relative humidity on autogenous shrinkage is also investigated.

2   Experimental procedures

2.1   Materials and mix proportion

In series 1, ordinary Portland cement was used. Hard-sandstone was used for fine aggregate(specific gravity: 2.56, absorption: 1.19 %, fineness modulus: 2.86) and coarse aggregate(specific gravity: 2.60, absorption: 1.00 %, maximum size: 20 mm). Modulus of elasticity of the original rock measured with ø50×100 mm core was 7.57 ×10⁴N/mm². Cement paste and concrete with water-cement ratio of 0.2, 0.3 and 0.5 were prepared. Volume concentration of aggregate of concrete with 0.2, 0.3 and 0.5 water-cement ratio was 0.53, 0.62 and 0.67 respectively.

In series 2, ordinary Portland cement (N) and low heat cement with high C_2S content (L) were used. Type of aggregates were different among different mixes and the properties of aggregate are given in published papers which are shown with a number in Table 1. Water-cement ratio ranges form 0.2 to 0.56.

In series 3, ordinary Portland cement, river sand (specific gravity: 2.60, absorption: 1.76 %, fineness modulus: 3.15) and gravel (specific gravity: 2.75, absorption: 1.26 %, maximum size: 25 mm) were used. Ten percent of cement by weight was replaced with Silica fume. Water-binder ratio was 0.2.

In series 4, ordinary Portland cement, river sand (specific gravity: 2.63, absorption: 1.39 %, fineness modulus: 2.68) and gravel (specific gravity: 2.56, absorption: 1.84 %, maximum size: 25 mm) were used. Concretes with water-cement ratio of 0.2, 0.3 and 0.5 were prepared.

In any series, a polycarboxylic acid type superplasticizer was used for concretes with water-cement ratio less than 0.40 and an air entraining agent was used for concretes with water-cement ratio more than 0.50. Mix proportions of and properties of flesh concrete are shown in Table 1.

Table 1. Mix Proportion and Properties of Flesh Concrete

Cement Series	w/c	s/a (%)	Unit content (kg/m³)				ad. (%)	Aggregate volume fraction	air (%)	Slump (slump flow) (cm)	Initial setting (hour)	Ref. No.
			W	C	S	G						
Series 1 N	0.20	38	170	850	505	856	2.2	0.526	2.0	2.0	2.3	
	0.30	41	165	550	646	966	0.6	0.624	7.0	7.0	6.3	
	0.50	43	170	340	722	996	0.03	0.665	6.0	6.0	5.1	
Series 2 N	0.20	30	160	800	434	1070	1.6	0.556	2.1	4.5	3.8	
	0.20	31	160	800	452	1044	0.6	0.567	3.3	(56.0x55.0)	-	10
	0.23	32	160	696	498	1120	1.3	0.599	1.7	8.5	4.5	
	0.28	40	160	569	645	997	2.45	0.620	2.5	(57.5x56.5)	9.0	14
	0.30	37	165	550	606	1092	1.0	0.630	2.0	17.5	7.0	
	0.30	37	170	567	590	1044	0.2	0.621	3.0	19.5	-	10
	0.31	43	160	510	714	976	2.2	0.639	2.0	(60.5x58.5)	10.3	14
	0.40	40	170	425	671	1064	0.04	0.645	4.7	5.0	-	
	0.40	40	180	450	655	1021	0.6	0.637	4.1	12.5	-	9
	0.50	43	170	340	752	958	0.03	0.638	5.4	15.0	6.7	
	0.56	46	160	286	850	1025	0.25	0.709	2.9	6.5	9.0	14
Series 2 L	0.23	32	160	696	498	1120	1.3	0.599	0.6	(74.5x71.0)	5.2	
	0.28	40	160	580	645	997	1.9	0.620	4.0	(59.0x58.5)	9.0	14
	0.30	37	165	550	611	1097	1.0	0.630	3.7	(67.0x65.0)	9.0	
	0.30	34	160	533	633	1120	1.0	0.656	1.9	7.0	-	4
	0.38	47	160	421	818	951	1.7	0.669	2.3	(60.5x59.0)	6.0	14
	0.65	48	156	240	902	1025	0.25	0.729	3.5	7.0	-	14
Series 3 N	0.20	30	160	800*	426	1053	1.7	0.547	2.5	21.0	-	
Series 4 N	0.20	30	160	800	439	998	2.3	0.552	3.0	20.0	2.5	
	0.50	37	162	667	667	1106	0.04	0.680	2.3	16.0	-	

*: Cement(C)+Silica fume(SF), SF/(C+SF)=0.1

2.2 Measurement of length change

In series 1, 2 and 4, 100×100×400 mm specimens were used for autogenous shrinkage test. The test was done in accordance with the JCI test method[7]. For the first 24 hours after casting, the measurements were done with mixtures in the mold by dial gauges(Fig 1). In order to eliminate the restraint by the mold, foamed styrol mold was used for cement paste, and Teflon plate 1 mm in thickness was put on the bottom of steel mold for concrete. The measurement was started at time of initial setting which was determined by Japanese Industrial Standard(JIS R 5201) for cement paste and JIS A 6204 for concrete. During the first day, temperature of specimens rose due to hydration. Therefore, thermal expansion was excluded from the measured strain on the assumption that thermal expansion coefficient of cement paste and concrete were $20 \times 10^{-6}/°C$ and $10 \times 10^{-6}/°C$ respectively. After specimens were demolded at the age of 24 hours, the specimens were sealed with aluminum tape in order to prevent from evaporation, and length change was measured at specified ages. The maximum change in mass of tested specimen was no more than 0.02% during the test periods(it is specified 0.05 % in JCI method), therefore influence of moisture movement to or from the specimens can be ignored.

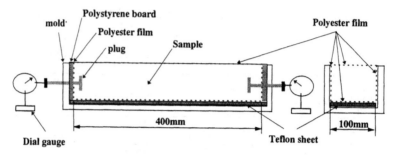

Fig. 1   Measurement of autogenous shrinkage of concrete during first 24 hours

In series 3, in order to study the effect of specimen size, both $100 \times 100 \times 400$ mm and $400 \times 400 \times 1600$mm specimens were used. Length change of specimens was measured from immediately after casting with embedded strain gauges, modulus of elasticity of which was 40N/mm^2. The specimens were sealed to prevent from evaporation throughout the test period.

In series 4, some of the specimens were exposed to different atmosphere at relative humidity of 40%, 60%, 80% and 90% after 7 days of sealed curing.

3   Results and discussions

3.1   Influence of volume concentration of aggregate (Series 1)

As autogenous shrinkage of concrete occurs in cement paste phase, it is less than that of cement paste itself because of the presence of aggregate particles. It is assumed that concrete can be expressed by a two-phase model, where aggregate particles are dispersing in cement paste matrix as inclusions. Hobbs' model[8], eq.(1), which was proposed for estimating drying shrinkage of concrete, is used to evaluate the effect of aggregate on autogenous shrinkage. It has been proved by the authors that influence of aggregate on autogenous shrinkage after the age of 24 hours can be estimated with Hobbs' model for mortar and concrete with different volume concentration of aggregate[9]. In this study, applicability of Hobbs' model to autogenous shrinkage before 24 hours is investigated.

$$\varepsilon_c / \varepsilon_p = \frac{(1 - V_a)(K_a / K_p + 1)}{1 + K_a / K_p + V_a(K_a / K_p - 1)} \tag{1}$$

where

ε_c is autogenous shrinkage of concrete

ε_p is autogenous shrinkage of cement paste

V_a is volume concentration of aggregate

K_p is bulk modulus of elasticity of cement paste

K_a is bulk modulus of elasticity of aggregate

$K = E / 3(1 - 2v)$, v is Poisson's ratio, E is modulus of elasticity

Poisson's ratio of the cement pastes, on which the calculated value of $\varepsilon_c/\varepsilon_p$ is hardly dependent, is assumed to be 0.16. The test results are shown in Fig 2, with estimated curves given by eq.(1), a parallel and a series models. In the calculations, the values of K_a and K_p are determined by experiments, as shown in Table 2. It can be seen from these figures that Estimation by Hobbs' model gives good agreement with the observed values.

Table 2. Modules of elasticity of aggregate and cement paste($\times 10^4 N/mm^2$)

water-cement ratio	age			aggregate
	1 day	7 days	28 days	
0.2	1.93	2.51	2.87	
0.3	1.49	2.37	2.47	7.57
0.5	0.94	1.47	1.42	

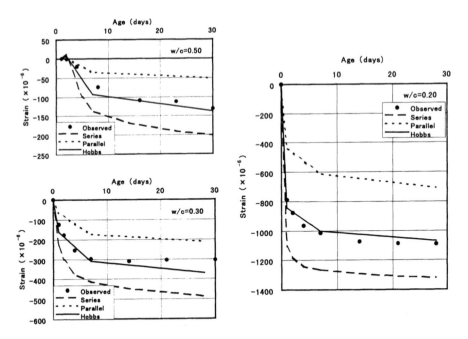

Fig. 2. Autogenous shrinkage of concrete and its estimation by composite models.

3.2 Influence of water cement ratio(Series 2)

Autogenous shrinkage of concrete is strongly dependent on water-cement ratio as shown in Fig. 3. It can be seen that the ultimate value of autogenous shrinkage increases with decrease in water-cement ratio. And autogenous shrinkage of concrete with low water-cement ratio grows more rapidly and reaches the ultimate value at earlier ages than that of concrete with high water-cement ratio.   From these points of view, in the prediction model proposed by Japan Concrete Institute (JCI), autogenous shrinkage is expressed as the product of the ultimate value ε_{c0}(w/b) and the coefficient β(t) which describes the development of autogenous shrinkage with time. The ultimate value ε_{c0}(w/b) and the development with time β(t) are obtained by the approximation of observed values.

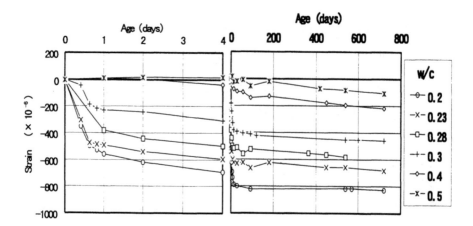

Fig. 3.   Autogenous shrinkage of concrete with different water-cement ratio.

$$\varepsilon_c(t)= \gamma\ \varepsilon_{c0}(w/b)\ \beta(t)\times 10^{-6} \tag{2}$$

For $0.2\leq w/b\leq 0.5$: $\varepsilon_{c0}(w/b)=3070\exp\{-7.2(w/b)\}$ (3)

For $0.5 < w/b$: $\varepsilon_{c0}(w/b)=80$ (4)

$\beta(t)=[1-\exp\{-a(t-t_0)^b\}\ ]$ (5)

where
$\varepsilon_c(t)$ is autogenous shrinkage of concrete at age t
γ is a coefficient to describe the effect of cement type ($\gamma =1.0$ for ordinary Portland cement)
$\varepsilon_{c0}(w/b)$ is the ultimate autogenous shrinkage
$\beta(t)$ is a coefficient to describe the development of autogenous shrinkage with time
w/b is water to binder ratio
a and **b** are constants which are given in Table 3
t is age in day
t_0 is initial setting time in day

If concrete temperature is not 20℃, t and t_0 are modified with eq. (6)

$$t, t_0 = \sum_{i=1}^{n} \Delta t_i \cdot \exp\left[13.65 - \frac{4000}{273 + T(\Delta t_i) / T_0}\right] \tag{6}$$

where

Δt_i is the number of days where a temperature T (℃) prevails

$T(\Delta t_i)$ is the temperature during the time period Δt_i $T_0 = 1$℃

As unit water content of concrete is generally ranging form 150 to 170 kg/m³, the volume concentration of aggregate generally ranges form 0.55 to 0.60 for concrete with 0.2 water-cement ratio, and 0.69 to 0.72 for concrete with 0.40 water-cement ratio. Difference in volume of aggregate results in only 5 % variation in autogenous shrinkage strain, when the composite model is used for the estimation. Therefore, the variation in aggregate volume concentration can be neglected without significant error for estimating autogenous shrinkage of concrete including common volume of aggregate.

The prediction model for autogenous shrinkage is valid for concretes with water-cement ratio ranging from 0.20 to 0.56, and with normal volume concentration of aggregate, at an environmental temperature ranging from 20℃ to 60 ℃.

Water-binder ratio (w/b) is taken as the main valuable in a function of the ultimate value of autogenous shrinkage ε_{co}(w/b), although ε_{co} is independent of w/b for concrete with w/b more than 0.50. In Fig.4, the calculated values are compared with the observed ones for concrete at the age of 91 days. It can be seen from this figure that the ultimate value of autogenous shrinkage can be properly predicted with the proposed equation.

The coefficients **a** and **b** in eq. (5), which describe the rate of autogenous shrinkage, are dependent upon water-cement ratio. Coefficients **a** and **b** obtained from observed values are shown in Table 3. The relationship between observed ε $_c$(t)/ ε_{co} and calculated one from eq. (5) is shown in Fig. 5. The calculated values have a good agreement with the observed ones.

Table 3.   Coefficients a and b in eq. (5)

w/c	a	b
0.20	1.2	0.4
0.23	1.5	0.4
0.30	0.6	0.5
0.40	0.1	0.7
more than 0.50	0.03	0.8

3.3   Influence of cement type(Series 2)

Autogenous shrinkage of concrete is strongly dependent on the type of cement. Medium heat Portland cement and belite rich cement result in lower autogenous

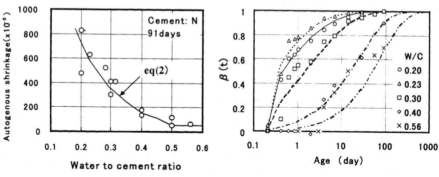

Fig. 4. Autogenous shrinkage versus w/c.   Fig. 5. β (t) and observed value.

shrinkage than ordinary Portland cement[6].   It has been reported that autogenous shrinkage is increased by using silica fume and blast furnace slag with high specific surface area and that it is slightly decreased by using fly ash[10][11].   In the proposed model, influence of the type of cement and both type and content of mineral admixture is taken into account by coefficient γ.   Autogenous shrinkage of concrete with belite rich cement (C_2S: $53.3\sim57.8\%$) is shown in Fig. 6 with the calculated value by the proposed equation, where γ is taken to be 0.6.

Table 4.   Example for mineral composition of cement(%)

Type	C_3S	C_2S	C_3A	C_4AF	$CaSO_4$	total
N	64.9	11.0	7.1	8.2	3.9	95.1
L	22.4	57.8	3.3	9.7	4.1	97.3

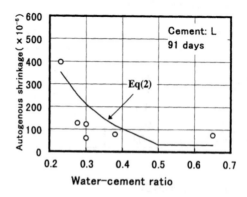

Fig. 6. Relation between water-cement ratio and autogenous shrinkage.

Following equations for predicting autogenous shrinkage has been proposed by

French Chapter of RELEM (AFREM) for the purpose of design of prestressed concrete structures[12](eqs.(7), (8) and (9)). In this study, observed autogenous shrinkage is also compared to the calculated value from the AFRM model.

For t<28days:

$$\varepsilon_{as}(t, f_{c28})=0 \qquad\qquad :f_c(t)/f_{c28}<0.1 \qquad\qquad (7)$$

$$\varepsilon_{as}(t, f_{c28})=(f_{c28}-20)\{2.2f_c(t)/f_{c28}-0.2\}10^{-6} \quad :f_c(t)/f_{c28}\geq0.1 \qquad (8)$$

For t≥28days:

$$\varepsilon_{as}(t, f_{c28})=(f_{c28}-20)\{2.8-1.1\exp(-t/96)\}10^{-6} \qquad\qquad (9)$$

where

$\varepsilon_{as}(t, f_{c28})$: autogenous shrinkage form the initial setting time to a certain age t
f_{c28}: compressive strength at 28 days
$f_c(t)$: compressive strength at a certain age t
If $f_c(t)$ is unknown, it may be obtained with the next equation.
$f_c(t)=\{t/(1.40+0.95t)\}f_{c28}$
t: age in day

In this equation, it is assumed that autogenous shrinkage before 28 days is related to the degree of hydration and is a functoin of $f_c(t)/f_{c28}$, and that autogenous shrinkage after the age of 28 days is a function of time. It is also assumed that autogenous shrinkage does not occur before the age corresponding to $f_c(t)/f_{c28}=0.1$.

For concrete with ordinary Portland cement, the AFRM model can give underestimation, as shown in Fig. 7. If the development of compressive strength is estimated by the AFRM model, the influence of water-cement ratio on the rate of autogenous shrinkage of concrete with wide range of water-cement ratio may not be precisely estimated, as shown in Fig. 8. Therefore, it can be said that compressive strength can not be used as a factor, and that influence of type of cement and water-

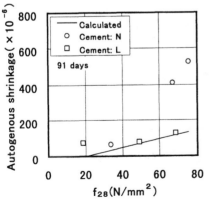

Fig. 7. ε_{as} by AFREM model.

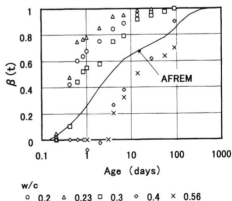

Fig. 8. β (t) by AFREM model.

cement ratio should be taken into account.

3.4   Influence of specimen size(Series 3)

Effects of specimen size on temperature history and autogenous shrinkage are shown in Fig. 9 and Fig. 10 respectively.   Thermal strain has been obtained by using a coefficient of thermal expansion of $12.0 \times 10^{-6}/\,°C$ which was experimentally determined with a 4 months old specimen.   It can be seen that autogenous shrinkage at early ages increases with specimen size, which is explained by temperature effect[4], and that ultimate value is not considerably influenced by specimen size.   In the JCI prediction model, effective age proposed by CEB-FIP model code[13] is used for considering the effect of temperature on the rate of autogenous shrinkage.   It is be seen from Fig. 11 that size effect can be considered as temperature effect and then size effect itself can be neglected, when autogenous shrinkage strain is estimated by the proposed model   It is important to establish the test method for coefficient of thermal expansion at early ages in order to estimate autogenous shrinkage precisely for the observed strain, as shown in Fig. 11.

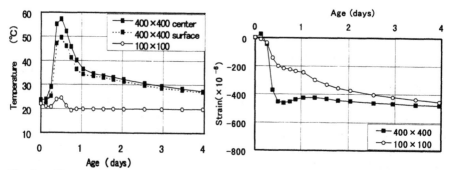

Fig. 9 .   Temperature history.   Fig. 10.   Autogenous shrinkage versus age.

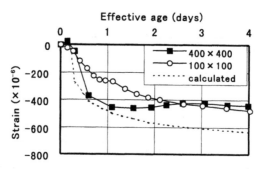

Fig. 11.   Autogenous shrinkage versus effective age.

3.4 Influence of ambient relative humidity(Series 4)

Effect of relative humidity of on shrinkage of concrete is shown in Fig.12. For the specimens with 0.5 water-cement ratio, shrinkage is increased by exposure to the surrounding atmosphere with relative humidity ranging form 40% to 90%. For specimen with 0.2 w/c, slight expansion is observed for the specimens at 80% and 90% of relative humidity. The intrinsic humidity in concrete with low water-cement ratio is decreased by self-desiccation, and may become lower than ambient humidity. Then swelling may occur due to moisture absorption under higher ambient humidity.

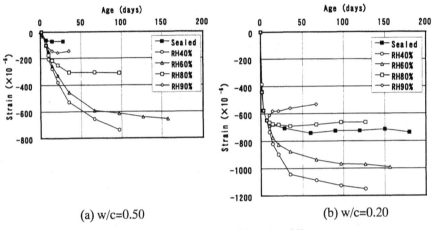

(a) w/c=0.50   (b) w/c=0.20

Fig. 12. Shrinkage of concrete at different ambient humidity.

As previously mentioned, the change in mass of the specimens during the tests was less than 0.02 %. It can be seen from Fig. 13 that this moisture transportation does not cause extra shrinkage or expansion to autogenous shrinkage. The strain predicted from the inclination of the curve is negligibly small.

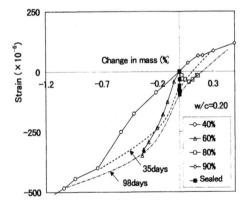

Fig.13. Relation between change in mass and shrinkage

4 Conclusions

The prediction model for autogenous shrinkage of concrete, which has been proposed by Japan Concrete Institute, is introduced. It is proved that the influence of aggregate volume concentration on autogenous shrinkage can be estimated by an existing composite model. Influence of ambient relative humidity on shrinkage of high-strength concrete is quite different from that of ordinary concrete, and a predicting model of shrinkage of high-strength concrete including autogenous shrinkage and drying shrinkage should be established.

5   References

1. Paillere, A.M. et al (1989) Effect of fiber addition on the autogenous shrinkage of silica fume concrete, *ACI Material Journal*, Vol..86, No.2, pp.139-144.
2. Tazawa, E. and Miyazawa, S. (1992) Autogenous shrinkage of cement paste with condensed silica fume, *Proceedings of the 4th CANMET/ACI International Conference on Fly Ash, Silica Fume, Slag and Natural Pozzolans in Concrete*, ACI, pp.875-894.
3. Miyazawa, S., Tazawa, E., Sato, T. and Sato, K. (1993) Autogenous Shrinkage Stress of Ultra-high-strength Concrete Caused by Restraint of Reinforcement, *Transactions of the Japan Concrete Institute*, Vol.15, pp.115-122.
4. Tazawa, E., Matsuoka, Y., Miyazawa, S. and Okamoto, S. (1994) Effect of autogenous shrinkage on self stress in hardening concrete, *Proceedings of RILEM International Symposium on Thermal Cracking in Concrete at Early Ages*, pp.221-228.
5. Japan Society of Civil Engineers (1996) *Standard Specification for Design and Construction of Concrete Structures* (in Japanese).
6. Tazawa, E. and Miyazawa, S. (1997) Influence of cement composition on autogenous shrinkage of concrete, *Proceedings of the 10th International Congress on the Chemistry of Cement*, 2ii071.
7. Japan Concrete Institute (1996) Test method for autogenous shrinkage and autogenous expansion of cement paste, mortar and concrete, *Report by Technical Committee on Autogenous Shrinkage of Concrete*, pp.195-198 (in Japanese).
8. Hobbs, D.W. (1974) Influence of Aggregate Restraint on the Shrinkage of Concrete, *Journal of ACI*, Vol.71,No.9, pp.445-450.
9. Tazawa, E., Miyazawa, S., Sato, T. and Konishi, K. (1993) Autogenous Shrinkage of Concrete", *Transactions of the Japan Concrete Institute*, Vol.14, 1992, pp.139-146.
10. Miura, T., Tazawa, E. and Miyazawa, S. (1995) Influence blast furnace slag on autogenous shrinkage of concrete, *Proceedings of the Japan Concrete Institute*, Vol.17, No.1, pp.359-364 (in Japanese).
11. Tazawa, E. and Miyazawa, S. (1992) Autogenous shrinkage caused by self desiccation in cementitious material, *Proceedings of the 9th International Congress on the Chemistry of Cement*, Vol.IV, pp.712-718.
12. Le Roy, R., De Larrard, F. and Pons, G. (1996) The after code type model for creep and shrinkage of high-performance concrete, *Proceedings of the 4th International Symposium on Utilization of High-strength/ High-performance Concrete*, pp.387-396.
13. CEB-FIP (1990) *model code CEB-FIP* 1990.
14. Miyazawa, S. and Matsumura, K. (1997) Shrinkage of concrete with low heat cement, *Proceedings of the Japan Concrete Institute*, Vol.19, No.1, pp.739-744, (in Japanese).

Discussions

Thermal dilation- Autogenous shrinkage: How to separate?
Ø. Bjøntegaard and E.J. Sellevold, The Norwegian University of Science and Technology, Norway

E. Tazawa	I appreciate your detailed research on the interaction between thermal strain and autogenous shrinkage. In the last slide, you showed different thermal coefficient of expansion for different way of experiment for three type. How did you determine thermal expansion coefficient for each time interval?
Ø.Bjøntegaard	The thermal dilation coefficient is determined by applying temperature steps ranging form 3 to 10 °C to the concrete. Temperature changes are provided by a refrigerated/heated bath circulating water within the concrete copper mould. Especially at early ages, the autogenous shrinkage may be significant and will interrupt the measurement of thermal dilation. Therefore, both before and after each temperature step there are isothermal periods of 2 hours or more. In this way, the autogenous shrinkage that occurs during each temperature step can be estimated and removed from the measured deformation (which is, of course, the sum of thermal dilation and autogenous shrinkage). The remaining thermal dilation is then simply divided by the measured temperature change giving the thermal dilation coefficient.
E. Tazawa	You controlled ambient temperature correctly but how do you determine the specimen has a certain temperature which is constant.
Ø.Bjøntegaard	A thermocouple is cast into the concrete specimen. All temperatures shown here are measured in the concrete.
E. Tazawa	May be you have some time lag for ambient temperature and actual temperature in the specimen. You have temperature distribution across the specimen. So you have to measure at the different points to get average temperature of the whole specimen. Specimen behaves by average temperature in a balanced way.
Ø.Bjøntegaard	We have tested with several thermocouples around in a sample and it's nice homogenous temperature distribution. No temperature differences were observed across the section along with the sample.

E. Tazawa	Your experiments were completely done with cement paste only?
Ø.Bjøntegaard	Everything is concrete.
K.van Breugel	Your finding is based on the relation between vapor pressure in the pore system and temperature. Have you ever thought about the possibility to look to thermal dilation coefficient of hardened concrete as a function of the degree of saturation of the pore system? We know that the thermal dilation of hardened concrete changes with the moisture content. Since the moisture content in the hardening concrete changes as well due to hydration, we may expect non-constancy of the thermal dilation coefficient of hardening concrete
Ø.Bjøntegaard	We have performed tests also on mature concrete (1 month old). It seems that the measured thermal dilation coefficient is dependent on both the temperature level and the applied temperature amplitude. So, even when the system does not change due to hydration, changes in vapor pressure/redistribution of water within the pore system interfere and make it difficult to estimate the "most valid" thermal dilation coefficient.
E. Tazawa	Cement paste has different thermal coefficient of expansion depending on water saturation. Many researchers have reported that the highest coefficient of thermal expansion is obtained at 70% of water saturation, and for higher and lower saturation it always decreases. Do you have any idea that your actual determination of thermal coefficient of concrete could be related to relative water content of the concrete in average?
Ø.Bjøntegaard	Not much work has been done on very young concrete. However, form the literature we might expect an increase of the thermal dilation coefficient of the cement paste due to the hydration process (Wittman performed tests on cement paste at 100% relative humidity (RH)) and due to reduced RH (Meyers preformed tests at different RH on mature cement paste). For concrete the RH effect is believed to be less but still significant. We know that our tested concrete self-desiccates form 100% RH at casting to approximately 80% RH after 1 month meaning that the thermal dilation coefficient should increase continuously when considering the results above. This is exactly what we measured when the temperature was continuously changed between 23 and 17 °C (see isothermal (saw-tooth) test in fig.10, page 242 in the proceedings).

E. Tazawa   For engineering purposes we can not control relative humidity.

T. Noguchi   In real structures, most of the members are subjected to external restraint for deformation to some extent. As Fig.3 in page 238 of the proceedings, the autogenous shrinkage after t_0 (the time when stress initiates) seems to be important.

May we neglect the autogenous shrinkage before the time t_0?

Is there any difference in the performance of hardened concrete between the restraint concrete and the restraint-free concrete?

Ø. Bjøntegaard   The time t_0 (in Fig.3, page 238) coincides approximately with the time of setting, that is, before t_0 the concrete has no measurable mechanical properties (E-modulus, tensile/compressive strength capacity, creep). Therefore, autogenous shrinkage (and any other strains) that occur before t_0 may be neglected in stress-calculations.

When it comes to the actual cracking tendency of the concrete, however, autogenous shrinkage is important also before t_0. In Norway, severe cracking has been experienced both before and during the time of setting and autogenous shrinkage is believed to play an important role. Thus, the cracking occurs before any stresses can be measured. The cracking problem before and during setting is very complicated and is possibly related to a kind of semi-liquid or "gel" behavior (thixotropy) involving a very low strain capacity at early ages (shown for instance by Y. Kasai in 1974, Kyoto). Therefore, during the time before t_0, we focus our research on strain/strain capacity and not on stresses.

We have not done any systematic studies on the effect of restraint stress and the performance of hardened concrete. I am sure the topic must be covered somewhere in the literature.

However, our tests of 1 to 3 weeks duration indicate that specimens that develop high tensile stresses (which may be the case during and after the cooling phase of the concrete) self-destruct (fail) at somewhat lower stress levels (5-10%) than specimens that develop lower tensile stresses, these specimens are manually loaded until failure after the end of test. This may indicate that high stress levels caused by the hydration process and restraint reduces the tensile stress capacity. However, the fact that the load intensity differ strongly within the two cases may also influence the result (a realistic stress development is slow, manually loading is fast).

Autogenous shrinkage measurement
P. -C. Aïtcin, *University of Sherbrooke, Canada*

T.A.Hammer
The effect of water curing on autogenous shrinkage (internal) is obviously dependent on size. What do you think the effect would be on the average autogenous shrinkage of a 1 m thick slab etc. ?

P.-C.Aïtcin
As I said, in our experiments, the vibrating wire gage was at 60 mm from the surface. I think that this is the most important part of the concrete to be cured because this cover protect the reinforcing bars. But I also think that if we can keep continuity of liquid phase within our sample, the water can go very deep in the sample. At Lund Conference, Prof. Guse and Hilsdorf also proposed to put a fleece all along the forms in order to have a source of water all along in the vertical surfaces. Yesterday, I was speaking of good curing practice, we have to develop good curing practice for high-performance concrete in order to control autogenous shrinkage. I think that we can get rid of most of the autogenous shrinkage even in thick element if we have proper water curing.

Ø. Bjøntegaard
The cement pastes at water-to-cement ratio of 0.35 were unaffected by curing method (water curing + different cover sheets). In the first 5 – 6 hours, the all specimens expanded equally. The cement pastes at water-to-cement ratio of 0.30 also showed expansion with the highest expansion for water cured specimens. Why do all specimens expand the first hours? My suggestion is that the expansion of sealed specimen is due to re-adsorption of bleeding water, and less expansion for the low cement pastes due to less bleeding and high density.

P. -C. Aïtcin
We just are starting our research in this field and we have comprehensive program and I hope that the next conference we can give you more satisfying explanation. Presently I don't want to give you an explanation.

Ø.Bjøntegaard
At least, I think the expansion is linked to suction of water. In the case of sealed specimen, it is hard to think that the chemical reactions within the cement changes fundamentally from a shrinking to an expanding mechanism only by changing the water-to-cement ratio.

P.-C.Aïtcin   Form the practical point of view, in between 0.30 and 0.40 there is a magic water-cement ratio where the micro structure is open enough to allow water to penetrate easily, but the drying shrinkage will not be too high and the autogenous shrinkage won't be so high. I think that we are trying to find this magic water-cement ratio, but, if we refine our curing technique specifically for high-performance concrete we will be able to cure high-performance concrete without any microcracking due to uncontrolled autogenous shrinkage.

Effect of constituents and curing condition on autogenous shrinkage of concrete
E. Tazawa, Hiroshima University, Japan
S. Miyazawa, Ashikaga Institute of Technology, Japan

Ø.Bjøntegaard   It seems that you have found the same as us in that the autogenous shrinkage is extremely temperature dependent within the first day of hydration. My conclusion was that the maturity concept fail due to this (temperature) effect. Your two autogenous shrinkage curves (one close to isothermal and one from a realistic temperature test (and time shifted with a chosen activation energy)) deviate up to abut 100% within a certain period of time. Do you consider this result satisfactory and does it justify the use of the maturity concept?

S.Miyazawa   In some cases, maturity concept doesn't work very well. So we have to study for many cases, different materials and different mix proportions. We don't have another method for estimation. In order to obtain thermal strain, we used a coefficient of thermal expansion which was measured with very old specimen. We don't have any method for measuring a coefficient of thermal expansion for very young concrete. I think this line should be continuously increasing. I think maturity concept is working for this particular concrete.

Ø.Bjøntegaard   The curves deviate something like 100 % at 1 day effective age (the absolute deviation is about 200×10^{-6}). It then seems very clear that the rate and magnitude of autogenous shrinkage is extremely important to take into account when calculating (so-called) "thermal stresses".

S.Miyazawa   It is very difficult to predict autogenous shrinkage precisely. This prediction model has an error at maximum 40 %, so we have sometime large difference.

19 MODELLING DIMENSIONAL CHANGES IN LOW WATER/CEMENT RATIO PASTES

E.A.B. Koenders and K. Van Breugel
Delft University of Technology, Delft, The Netherlands

Abstract
Hydration processes and the associated microstructural development of cement-based systems is accompanied by a number of volume changes. One of these volume changes originates from changes in the pore system itself and of the state of water in the pore system, i.e. a decrease of the vapour pressure and the associated role of surface tension in the pore system. In this contribution it will be shown how the hydration process, the microstructural development and the associated volume changes, which are supposed to be correlated to the relative humidity in the pore system, can be modelled numerically. First the basic principle of the model for hydration and microstructural development is briefly explained. Subsequently it is explained how this model is used as a basis for numerical modelling of volume changes associated with the changes in the state of water in the pore system. It is briefly shown how thermodynamic equilibrium in the continuously changing pore system is modelled. Changes in the relative humidity and in the capillary water and associated volume changes are calculated for isothermally cured mixes. Numerical and experimental results obtained for pastes and for one concrete mix are compared.
Keywords: Hydration, microstructure, self desiccation, volume changes, numerical modelling.

1 Introduction

The pore volume of cement paste is generally defined as the initial paste volume minus the volume of the solid material. The ratio between the pore volume and the initial paste volume is defined as the porosity. For cement paste, it is generally assumed that the pore diameters exhibit a continuous pore size distribution. The

Autogenous Shrinkage of Concrete, edited by Ei-ichi Tazawa. Published in 1999 by E & FN Spon,
11 New Fetter Lane, London EC4P 4EE, UK. ISBN: 0 419 23890 5

porosity and the pore size distribution depend on the degree of hydration, the water/ cement ratio, the temperature, the cement composition and the particle size distribution of the cement. With progress of the hydration process the pore structure changes and so does the state of water in the pore system. From the moment that the microstructure has gained a certain strength, the reduction of the volume of the reaction products relative to the volume of the constituents, i.e. cement and water, will result in a gradual "emptying" of the larger pores of the pore system. The "empty" pores are filled with vapour, whereas at the surface of the wall of the empty pores a layer of water is adsorbed. The process of gradual emptying of the pore system, known as self desiccation, is associated with a decrease of the relative humidity and a change in the effective surface energy caused by changes in the adsorption layer. The relationship between the pore structure, the amount of water in the paste, the relative humidity and the change in effective surface energy follow from the thermodynamic equilibrium in the pore system. Moreover, the volumetric changes of the paste have been shown to be correlated with changes of the relative humidity [1] or, alternatively, to changes in the effective surface energy [2].

In a numerical model, with which quantification of volume changes in low water/ binder ratio pastes at early ages is aimed at, the aforementioned quantities and mutual interrelationships have all to be dealt with. At first the evolution of the microstructure has to be described and modelled. For this purpose the numerical simulation program HYMOSTRUC [3] will be used. This program has the potential to simulate and predict the microstructural development as a function of the particle size distribution and chemical composition of the cement, the water/cement ratio and the temperature. For the determination of the volume changes caused by self desiccation a modified Bangham equation is used.

2 Numerical simulation of microstructural development

2.1 Modelling approach
Only few models exist with which the development of the microstructure of hardening cement-based systems can be quantified numerically [4]. One of the available

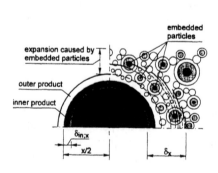

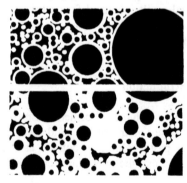

Fig. 1. Formation of interparticle contacts during hydration as considered in HYMOSTRUC [3].

Fig. 2. Result of numerical simulation of a microstructure. wcr = 0.3.
Upper part: α = 10%. Lower part: α = 50%.

models is the simulation model HYMOSTRUC, developed at the TU-Delft in the late eighties [3]. The "core" of the model is shown diagrammatically in Fig. 1. On contact of cement particles with water reaction products are formed. These products are assumed to precipitate, in part, at the surface of the hydrating grain. Hydrating cement grains can thus be considered as growing spheres. These spheres will make contact with adjacent particles. The process of external growth and formation of contacts with adjacent particles has been developed in a mathematical series. The original model, in which cement grains of the same size are considered to be positioned at equal distances, has been modified by Koenders [5] in order to allow for the actual randomness of the cement grains in the paste. For a spherical paste volume, with a diameter of the sphere of 200 µm, the result of a numerical simulation is shown in Fig. 2. The simulation concerns a paste made with a cement with Blaine fineness 420 m²/kg and a water/cement ratio 0.3. Two stages of the hydration process are presented, viz. for a degree of hydration $\alpha = 10\%$ and $\alpha = 50\%$.

With progress of the hydration process the volume of the capillary pores, V_{por}, decreases. A part of the pore system will be filled with water, V_{wat}, whereas the remaining part is filled with vapour. The water is accumulated in the smaller pores. Diagrammatically this is shown in Fig. 3. From measurements on pore size distributions it is known that the angle of the curve with which the cumulative pore size distribution can be described does not changes significantly with progress of the hydration process. In other words, the pore structure constant "a", i.e. the angle of the curve indicating the cumulative pore size distribution, is almost constant. In [5] is has been explained how the pore structure constant can be determined from the calculated pore volume V_{por} and the hydraulic radius R_H. For the hydraulic radius it holds:

$$R_H(\alpha) = \frac{V_{por}(\alpha)}{A_{por}(\alpha)} \tag{1}$$

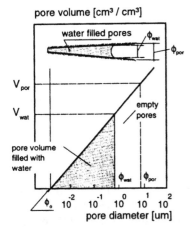

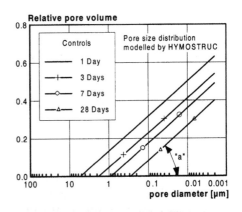

Fig. 3. Representation of (capillary) pore size distribution. Smaller pores filled with capillary water [5].

Fig. 4. Pore size distribution as used in numerical modelling. Constant value of the pore structure constant a [5].

where $A_{por}(\alpha)$ is the total pore wall area at a degree of hydration α. The pore wall area can be determined from the simulated pore structures like those presented in Fig. 2. With progress of the hydration process the calculated hydraulic radius decreases.

The measured pore size distribution can be approximated by curves as shown in Fig. 4. In this figure the relative pore volume is presented as a function of the pore diameter for different ages of a cement paste. The pore structure constant "a" appeared to vary between 0.08 and 0.12, depending on (mainly) the fineness of the cement and the presence of fines like silica fume and fly ash. For individual mixes the appropriate values of "a" were deduced from published experimental data. In a parameter study it was also found that the change of the hydraulic radius as deduced from measured pore size distributions did not differ much from those deduced from the numerically simulated pore structures [5]. Based on these findings it was considered justified to use the calculated changes of the hydraulic radius, R_H, together with the calculated changes of the pore volume V_{por}, as a basis for the determination of the changes in the pore size distribution with progress of the hydration pro-cess.

3 Thermodynamic equilibrium in the pore space

With progress of the hydration process the total capillary pore volume V_{por} and the maximum pore diameter ϕ_{por} decrease (Fig. 3). The same holds for the volume of the capillary water V_{wat} and the maximum pore that is still filled with water ϕ_{wat}. The "empty pores" are partly filled with vapour with a gas pressure p_g and partly with water that is adsorbed at the pore walls. The thickness of the adsorption layer, Γ, decreases with decreasing relative humidity. For the interrelationship between the thickness of the adsorption layer Γ, the change in relative humidity and the change in effective surface tension $\Delta\Phi$ it holds (expression after [5]):

$$\Delta\Phi = R.T \int_{p_g/p_0}^{p_g/p_0=1} \Gamma(p_g/p_0) \; d(\ln(p_g/p_0)) \tag{2}$$

in which p_g/p_0 is the relative humidity in the empty pore space, R the universal gas constant and T the absolute temperature.

In the well-known Kelvin equation the relative humidity and the size of the largest pore which is still completely filled with water, i.e. ϕ_{wat} (Fig. 3), are linked up with each other according to (after [5]):

$$\ln RH(\alpha) = \ln \frac{p_g(\alpha)}{p_0} = \frac{-4 \; \Delta\Phi}{R \; T \; \rho_w \; \phi_{wat}(\alpha)} \tag{3}$$

with ρ_w is the specific mass of water. The value of $\phi_{wat}(\alpha)$ follows from the pore size distribution and the amount of water in the pore system. These two quantities are calculated step-wise with HYMOSTRUC. The two equations (2) and (3) constitute the core of the calculation procedure for determination of the changes in the state of water and in the surface energy. For a given mix the changes in the parameter values, with which the pore structure is quantified and the changes in the state of water are determined, are all described as functions of the degree of hydration as calculated with HYMOSTRUC.

4 Towards simulation of autogenous deformations due to self desiccation

4.1 Modified Bangham concept

The volume changes of a hydrating cement paste are the result of the reduction of the relative humidity in the pore space (self-desiccation). It was Bangham [7] who found that deformational changes of coal could be related linearly to the change in the effective surface tension. This approach has been shown to be valid for hardened cement paste at low relative humidity by, among others, Wittmann [8]. It is noticed that the Bangham equation has a semi-phenomenological basis and should not be considered a physical law. This should be borne in mind when we propose to apply a modified Bangham equation for quantification of the deformational behaviour of hardening cement-based systems in which the relative humidity is beyond the range considered by Bangham, i.e. up to about 40%.

For hardened cement paste, the material properties are totally developed. As far as hardening cement paste is concerned, the material properties and the state of water change continuously with increasing degree of hydration. Therefore, the relationship between the external deformation and the change in effective surface tension must be considered incrementally. For strain increments associated with changes in the effective surface tension it is proposed:

$$\frac{\partial \varepsilon_a(\alpha)}{\partial t} = \lambda(\alpha) \cdot \frac{\partial(\Delta \Phi(\alpha))}{\partial t} \tag{4}$$

where $\partial \varepsilon_a(\alpha)$ is the strain increment that describes the microstructural deformation of the hardening paste (autogenous deformation due to self desiccation), $\partial(\Delta \Phi(\alpha))$ the surface tension increment and $\lambda(\alpha)$ a proportionality factor. For this proportionality factor an expression is proposed similar to the one used by Hiller [9]:

$$\lambda(\alpha) = \frac{\Sigma(\alpha) \cdot \rho_{pa}}{3 \, E_{pa}(\alpha)} \tag{5}$$

where:
$\Sigma(\alpha)$ = $A_{por}(\alpha)$-$A_{wat}(\alpha)$ = adsorption area
$A_{por}(\alpha)$, $A_{wat}(\alpha)$ = total pore wall area and pore wall area of filled pores
ρ_{pa} = specific mass of the cement paste
$E_{pa}(\alpha)$ = modulus of elasticity of the cement paste

As can be noticed from this relation, the pore wall area, the specific mass of the cement paste and the modulus of elasticity play an important role in the simulation of the volume changes of the microstructure (autogenous shrinkage) during the hardening phase. Thereby it is worthwhile to notice that the pore wall area of the empty pores and the modulus of elasticity are all functions of the degree of hydration. A modification of the original Bangham concept as outlined in the foregoing has been suggested tentatively by Setzer [12] when proposing to consider the change in

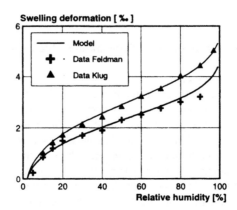

Fig. 5. Length changes of cement pastes as a function of the relative humidity in the pore system. Water/cement ratio 0.5 (Feldman) and 0.4 (Klug). Simulation with HYMOSTRUC [5].

effective surface tension of the adsorption area $\Sigma(\alpha)$ as adopted in eq. (5). He emphasized that, although an improvement, this modification must be considered as an approximation and is, in fact, still a simplification of reality. This should be borne in mind when interpreting the results on measurements and numerical simulations.

4.2 Preliminary verification of the modified Bangham concept

In order to have a first check of the reasonableness of the proposed modification of the Bangham concept, experimental data published by Feldman [13] and Klug [14] have been used. They have measured length changes of hardened pastes with water/cement ratio's of 0.5 and 0.4, respectively. The measured and simulated length changes are presented in Fig. 5 as a function of the relative humidity. In the simulation the proportionality factor $\lambda(\alpha)$ has been determined iteratively in order to allow for the increase of the adsorption area with decreasing relative humidity.

5 Measurements versus numerical simulations

5.1 Shrinkage of hardening cement paste

In Fig. 6 the autogenous shrinkage strains due to self desiccation are presented as a function of time for a cement paste made with a fine cement, Blaine 550 m²/kg. Results of the simulations and measurements are shown for pastes with water/cement ratio of 0.3, 0.4 and 0.5. All tests were performed at 20°C in order to avoid temperature effects which are expected to complicate the interpretation of experimental results. Shrinkage strains increase with decreasing water/cement ratio. The largest shrinkage strains, ca. 2‰, are measured after about 168 hours hydration for

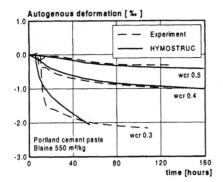

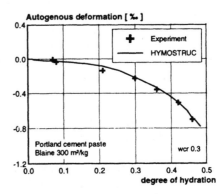

Fig. 6. Calculated autogenous shrinkage strains of cement paste caused by self desiccation, compared with measured strains. Water/cement ratios 0.3, 0.4 and 0.5.

Fig. 7. Calculated shrinkage due to self desiccation and observed shrinkage as a function of the degree of hydration. Water/cement ratio 0.3.

the paste with water/cement ratio of 0.3. For a paste with water/cement ratio 0.4 shrinkage strains of about 1‰ were reached after 168 hours hydration. This is a reduction of about 50% in comparison with the paste with a water/cement ratio of 0.3. A further increase of the water/cement ratio to 0.5 reduces the shrinkage strains by another 50%.

For a paste made with a coarse cement, Blaine 300 m²/kg and water/cement ratio 0.3, the calculated and measured shrinkage deformation due to self desiccation is shown in Fig. 7 as a function of the degree of hydration. The latter quantity is approximated by the calculated amount of heat of hydration divided by the amount of heat liberated at complete hydration of the cement according to the procedure

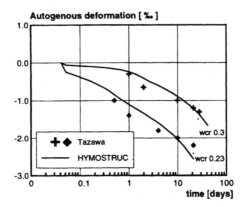

Fig. 8. Calculated shrinkage due to self desiccation and measured shrinkage. Blaine 350 m²/kg). Experiments: Tazawa [10]. Simulations with HYMOSTRUC.

proposed by van Beek [15]. The figure shows that also for a coarse cement the proposed model is able to predict the deformation of the hardening paste fairly well.

Tazawa [10] has presented results of an extensive experimental study on autogenous shrinkage of cementitious material. Several experiments performed with plain cement pastes have been simulated with the proposed numerical model. Two of these simulations is presented in Fig. 8. Good agreement was obtained between the experimental results and the numerical simulations. This holds true for pastes with a water/cement ratio 0.3 and 0.23. The shrinkage strains increase to about 1.5‰ and 2.5‰ for pastes with a water/cement ratio 0.3 and 0.23, respectively. With progress of the hydration process the shrinkage strains further increase.

5.2 Shrinkage strains in hardening concrete

In case shrinkage strains in hardening pastes are restrained, stresses will occur. Shrinkage deformations are restrained if stiff aggregate particles are enclosed in the paste. The aggregate particles were assumed not to absorb or contain any water, which could affect the hydration process. For a quantitative evaluation of shrinkage-induced stresses and of resulting deformations of the concrete, the so called "lattice model" has been applied as proposed by Schlangen [11]. In the lattice model the matrix is represented as a frame-work of short "beams", or lattices, with which the individual aggregate particles are connected. Fig. 9 shows the top half of a concrete sample with dimensions 40x40x160 mm³. The numerically determined shrinkage properties of an arbitrary paste, see Fig. 9c, are assigned to the "beams" of the lattice system. With progress of the hydration process the beams intend to shrink, causing deformations of the concrete sample. At the same time stresses are generated in those beams of which the deformations are prevented in whole or in

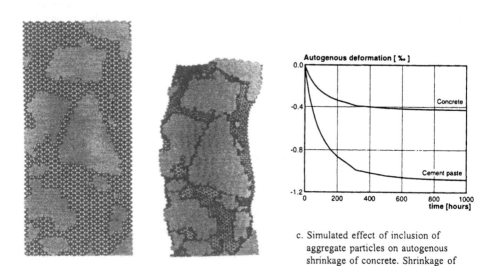

a. t = 0 hrs   b. t = 100 hrs

c. Simulated effect of inclusion of aggregate particles on autogenous shrinkage of concrete. Shrinkage of paste is input for the simulations.

Fig. 9. Numerical simulation (2-dimensional) of autogenous shrinkage deformation of a concrete sample after t = 100 hours hardening (b) and as a function of time (c).

part. The stresses are highest in areas with a high degree of restraint. In Fig. 9b the calculated deformation of the concrete sample is shown after 100 hrs hydration. Due to the inclusion of rigid particles in the paste, the (calculated) shrinkage of the concrete specimen is about 70% less than that of the plain paste (Fig. 9c). The areas where high stresses occur are susceptible to microcracking.

6 Discussion

Like drying shrinkage, shrinkage due to self desiccation is caused by changes in the state of water in the pore system. In hardening cement-based systems changes of the state of water are caused by the hydration process. With progress of the hydration process a porous microstructure is formed. The pores in the hardening system are partly filled with water and partly empty, i.e. filled with water vapour, while an adsorption layer is formed at the pore walls. The relative humidity in the pore system decrease with increasing degree of hydration. At the same time the effective surface tension of the adsorption area changes. With increasing surface tension and an increasing surface on which the effective surface tension is operational, the paste will be put under stress and deforms.

In this paper the development of the pore system and changes in the state of water in the pore system have been determined with the numerical simulation program HYMOSTRUC. The volume changes are related to changes in the effective surface tension with a modified Bangham equation. The most important modification concerned the introduction of a variable proportionality factor λ for hardening pastes. This factor was adjusted step-wise, making allowance for the gradual change of the pore wall area covered with an adsorption layer and of the modulus of elasticity. Both quantities were calculated with HYMOSTRUC as functions of the degree of hydration. For a number of cement pastes numerical simulations of the shrinkage strains during hardening have been compared with measurements. The simulated tests were exclusively isothermal tests, mainly on portland cement pastes. Allowance for temperature effects is a next step towards a more comprehensive numerical model for evaluation of autogenous shrinkage. In this respect it is noticed that curing at higher temperatures affects the rate of hydration, but will also cause microstructural changes and associated changes in the thermodynamic equilibrium in the pore system.

On the average the agreement between simulations and measurement were promising. Depending on the type of cement some pastes exhibited expansion in the very early stage of hydration. In the simulations, in which only sealed curing was considered, the simulated self desiccation process can not yield expansion. It is noticed that possible expansion mechanisms, whatever the cause of expansion could be, complicate the interpretation of measurements of strains at early ages. In tests we will always measure the combined effect of (eventual) expansion mechanisms and shrinkage mechanisms. Strictly speaking a comparison of the measured and calculated strains would only be justified if the presence of any expansion mechanism could be excluded.

A numerical simulation of the effect of aggregate particles on the autogenous deformation of a concrete sample was presented here as an example of the potentialities of presently available simulation programs. In this simulation the aggregates particles were considered to behave as rigid inclusions. No moisture exchanges with the surrounding paste was considered. In case of lightweight aggregate concrete, moisture transport from the aggregate to the paste and vice versa will have a major effect on the autogenous deformation. In future modelling work this is one of the topics that has to be considered in more detail.

7 References

1. BaroghelBouny, V. (1996) Texture and Moisture Properties of Ordinary and High Performance Cementitious Materials, in *Proc. RILEM-seminar "Béton: du Matériau à la Structure"*. Arles, (Fr). 22 p.
2. Bangham, D. H. and Fakhoury, N. (1931) *The Swelling of Charcoal.* Royal Society of London CXXX (Series A). pp. 81-89.
3. Breugel, K. van (1991) *Simulation of hydration and formation of structure in hardening cement-based materials.* PdD, Delft, p. 295.
4. Garboczi, E.J., Bentz, D.P. (1992) *Computer-based models of the micro-structure and properties of cement-based materials.* 9th Int. Conference on the Chemistry of Cement, Vol. VI, 3-15.
5. Koenders, E.A.B. (1997) *Simulation of volume changes in hardening cement-based systems.* PhD, Delft University Press, p. 171.
6. Setzer, M., J. (1978) *Einfluss des Wassergehalts auf die Eigenschaften des erhärteten Betons.* Deutscher Ausschuss Für Stahlbeton DAfSt H. 280: 43-79.
7. Bangham, D.H. and Maggs, F.A.P. (1944) *The Strength and Elastic Constants of Coal in Relation to their Ultra-fine Structure.* The British Coal Utilisation Research Association, The Royal Institution, London.
8. Wittmann, F. (1968) *Physikalische Messungen an Zementstein.* München, TU München.
9. Hiller, K.H. (1964) *Strength Reduction and Length Changes in Porous Glass Caused by Water Vapor Adsorption.* J. Applied Physics 35(5): 1622-1628.
10. Tazawa, E. and Miyazawa, S. (1992) *Autogenous shrinkage caused by self desiccation in cementitious material.* 9th Int. Conf. on Chemistry of Cement, New Delhi.
11. Schlangen, E. (1993) *Experimental and numerical analysis of fracture processes in concrete.* PhD, Delft University of Technology, 121 p.
12. Setzer, M.J. (1972) *Oberflächenenergie und Mechanische Eigenschaften des Zementsteins.* PdD., München, 113 p.
13. Feldman, R.F. (1968) *Sorption and Length-Change Scanning Isotherms of Methanol and Water on Hydrated Portland Cement.* Fifth Int. Symposium on the Properties of Cement Paste and Concrete. Tokyo.
14. Klug, P. (1973) *Creep, Relaxation and Swelling of Hardened Cement Paste.* PdD, Munich, 69 p.
15. Beek, A. van, Lokhorst, S.J., Breugel, K. van (1996). *On-site determination of degree of hydration and associated properties of hardening concrete".* Proc.3rd. Conf. on Non-destructive evaluation of civil structures and materials. Ed. Schuller et al., Boulder, Colorado, pp. 7-22.

20 A STUDY ON THE HYDRATION RATIO AND AUTOGENOUS SHRINKAGE OF CEMENT PASTE

K.B. Park, T. Noguchi and F. Tomosawa
*Department of Architecture, Faculty of Engineering,
The University of Tokyo, Tokyo, Japan*

Abstract
This study aims to clarify the relationship between autogenous shrinkage and hydration by investigating the effects of hydration reaction on the autogenous shrinkage phenomenon and the relationship between the autogenous shrinkage and ratio of hydration, in order to set course for devising a model of autogenous shrinkage. The authors therefore measured autogenous shrinkage strain of cement paste with a W/C of 25 and 30% using an experimental laser measuring device. The pore size distribution and ignition loss were also measured to investigate the relationship between autogenous shrinkage and the microstructure in the hardened paste and hydration reaction. As a result, it was found necessary to quantify the time-related changes in the ratio of hydration and the intermolecular forces and surface forces resulting from the changes in the microstructure of hardened paste.
Keywords: Autogenous shrinkage, Chemical shrinkage, Hydration ratio, Ignition loss, Pore size distribution

1 Introduction

Autogenous shrinkage of hardening cement is said to be a phenomenon in which water in newly formed voids is lost by the reaction of cement with water and the low relative humidity in the voids causes a capillary tension, which causes a reduction in the volume of cement. This phenomenon can be considered from three levels of scale: a microscopic level of hydration of cement, intermediate level of pore structure, and macroscopic level of autogenous shrinkage development. Grasping such a macroscopic phenomenon from different levels of scale will lead to the formulation of a

Autogenous Shrinkage of Concrete, edited by Ei-ichi Tazawa. Published in 1999 by E & FN Spon, 11 New Fetter Lane, London EC4P 4EE, UK. ISBN: 0 419 23890 5

more realistic and reliable evaluation model.

Hydration of cement is impossible to explain with a simple parameter. However, from the standpoint of the final goal of this study, "to explain the early-age physical properties of hardening cement (hydration heat generation, autogenous shrinkage, strength development, and creep) with a single parameter," it is considered most appropriate to evaluate these properties with the ratio of hydration, which can well explain hydration of cement. If a model for cement hydration is formulated on a more microscopic level using more parameters, the applicability of the model will conversely be limited.

2 Research significance

Attempts have been made for a long time to devise a model for reactions or phenomena occurring in cement concrete, but the focus has been mostly on the prediction of the yield strength, deformation under loading, and cracking of sufficiently hardened concrete. They have not yet reached the stage of explaining the changes in the physical properties of concrete at very early ages. However, the wide varieties of materials and requirements for structures in recent years have led to the extensive use of high strength and high performance concrete. This has cast light on the significant effects of early age physical properties in the process of hardening, such as hydration heat, autogenous shrinkage, creep and strength development, on the cracking at very early ages and durability after hardening of concrete. Moreover, these early-age phenomena affect one another and synergistically affect concrete. It is therefore difficult to explain the stress and cracking tendency of actual concrete structures with a model that describes a single physical property.

In order to solve this problem, a universal model is necessary, with which complicated early-age behavior of concrete is comprehensively described using a single parameter.

Our laboratory proposed an integrated system for predicting the adiabatic temperature rise of concrete and temperature history and strength development of real-scale concrete members using Tomosawa's hydration model [1][2], and has conducted analysis of cracks due to temperature changes [3].

As a result, temperature changes and strength development of concrete have become predictable by the hydration ratio. The laboratory therefore intended a more comprehensive model for predicting other early-age physical properties of concrete as well, including autogenous shrinkage.

3 Experimental details

3.1 Materials and specimens
Normal portland cement containing no mineral admixtures (NO) and belite-rich cement (HF) were used as the cements. The mineral composition and physical properties of these cements are listed in Table 1.

composition

pecific	Mineral Composition (%)[1]				
ravity	C3S	C2S	C3A	C4AF	CaSO4
15	57	18	10	8	3.2
ɔ.20	35	47	3	8	4.6

1) by Bogue's eq.

Table 2. Mix design of the paste and set time

Cement	W/C	Super-	Time of Setting (hr.)	
	(%)	plasticizer(%)	Initial Set	Final Set
NO	25	0.6	3-20	5-13
	30	0.3	4-26	6-30
HF	25	0.5	2-52	4-44
	30	0.2	3-02	4-57

Cement paste was proportioned to have a water-cement ratio of 25 and 30. A polycarboxylate-type air-entraining and high-range water-reducing admixture was added to attain the target flow value of 150mm (Table 2). The paste was mixed using a mechanical mixer in an air-conditioned room at 20℃ and 60% RH for 90 seconds at a low speed and then for another 90 seconds at a high speed after an interval of 30 seconds. The paste was then placed in molds 40 by 40 by 160 mm and seal-cured up to the specified ages. The autogenous shrinkage and other physical properties were measured at the time of initial setting for datum measurements and at 6, 8, 12, 18, 24, 48, 72, 120, 168, and 336 hours.

3.2 Test methods

3.2.1 Setting test
Setting of cement was measured in accordance with the test method for setting provided in JIS R 5201 (Physical testing methods of cement). The results are given in Table 2.

3.2.2 Ignition loss
After measuring at the specified age the compressive strength of paste seal-cured for a certain period, the remaining hardened paste was crushed into particles of approximately 5 mm in size. The specimens were D-dried for 2 weeks after terminating the hydration with acetone. The ignition loss was then determined by the method of ignition loss determination.

3.2.3 Pore size distribution
Pore radii were measured with a mercury porosimetry. The samples used for measurement were the same as those for ignition loss.

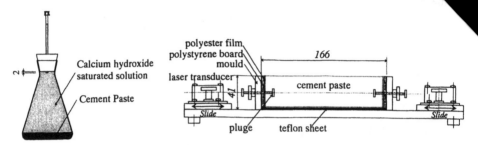

Fig.1. Chemical shrinkage test   Fig.2. Experiment set up for autogenous shrinkage

3.2.4 Chemical shrinkage
The chemical shrinkage was measured in accordance with the method proposed by the JCI Committee for Autogenous Shrinkage shown in Fig. 1. This is a method in which a saturated solution of calcium hydroxide is allowed to permeate through voids resulting from hydration of cement paste placed in a conical flask.   The chemical shrinkage is determined by measuring the reduction in the volume of the solution by the changes in the solution level of the measuring pipette mounted on top of the conical flask.

3.2.5 Autogenous shrinkage
The autogenous shrinkage strain from initial setting up to an age of 2 days was measured using an experimental device shown in Fig. 2.   For measurements later than 2 days, contact tips for measuring the length change were glued with an epoxy adhesive to sides of three specimens prepared under the same conditions, and the entire surfaces of the specimens were then sealed with aluminum foil adhesive tape 0.05mm in thickness.   The length change was measured with contact gages.

4   Results and discussion

4.1   Hydration ratio
The percentage of bound water contained in the hydrates resulting from the reaction of cement with water was assumed to be the hydration ratio.   The hydration ratio was determined by assuming that the quantity of bound water was the same as the ignition loss.

$$\text{Hydration ratio } (\alpha) = (\text{bound water})/(\text{bound water at 100\% hydration}) \quad (1)$$

The quantity of bound water at 100% hydration of cement paste was determined by obtaining the amount of clinker-hydrate and $CaSO_4$ according to the following chemical equations, which are based on Bogues's method [4].

$$2C_3S + 6H_2O \longrightarrow C_3S_2H_3 + 3Ca(OH)_2 \tag{2}$$

$$2C_3S + 4H_2O \longrightarrow C_3S_2H_3 + Ca(OH)_2 \tag{3}$$

$$C_3A + 3CaSO_4 + 32H_2O \longrightarrow C_3A \cdot 3CaSO_4 \cdot 32H_2O \tag{4}$$

$$2C_3A + C_3A \cdot 3CaSO_4 \cdot 32H_2O + 4H_2O \longrightarrow 3[C_3A \cdot CaSO_4 \cdot 12H_2O] \tag{5}$$

$$2C_3A + 21H_2O \longrightarrow C_2AH_8 + C_4AH_{13} \tag{6}$$

In principal, the ferrite phase (C4AF) will form the same reaction as C3A, but since it is much less reactive it is likely to combine with very little of the gypsum. So unless the C3A content is low, a better representation of the C4AF reacting is [5]:

$$C_4AF + 2CH + 14H_2O \longrightarrow C_4(A,F)H_{13} + (A,F)H_3 \tag{7}$$

In addition, because the mole composition rate of C3A and CaSO4 in belite-rich cement is 1 to 3, only the reaction by the equation 4 is possible, whereas the reaction by either of the equation (5) or (6) is impossible, under the assumption that the final hydrate of normal portland cement becomes C3S2H3, CH, C2AH8, C4AH13, C4FH13 and C3A(CS)3H12. Thus, if we assume that the full hydrate of belite-rich cement becomes C3S2H3, CH, C4AH13, C4FH13 and C3A(CS)3H32, we obtain the quantity of combined water according to the equation (2), (3), (4) and (7). As a result, when 1 gram of cement fully reacts with water, the determined quantities of bound water for normal portland cement (NO) and belite-rich cement (HF) were 0.284 and 0.288 respectively.

Figure 3 shows the changes in the hydration ratio with age. It increases with age and becomes higher as the W/C increases. Normal portland cement containing more C3A and C3S minerals that react quickly with water is found to lead to much higher hydration ratios at early ages. Whereas no effect of W/C is observed before an age of 24 hours, a lower W/C tends to lead to a lower increase in the hydration ratio thereafter. This phenomenon may be because the hydrates generated in abundance fill the voids between particles, reducing the spaces for subsequent hydrates, thereby inhibiting hydration reaction.

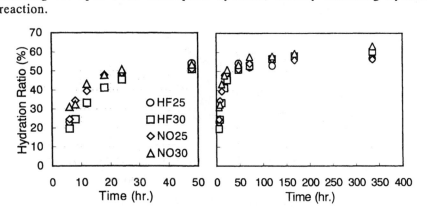

Fig.3.   Rate of hydration of cement paste

4.2   Chemical shrinkage

Measurements of the chemical shrinkage factor of cement paste are shown in Fig. 4.   This figure reveals that a higher chemical shrinkage factor results from a lower W/C or a cement with a higher content of minerals that react quickly with water.

Chemical shrinkage is caused by the fact that the volume of hydrates formed by the reaction of cement with water is smaller than the total volume of the original cement and water. From the results of this experiment, the relationships between the hydration shrinkage and the bound water content and void content were examined, and the hydration shrinkage phenomenon was assumed to consist of two mechanisms.   In the first stage, water between dispersed particles is consumed by the chemical reaction of cement, and the distances between cement particles radically decrease, causing shrinkage.   In the second stage, particles are entangled in hydrates and in full contact, and shrinkage proceeds slowly with balanced chemical shrinkage forces and rigidity.

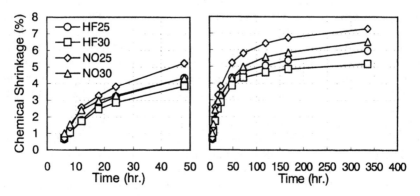

Fig.4.   Chemical shrinkage of cement paste

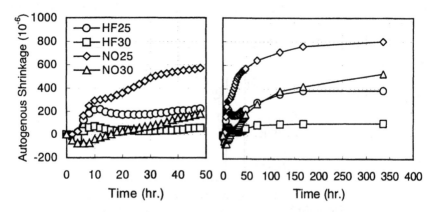

Fig.5.   Autogenous shrinkage of cement paste

4.3 Autogenous shrinkage

Figure 5 shows the autogenous shrinkage measurements of various cement pastes. The autogenous shrinkage phenomenon is roughly divided into 3 stages: the first shrinking stage followed by the first relaxing stage, and the second shrinking stage.

In the first shrinking stage, C3S in cement compound reacts with water, and the amount of Ca^{2+} in liquid phase becomes maximal. Then, the ion density of Ca^{2+} becomes higher than SiO^-; the penetration pressure increase in particles. At this phase, the layer of hydrate which surrounds the cement particles is destroyed either by the penetration pressure of ion or Maxwell force. Again, the inner side of unhydrated parts will begin to be hydrated violently[6]. By chemical shrinkage following this hydration, and by cohesiveness of cement hydrates in free pores, cement paste get massively shrinked[7].

We think that shrinkage relaxation in the first relaxing stage is due to the fact that the expansive forces of ettringite account for the slight expansion in this stage. Also, the expansion of the volume when Ca(OH2) is created by the reaction between CaO and water could be the possible cause [8][9].

On the other hand, autogenous shrinkage in the second shrinking stage is considered to occur by the following two phenomena: 1) capillary pressure produced by surface tension of water in the space of xerogel, as the hydration of cement proceeds further, and 2) the decrease of distance between layers, as the laminar void water and the zeolite water in cement paste decrease[10].

Figure 6 shows the relationship between the autogenous shrinkage and chemical shrinkage of various cements. Belite-rich cement with a lower C3A content led to a smaller autogenous shrinkage when the chemical shrinkage factors are the same. This suggests that autogenous shrinkage closely relates to the formation of hydrate structure in each mineral in cement. A period was also found in the first relaxing stage in autogenous shrinkage, in which

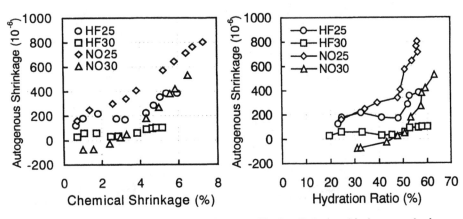

Fig.6. Relationship between chemical shrinkage and autogenous shrinkage

Fig.7. Relationship between hydration ratio and autogenous shrinkage

autogenous shrinkage does not proceed while chemical shrinkage proceeds.

Figure 7 shows the relationship between the autogenous shrinkage and the hydration ratio. This figure suggests that from a certain age autogenous shrinkage becomes no loner explainable by the hydration ratio of cement. In other words, Fig. 7 shows an age from which autogenous shrinkage radically increases while hydration of cement remains the same. In this light, autogenous shrinkage is found to be subject to two different mechanisms: hydration of cement and changes in the pore structure of hardened cement. A single parameter to evaluate these would realize a simple and clear model of autogenous shrinkage.

4.4   Pore structure

Pores in hardened cement are roughly classified into three types: pores that have been occupied by water, gel pores, and pores due to air entrapped during mixing. Capillary pores are those occupied with water used for mixing yet to be replaced with hydrates. These are filled with gel resulting from hydration and are reduced as hydration proceeds with age. Gel pores are those exist within the laminar structure of C-S-H gel. The total volume of these pores varies depending on the amount of C-S-H formation. There have been a number of studies that intend to explain autogenous shrinkage by the theory of capillary tension within the voids using Kelvin equation, Laplace equation, or Bingham equation [11][12][13]. However, how such pores work and affect the autogenous shrinkage by what reaction is still unknown.

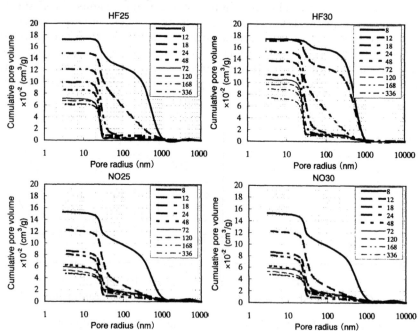

Fig.8.   Pore structure of transition zone in cement paste

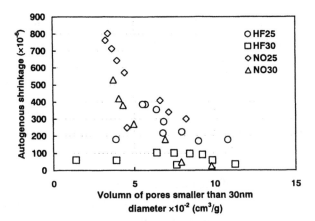

Fig.9. Relationship between volume of pores
and autogenous shrinkage

In this study, the author experimentally determined the time of threshold pore radius, at which the pore volume drastically increases as the pore radius decreases, and investigated its relationship with autogenous shrinkage. As a result, a threshold pore radius of 30 nm was observed at 24 hours as shown in Fig. 8. This is found to coincide with the time when autogenous shrinkage does not depend on the hydration ratio (Fig. 7).

Figure 9 shows the relationship between the pore volume (pores 30 nm or less in radius) and the autogenous shrinkage. This figure reveals that the volume of voids 30 nm or less in radius decreases with age and that the smaller the pore volume, the larger the autogenous shrinkage.

5 Conclusions

The relationships between the autogenous shrinkage and the hydration in cement paste and changes in the microstructure were investigated to devise a model for autogenous shrinkage based on the hydration ratio of cement. The following conclusions were obtained:

Autogenous shrinkage was found unexplainable solely by the hydration shrinkage of cement. In other words, there is a period during which autogenous shrinkage radically increases while hydration of cement scarcely proceeds. In this period, autogenous shrinkage is considered to occur because of capillary pressure produced by surface tension of water in the space of xerogel, and   the decrease of distance between layers.

From this standpoint, autogenous shrinkage can be considered by two mechanisms: hydration of cement and changes in the pore structure of hardened cement. Explaining these mechanisms using a single parameter would realize a model for autogenous shrinkage.

6   References

1. Tomosawa, F. (1970) A study on strength development of concrete in view of hydration rate, (Japanese), Proc. of Japan Arch. Inst. Kanto Branch Conf., No.41, pp.229-240.
2. Tomosawa, F. (1997) Development of a kinetic model for hydration cement, Proc. of 10th ICCC, Vol.2, 2ii051, 8pp.
3. Tomosawa, F., Noguchi, T., Hyeon, C. (1997) Simulation model for temperature rise and evolution of thermal stress in concrete based on kinetic hydration model of cement, Proc. of 10th ICCC, Vol.4, 4iv072, 4pp.
4. Taylor, H.F.W (1990) Cement Chemisry
5. Young, J.F., Hindess, S., Gray, R.J., Bentur, A. The science and Technology of Civil Engineering Materials
6. Group of chemist in Tokyo (1995) Colloid Science I , (Japanese)
7. Pracevaux, P. (1984) Pore size distribution of portland cement surries at very early stages of hydration, Cement and Concrete Research, Vol.14, No.3, pp.419-430
8. Takahashi, T., Nakata, H., Yoshida, K., Goto, S. (1996) Influence of hydration on autogenous shrinkage of cement paste, (Japanese), Concrete Research and Technology, Vol.7, No.2, pp.137-142.
9. Kasai, J. (1984) The chemical of cement, (Japanese), Concrete Journal, Vol.22, No2, pp.50-55.
10. Wittman, F.H. (1983) Structure and Mechanical Properties of Concrete, (Japanese), Concrete Journal, Vol.21, No.3, pp.19-30.
11. Breugel, K., Koenders, EddieA.B. (1997) Modeling dimensional changes in low water/cement ratio pastes, Proc. Int.Res.Sem. Lund, pp.158-173.
12. Ishida, T., Chaube, R.P., Kishi, T. (1998) Micro-physical approach to coupled autogenous and drying shrinkage of concrete, Proc. of 6th EAPC, Vol.3, pp.1905-1910.
13. Hiller, K.H. (1964) Strength Reduction and Length Changes in Porous Glass Caused by Water Vapor Adsorption, Journal of Applied Physics, Vol.35, Number5, pp.1622-1628.

21 SOLIDIFICATION MODEL OF HARDENING CONCRETE COMPOSITE FOR PREDICTING AUTOGENOUS AND DRYING SHRINKAGE

R.T. Mabrouk, T. Ishida and K. Maekawa
The University of Tokyo, Tokyo, Japan

Abstract
In this study, the prediction of hardening young concrete shrinkage and autogenous shrinkage is attempted using a solidification model based on microphysical information of temperature, hydration ratio, porosity, saturation, isotherm and others. The micro-scale surface tension is treated as a driving force for the shrinkage associated with and without water migration. The solidification model is expected to offer the deformability of hardening concrete under estimated capillary stress by the thermo-dynamic computation. The solid model deals with cement paste as the solidified finite fictitious clusters having each creep property. Aggregates are idealized as suspended continuum media of perfect elasticity. The combination of both phases may create overall features of concrete composite under unstable transient situations at early age. Experiments are performed to check the properties of cement paste matrix and a parametric study is conducted to verify the functionality of the model proposed. The predicted shrinkage strain, obtained from the model, is compared with experimental data.
Keywords: Autogenous shrinkage, microphysics, multiphase material, shrinkage, solidification, young concrete

1 Introduction

Needs of predicting behaviors of early aged concrete have always been encountered, especially, in the case of massive concrete such as dams or raft foundation. These types of structures are always subjected to rather huge temperature gradients leading to thermal cracking. This makes it of great importance to be able to predict correctly the early age behavior of these structures such as drying shrinkage and creep.

Autogenous Shrinkage of Concrete, edited by Ei-ichi Tazawa. Published in 1999 by E & FN Spon, 11 New Fetter Lane, London EC4P 4EE, UK. ISBN: 0 419 23890 5

Another point of interest is the autogenous shrinkage. The use of self-compacting high performance concrete (HPC) [1] has become a means of avoiding uncertainties rooted in construction methods thus producing a reliable structural concrete in terms of crack control or durability design. However, to obtain the super fluidity and segregation resistance of HPC, a small amount of free water and low water to cement ratio need to be specified. According to this, it has been reported that a large amount of autogenous shrinkage occurs. Thus, the autogenous shrinkage, which can be neglected for ordinary concrete, has to be considered in durability design of structures using HPC concrete.

In the past, it was somehow difficult to treat both shrinkage and creep of young concrete based on the unified approach of material science. Now, more microphysical information is available like temperature, hydration ratio, porosity, pore water pressure, transient water content, isotherm and others. Based on this, the prediction of both shrinkage and autogenous shrinkage as well as creep of hardening young concrete is attempted. One advantage of this research is that it attempts to deal with these behaviors using only one microphysical model in a unified manner without any separation of creep and shrinkage.

2   Analytical model

The total deformation of concrete is divided into two components, that is to say, volumetric component and deviatoric component.

2.1   Volumetric deformation

Concrete is idealized as a two-phase solid dispersion system, namely cement paste and aggregate. Aggregate is assumed as a linear elastic material suspended in the cement paste. The solidification concept [2], [3] is used to represent the behavior of cement paste. The growth of cement paste is idealized by the formation of finite fictitious clusters or layers. If the total volume of cement paste is V_{cp} and the effective volume solidified at time t is $V(t)$, then the ratio of hydration controlling the solidification of layers is $\psi(t) = V(t) / V_{cp}$. New layers solidify one by one

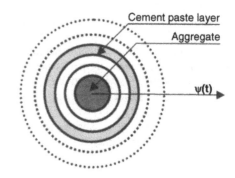

Fig. 1   Schematic representation of aggregate and cement paste

according to the hydration development. It is assumed that the new layers solidified join the existing ones in parallel coupling. Fig. 1 shows a schematic representation of the system at an arbitrary stage.

Concerning the coupling of aggregate and cement paste, virtual work principle (basically, Green's formula for conversion of volumetric divergence to the surface outer stream integral) was utilized for theoretically deriving the mean stress and strain on the aggregate and cement paste. According to this, the following two equations of phase requirements and two constitutive laws can be obtained.

$$\overline{\sigma}_o = \rho_{ag}\overline{\sigma}_{ag} + \rho_{cp}\overline{\sigma}_{cp} \qquad (1)$$

$$\overline{\varepsilon}_o = \rho_{ag}\overline{\varepsilon}_{ag} + \rho_{cp}\overline{\varepsilon}_{cp} \qquad (2)$$

$$\overline{\varepsilon}_{cp} = f(\overline{\sigma}_{cp}) \qquad (3)$$

$$\overline{\varepsilon}_{ag} = \frac{1}{3K_{ag}}\overline{\sigma}_{ag} \qquad (4)$$

Where $\overline{\sigma}_o$, $\overline{\sigma}_{ag}$ and $\overline{\sigma}_{cp}$ are the average volumetric stresses on concrete, aggregate and cement paste, respectively, and $\overline{\varepsilon}_o$, $\overline{\varepsilon}_{ag}$ and $\overline{\varepsilon}_{cp}$ are the average volumetric strains. ρ_{ag} and ρ_{cp} are the volume fractions of aggregate and cement paste, respectively. K_{ag} is the volumetric stiffness of aggregate.

Capillary tension has been introduced by many of the past researchers as the source of the drying shrinkage behavior of concrete [4]. Under this assumption, the volume change and the deformation of the cement paste will occur due to the surface tension force of capillary water across the curved meniscus. In addition, due to the hydration process, the volume of the hydrated cement paste is smaller than the sum of both unhydrated powder and free water for hydration. Therefore, the pore space in the hydrated cement paste can not be filled with water and the relative humidity in the pore structures would decrease as the hydration proceeds. This mechanism will be the cause of the autogenous shrinkage of concrete [5].

In this study, the combined effect of external loads and pore water pressure created by the micro-scale surface tension is treated as the driving force for the deformation of concrete. This deformation includes autogenous and drying shrinkage together with creep. Accordingly, the total average volumetric stress on concrete $\overline{\sigma}_{ot}$ is calculated by using eq. (5) as,

$$\overline{\sigma}_{ot} = \overline{\sigma}_o + \eta \cdot \sigma_s \qquad (5)$$

where σ_s denotes capillary stress computed by the thermo-hygro-physical approach [5] such that $\sigma_s = \sigma_s$ (pore water radius, surface tension of liquid water) and η is a factor depending on the saturation degree of the paste. More investigation is required in order to determine this factor η. For this analysis, η is taken as 1.0.

Here, there are 5 independent variables and 4 equations. Thus, one more governing equation is required to specify the multi-phase system of interaction. As expressed by Maxwell and Kelvin chains of viscous continuum, there can be a parallel system and a serial one. If the cement paste matrix would be a perfect liquid losing resistance to the shear deformation, then $\overline{\sigma}_{ag} = \overline{\sigma}_{cp}$, where, shear stiffness of cement paste G_{cp} becomes null. This case corresponds to the Maxwell chain of a serial system of elements. On the contrary, if the shear stiffness G_{cp} is infinitely large, it brings no shear deformation at all, and the deformed shape with volumetric expansion or contraction is perfectly similar to the referential shape at the initial time. Thus, aggregate and cement paste deform with perfect proportionality or, $\overline{\varepsilon}_{ag} = \overline{\varepsilon}_{cp}$. This case corresponds to the Kelvin chain of a parallel system of elements. The actual case is somehow in between these two cases. According to this, the fifth equation of concrete composite isotropy is associated with paste shear rigidity and related degree of parallelism. One of the possible formulas is the Lagragian method of linear summation.

$$\left(\frac{\overline{\sigma}_{ag} - \overline{\sigma}_{cp}}{G_{cp}}\right) + (\overline{\varepsilon}_{ag} - \overline{\varepsilon}_{cp}) = 0 \tag{6}$$

This equation satisfies the above stated extreme cases in the parallel and serial figures of composite. By adding this last equation to the other 4 equations, the simultaneous equations get mathematical completeness. That is, the number of variables is equal to the number of equations.

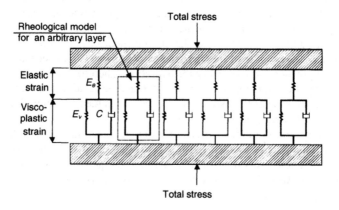

Fig. 2 Rhelogical model for cement paste layers

As shown in Fig. 2, the layers of cement paste are assumed to join in parallel where each layer is represented using two springs and a dashpot. Thus, the total average stress in cement paste at a certain time is the summation of the stress in all individual layers found at that time and the average strain in cement paste is equal to the strain induced in each layer. It should be noted here that when a new layer is just formed it should be stress free. Hence, the stress developed in each infinitesimal layer should be a function of the time when this layer is solidified. Let S_{cp} denote the average stress in a general layer, t is the time and t' is the time when this layer is solidified. Then we have, $S_{cp} = S_{cp}(t,t')$ and $S(t,t) = 0$. The constitutive relation for each layer can be obtained using the rate type creep law as follows.

$$(1 + \frac{E_v}{E_e})S_{cp}(t) + \frac{C}{E_e}\frac{dS_{cp}(t)}{dt} = E_v\overline{\varepsilon}'_{cp}(t) + C\frac{d\overline{\varepsilon}'_{cp}(t)}{dt} \tag{7}$$

where, $\overline{\varepsilon}'_{cp}(t)$ is the strain in a general layer, E_e and E_v are the stiffness of the elastic and the plastic springs, respectively. C is the constant of the dashpot fluid and it is related to the water motion through pores associated with thermo-hygro-physical requirement. The average stress in the cement paste is calculated from,

$$\overline{\sigma}_{cp}(t) = \int_{t'=0}^{t} S_{cp}(t',t)d\psi(t') \qquad (8)$$

The stiffness of each layer is calculated such that the summation of the stiffness over all layers at a certain time is equal to that of cement paste at this time. The plastic stiffness of each layer, E_v, is taken as 1/3 of the total stiffness of the layer. The dashpot constant, C, is assumed to be a function in water content, viscosity of paste water and porosity as shown in Fig. 3.

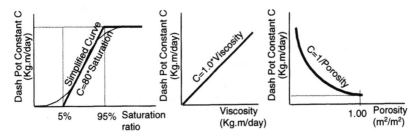

Fig. 3 Variation of dash pot constant against water content, viscosity and porosity

2.2 Deviatoric deformation

The same procedure used for the volumetric component is used. However, neglecting the shear strain of the aggregate component and assuming free rotation of the suspended particles reduces the previously used system of simultaneous equations to,

$$\overline{\gamma}_o \cong \overline{\gamma}_{cp} \quad \text{and} \quad \overline{\tau}_o \cong \overline{\tau}_{cp} \qquad (9)$$

$$\overline{\gamma}_o = f(\overline{\tau}_o) \text{ or } \overline{\gamma}_{cp} = f(\overline{\tau}_{cp}) \text{ or } \overline{\tau}_{cp}(t) = \int_{t'=0}^{t} \tau_{cp}(t',t)d\psi(t') \qquad (10)$$

$$(1+\frac{G_v}{G_e})\tau_{cp}(t) + \frac{V}{G_e}\frac{d\tau_{cp}(t)}{dt} = G_v\overline{\gamma}'_{cp}(t) + V\frac{d\gamma'_{cp}(t)}{dt} \qquad (11)$$

where $\overline{\tau}_o$ and $\overline{\tau}_{cp}$ are the average deviatoric stresses on concrete and cement paste, respectively, $\overline{\gamma}_o$ and $\overline{\gamma}_{cp}$ are the average deviatoric strains, τ_{cp} and $\overline{\gamma}'_{cp}$ are the deviatoric stress and strain in a general layer, G_e and G_v are the stiffness of the elastic and the plastic springs, respectively and V is the constant of the dashpot fluid.

2.3 Post cracking behavior

The cracking criterion is set based on the value of the maximum principle stress. At a certain age, cracking occurs if this value exceeds the tensile strength of concrete at that time. After cracking has occurred, the strain vector is transformed into the local co-ordinates representing the direction of the crack. The stress-strain relation in the plane perpendicular to the crack is assumed to follow the tension-softening curve [6] shown in Fig. 4. The descending part of the curve is represented using eq. (12).

$$\sigma = f_t \left(2 \cdot \varepsilon_{tu} / \varepsilon\right)^2 \qquad (12)$$

where σ and ε are the stress and strain perpendicular to the crack plane, f_t is the tensile strength of concrete and ε_{tu} is the strain at cracking.

The stress–strain relation in the other plane directions is calculated using the same modeling as before. For this analysis, the model parameters, before and after cracking, are defined using the same procedure. The flow chart of the routine used is shown in Fig. 5.

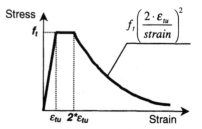

Fig. 4 Tension-softening curve perpendicular to crack plane

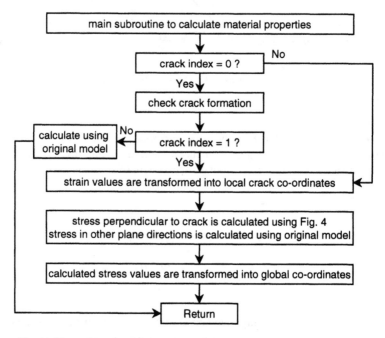

Fig. 5 Flow chart for post cracking calculation

3   Analytical study by the proposed model

An analytical study of creep and shrinkage of concrete was done using the proposed model. The microphysical properties needed are calculated using a thermo-dynamical program DuCOM [7]. The model is implemented in a finite element program COMT3 that is dynamically linked with DuCOM. For verification, some of the experimental works reported in the literature are simulated using the proposed model. The analytical results were compared with the experimental data to check the validity of the model. A parametric study was also conducted to decide the values of the model parameters.

4 Verifications

4.1 Autogenous and drying shrinkage of cement paste

One experiment was conducted to check the cement matrix properties. A pure cement paste specimen of size 30*30*30 cm was sealed for one day then tested under drying conditions. The conditions of temperature and relative humidity applied to the specimen are shown in Fig. 6a.

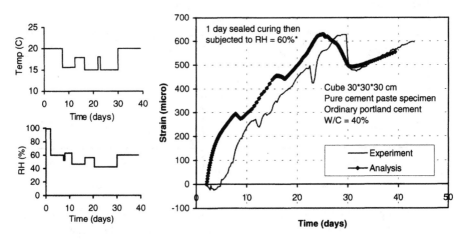

Fig. 6a Environmental
conditions applied

Fig. 6b Comparison between analytical and
experimental results – Drying shrinkage

The same problem is analyzed using the proposed model. It was observed that microcracking started to occur at the age of 2 days in both experiment and analysis. Figure 6b shows the results of the analysis compared with the experimental readings. The figure shows reasonable agreement between the two results.

Comparison is also made with Tazawa et al. [8] where he conducted autogenous shrinkage experiments on pure cement paste specimens. Figure 7 shows the analytical results compared with the experimental data. It can be seen that the analytical results are a little underestimated especially at the early age.

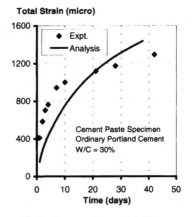

Fig.7 Comparison with Tazawa
et al. [8] – Autogenous shrinkage
of cement paste

4.2 Autogenous and drying shrinkage of concrete

First autogenous shrinkage of concrete was studied. The experimental data by Tazawa et al. [9] was used as shown in Fig. 8 and Fig. 9. These figures show two cases of autogenous shrinkage for two different W/C ratios of 30% and 40%. The figures show that the analytical results are underestimated in the first case where W/C is 30%. However, better agreement between analysis and experiment is shown in the case of the 40% water to cement ratio. Concerning the early swelling of concrete that appears in Fig. 9; the proposed model is only effective after the formation of microstructure starts to take place. No output results are available at the very early stage when cement paste has not started to solidify.

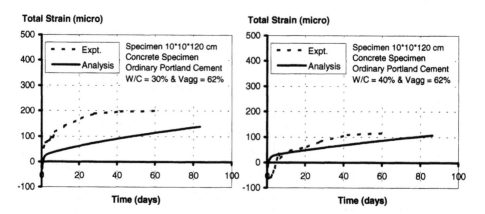

Fig. 8 Comparison with Tazawa et al. [9] – Autogenous shrinkage

Fig. 9 Comparison with Tazawa et al. [9] – Autogenous shrinkage

Next, comparison is made with Khan et al. [10] for both autogenous and drying shrinkage as shown in Fig. 10. One specimen was sealed for 0.53 days then subjected to drying under RH 50% and the other was sealed throughout the experiment. It should be noted here that microcracking appeared in the analysis of the first specimen 1.2 days after drying started. Figure 11 shows another comparison with Tazawa et al. [9] for drying shrinkage. Both figures show reasonable agreement between the computed results and the experimental data.

It has been reported that the autogenous shrinkage behavior of concrete with normal W/C ratio and that of low W/C ratio is quite different from each other. In the case of normal concrete, autogenous shrinkage is rather small compared to the drying shrinkage strain and thus can be neglected. While, in the case of low W/C ratio, the autogenous shrinkage has a substantial value and can not be ignored. Using the model proposed in this study, a qualitative analysis is performed using two specimens sealed for 28 days then subjected to RH of 60% but having quite different W/C ratios. The analytical results are shown in Fig. 12 which shows the above mentioned behavior clearly.

4.3 Creep behavior of concrete

Comparison is made with Khan et al [10]. The experimental conditions are the same as in the case of the drying and autogenous shrinkage mentioned before. Except that the specimens are loaded at the age of 16 hours with a stress level of 0.185. This means that the applied stress is 0.185 of the compressive strength at the time of the loading. The results are shown in Fig. 13, which shows reasonable agreement between the computed results and the experimental data.

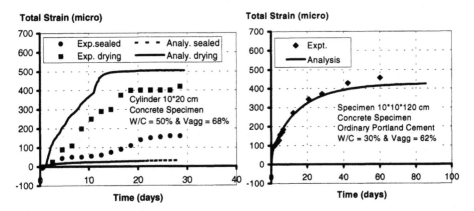

Fig. 10 Comparison with Khan et al. [10] – Autogenous and drying shrinkage

Fig. 11 Comparison with Tazawa et al. [9] – Drying shrinkage

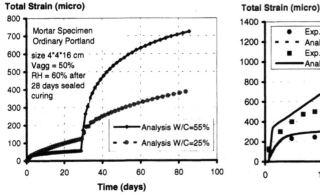

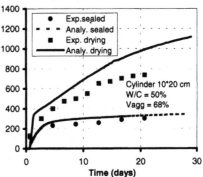

Fig. 12 Analysis of autogenous and drying shrinkage for different W/C ratios

Fig. 13 Comparison with Khan et al. [9] – Basic and drying creep

5 Conclusions

A solidification model based on microphysical information for the prediction of both creep and shrinkage of hardening young concrete is presented in this study. The solidification model deals with cement paste as the solidified clusters having each creep property. Aggregates are idealized as suspended continuum media of perfect elasticity. The combination of both phases is used to represent the overall composite under unstable transient situations at early age. The combined effect of external loads and pore water pressure is treated as the driving force for concrete deformation. Comparison with some experimental data form the literature proved that the proposed model shows reasonable agreement with experiments. This shows that this framework can be a solid method for the determination of autogenous shrinkage, drying shrinkage, basic creep and drying creep. With some enhancement of the model parameters this model shows good future prospects.

6 References

1. Okamura, H., Maekawa, K. and Ozawa, K. (1993) *High performance Concrete*, Gihodo, Tokyo.
2. Bazant, Z. P. (1977) Viscoelasticity of Solidifying Porous material-Concrete. *Journal of Engineering Mechanics ASCE*, Vol. 103, pp. 1049-1067.
3. Carol, I. and Bazant, Z. P. (1993) Viscoelasticity with Aging Caused by Solidification of Nonaging Constituent. *Journal of Engineering Mechanics ASCE*, Vol. 119, pp. 2252-2269.
4. Shimomura, T. and Maekawa, K. (1996) Analysis of the Drying Shrinkage Behavior of Concrete Using a Micromechanical Model Based on the Micropore Structure of Concrete. *Concrete Library of JSCE*, No. 27, pp. 121-143.
5. Ishida T. and et al. (Under Publication 1998) Analytical prediction of Autogenous and Drying Shrinkage of Concrete based on Microscopic Mechanisms. *CONSEC*.
6. Okamura, H. and Maekawa, K. (1991*) Nonlinear Analysis and Constitutive Models of Reinforced Concrete*, Gihodo Publishing, Tokyo.
7. Maekawa, K., Chaube, R. P. and kishi, T. (1995) Coupled Mass Transport, Hydration and Structure Formation Theory for Durability Design of Concrete Structures. *Proceedings of the International Workshop on Rational Design of Concrete Structures under Severe Conditions*, pp. 263-274.
8. Tazawa, E. and Miyazawa, S. (1995) Influence of cement and admixture on autogenous shrinkage of cement paste, *Cement and Concrete Research*, Vol. 25, No. 2, pp. 281-287.
9. Tazawa, E. and Miyazawa, S. (1994) Influence of binder and mix proportion on autogenous shrinkage of cementitious materials, *Proc. Of JSCE*, Vol. 25, No. 502, pp. 43-52.
10. Khan, A. A., Cook, W. D. and Mitchell, D. (1997) Creep, Shrinkage and Thermal Strains in Normal, Medium and High-Strength Concretes during Hydration. *Material Journal, ACI*, March-April, pp. 156-163.

22 MICRO-PHYSICAL APPROACH TO COUPLED AUTOGENOUS AND DRYING SHRINKAGE OF CONCRETE

T. Ishida, R.P. Chaube, T. Kishi and K. Maekawa
The University of Tokyo, Tokyo, Japan

Abstract
The proposed autogenous and drying shrinkage model is derived from micro-mechanical physics of water in pore structure in concrete. In the modeling, the shrinkage of concrete due to self-desiccation, known as autogenous shrinkage, is idealized to be driven by the surface tension force of capillary water across curved meniscus, similar to drying shrinkage processes. The overall material properties of concrete are evaluated considering the inter-relationship of hydration, moisture transport and pore-structure development processes based upon fundamental thermo-physical models. By implementing the shrinkage model to this system, the volumetric change of concrete can be predicted for different water to cement ratio, curing and mix proportions satisfactorily.
Keywords: Autogenous shrinkage, capillary tension, drying shrinkage, hydration, pore structure, water content.

1 Introduction
Self-compacting high performance concrete (HPC)[1] can avoid uncertainties rooted in construction methods (especially concreting works) and produce a reliable structural concrete with high durability. Owing to the requirements of super fluidity and segregation resistance of self-compacting HPC, a small amount of free water and low water to powder ratio are usually specified. For such a concrete, it has been reported that large amount of autogenous shrinkage occurs, and that the volume change due to autogenous shrinkage, which might be avoided for ordinary concrete, should be considered in crack control or durability check of RC structures [2].

In the hydration process of cementitious powder materials, the total volume of hydrate products will be smaller than the sum of the volumes of unhydrated powder

Autogenous Shrinkage of Concrete, edited by Ei-ichi Tazawa. Published in 1999 by E & FN Spon, 11 New Fetter Lane, London EC4P 4EE, UK. ISBN: 0 419 23890 5

materials and free water for hydration. Therefore, pore space in hydrated cement paste can not be filled with liquid water, and relative humidity in pore structures would decrease as the hydration proceeds. This phenomenon is called as self-desiccation and this mechanism would cause the autogenous shrinkage of concrete [2].

Many of the past researches have introduced capillary tension theory to explain drying shrinkage behavior [2] [5]. Under this assumption, the volume change and the deformation of cement paste would occur due to the surface tension force of capillary water across curved meniscus. In this paper also, it is assumed that capillary tension force caused by a drop of relative humidity in pore structures would lead to shrinkage stress, and this mechanism would cause autogenous shrinkage, as well as the shrinkage caused by drying out of concrete associated with water migration. As for the constitutive law, which describes the stress-strain relationship in terms of shrinkage behavior, basically the model proposed by Shimomura's research was applied in this study [5]. Here, shrinkage stress is determined by pore pressure, pore distribution and water content in hydrated cement paste. The deformability of the microstructure against shrinkage stress would be modeled based on computationally evaluated porosity of hydrated cement paste. These material properties of early aged concrete are evaluated by analytical method coded **DuCOM** considering inter-relationship of hydration, moisture transport and micro-pore structure development process [3]. In this system, autogenous shrinkage need not be distinguished from drying shrinkage caused by water loss. By the proposed system, the authors attempt to predict the volume change of concrete in a unified manner for arbitrary atmospheric conditions.

2 Analytical model

2.1 Pore structure development [3]

As a physical basis for pore-structure development computation at early ages of hydration, the overall pore-space is broadly divided into interlayer, gel and capillary porosity. The powder material is idealized as consisting of uniform sized spherical particles of same radius. Fig.1 shows a schematic representation of various phases at any arbitrary stage of hydration. Precipitation of the pore solution phase on the outer surfaces of particles leads to the formation of outer products whereas so called inner products are formed inside the original particle boundary. Inner product porosity and properties are assumed to be constant throughout the process of pore structure formation. It is assumed that the capillary porosity exists primarily in the outer product whereas the CSH hydrate crystals account for gel and interlayer porosity. Characteristic porosity of the CSH mass ϕ_{ch} is assumed to be constant throughout the progress of hydration. In this study a value of 0.28 is assumed [4].

Undertaking these assumptions, weight W_s and volume V_s of gel solids can be computed, provided average degree of hydration α and the amount of chemically combined water β per unit weight of powder material are known. The total volume of hydration products would decrease compared with the summation of both unhydrated powder and water volume. In this paper, this volume change during hydration is accounted by changing the water density for the sake of analytical convenience, that is, the density of free water ρ_L is 1.0 whereas the density for combined water ρ_w is assumed to be 1.25[4]. Overall volume balance thus gives interlayer (ϕ_l), gel (ϕ_g) and capillary (ϕ_c) porosity. These parameters are computed as,

$$\phi_l = \frac{t_w s_l \rho_s}{2} \qquad \phi_c = 1 - V_s - (1-\alpha)\frac{W_p}{\rho_p} \qquad (1)$$

$$\phi_g = \phi_{ch} V_s - \phi_l \qquad V_s = \frac{\alpha W_p}{1-\phi_{ch}}\left(\frac{1}{\rho_p} + \frac{\beta}{\rho_w}\right) \qquad (2)$$

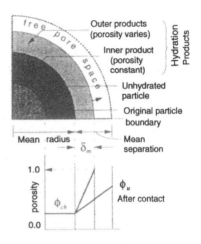

Fig.1 Micro pore-structure development model

where, t_w : interlayer thickness, s_l : specific surface area of interlayer, W_p : weight of the powder materials per unit volume, ρ_p : density of the powder material, ρ_s : dry density of solid crystals. The outer product of thickness δ_m is computed by assuming a bulk porosity variation which increases from the characteristics porosity of solid mass ϕ_{ch} at the particle surface to unity or ϕ_u at the external boundary of outer product. From these parameters, the surface areas of capillary and gel pores can be computed.

In the computational model a bi-model R-R porosity distribution $\phi(r)$ which gives the total porosity function is assumed [3][5]. Thus, we have,

$$\phi(r) = \phi_l + \phi_g\{1 - \exp(-B_g r)\} + \phi_c\{1 - \exp(-B_c r)\} \qquad (3)$$

where, r : pore radius. Distribution parameters B_c and B_g, which represent the peak of porosity distribution on logarithmic scale, correspond to the capillary and gel porosity components respectively. If we assume a cylindrical shape of the pores in such a distribution, these parameters can be easily obtained from the computed porosity and surface area values for the capillary and micro-gel region.

2.2 Multi-component hydration model [6]

Hydration process is simulated by using multi-component model for heat of hydration of cement proposed by *Kishi et al.* Specified chemical components of clinker minerals are treated as the characteristic parameter and the influence of variable moisture content and powder material proportions in the mix is taken into consideration. Total heat generation rate H per unit volume is described as eq. 4.

$$H = C\sum p_i H_i \qquad H_i = H_{i,T_o}\exp\left[-\frac{E_i}{R}\left(\frac{1}{T} - \frac{1}{T_o}\right)\right] \qquad (4)$$

where, C : the cement content per unit volume, p_i : corresponding mass ratio in the cement, H_i : the specific heat generation rate of individual clinker component, computed using Arrhenius's law, E_i : the activation energy of i-th component, R : gas constant, H_{i,T_o} : the referential heat rate of i-th component when temperature is T_o. The referential heat rate of each reaction embodies the probability of molecular collision with which hydration proceeds. In this model, it is taken as a function of the amount of free water, the thickness of the cluster made by already hydrated products and unhydrated chemical compound, and on the total accumulated heat of each clinker component. The total amount of free water in the model is in fact the total condensed water in micro-pore structure, which is obtained by moisture isotherm and transportation model, and pore structure development model.

In the above model, the average degree of hydration α and the chemical combined water β per unit weight of powder material are obtained at arbitrary stage of hydration. These parameters are essential for describing pore structure development process, and the total amount of water consumed per unit volume of concrete due to hydration is dynamically coupled with overall moisture balance.

2.3 Moisture transport formulation [7][8][9]

Ingress of moisture into the pores of concrete is a thermodynamic process, driven by the pressure and temperature potential gradients. In this study, total water present in matrix pores is subdivided into interlayer, adsorbed and condensed water. Interlayer water is probably the water which is under the influence of strong surface forces and which perhaps does not move under the application of pore pressure potential gradients. An absorption-desorption characteristic of the interlayer water is modeled based on Feldman and Sereda interlayer model [14]. Amount of water disposed in the remaining microstructure as condensed and adsorbed phases is obtained by integrating the degree of saturation of individual pores, considering local thermodynamical equilibrium and modified B.E.T. theory [7]. Furthermore, the hysteresis behavior, which shows that the water content in concrete is different under drying and wetting stages, is described by considering the geometrical characteristics of random pore structures [9]. Overall moisture balance which also takes into account the consumption of water during early age hydration can be obtained as,

$$\rho_L\left(\sum\phi_i\frac{\partial S_i}{\partial P_L}\right)\frac{\partial P_L}{\partial t}-div(K(P_L,T)\nabla P_L)+\rho_L\sum S_i\frac{\partial\phi_i}{\partial t}-W_p\frac{\partial\beta}{\partial t}=0 \tag{5}$$

where, ϕ_i : porosity of each component(interlayer, gel and capillary), S_i : degree of saturation of each component, P_L : equivalent liquid pore pressure, ρ_L : density of pore water. Moisture conductivity K is obtained from the flux models proposed by *Chaube et al*, using the computed R-R distribution function [8].

2.4 Autogenous and drying shrinkage model based on micromechanism [5].

Relative humidity in pore structures decreases by drying due to the gradient of RH between ambient conditions and inside concrete, or by self-desiccation even if the egress of moisture to outside does not occur. Considering local thermodynamic and interface equilibrium, vapor and liquid interfaces in pore structure would be formed due to the pressure difference by capillarity. When the interface is a part of an ideal spherical surface, this relation can be described by Kelvin's equation as,

$$\ln\frac{P_v}{P_{vo}} = -\frac{2\gamma M_w}{RT\rho_L}\frac{1}{r_s} \tag{6}$$

where, P_v/P_{vo} : Relative humidity of vapor phase, γ : surface tension of liquid water [N/m], M_w : molecular mass of water [kg/mol], R : universal gas constant [J/mol·K], T : absolute temperature [K], ρ_L : density of liquid water [kg/m³], r_s : the radius of the pore in which the interface of liquid and vapor is created [m]. Due to the surface tension of liquid water, the pressures of gas and liquid are not equal. This pressure difference ΔP would be described by Laplace's equation as,

$$\Delta P = P_G - P_L = \frac{2\gamma}{r_s} \tag{7}$$

where, P_G, P_L : pressure of gas and liquid phase respectively [Pa]. From eq.7 we can find clearly that the pressure of the liquid phase is lower than that of gas phase, and tensile stress would be applied on the pore walls where contact is made with liquid phase. In this paper, the volume change at a microscopic level due to this stress is assumed to cause both the autogenous shrinkage and drying shrinkage at macroscopic scale. The total intensity of the tensile stress per unit concrete volume will depend on both the magnitudes of the tension and the area where it is applied. We adopt the Shimomura's formulation for capillary stress which causes drying shrinkage as [5],

$$\sigma_s = A_s\frac{2\gamma}{r_s} \tag{8}$$

where, σ_s : capillary stress [Pa], A_s : area factor [m³/m³]. The capillary stress is applied on pore walls where liquid water exists. Therefore, in this paper, A_s is defined as the total liquid water content per unit concrete volume, which can be obtained by hysteretic isotherms, moisture transport and pore structure development model. The stress-strain relationship, which describes the micro deformation of cement paste due to capillary stress, might show a nonlinearity. However, at this stage, it is difficult to take account of the nonlinear behavior of C-S-H crystals at micro level. In this study, using elastic modulus for capillary stress E_s, which includes various aspects of nonlinear deformational behaviors, the stress-strain relationship is assumed as [5],

$$\varepsilon_{sh} = \frac{\sigma_s}{E_s} \tag{9}$$

where, ε_{sh} is unrestrained macroscopic shrinkage strain.

3. Analytical study of the shrinkage behavior by the proposed model

3.1 Outline of the computational scheme of solutions for material properties

An analytical study of the autogenous and the drying shrinkage behavior of mortar and concrete was done by using the proposed shrinkage model. To treat both autogenous, drying behavior and their coupling in the same scheme, the various material properties, namely, hydration, pore-structure development, moisture distribution and relative humidity in pores, or deformability for capillary stress should be predicted with time and space. In this study, to verify the shrinkage behavior of concrete for arbitrary mix proportions, curing and environmental conditions, the proposed shrinkage model was implemented into a finite-element computational program **DuCOM** (Fig.2) [3]. This system can deal with the coupled pore-structure development, hydration and moisture transport behaviors of early aged concrete and is accessable on WWW addressed by **http://concrete.t.u-tokyo.ac.jp/index.html**. This program can give the solutions of temperature, pore pressure, pore distributions and other material properties in 3-D space and time domain. In this computational system, it is not necessary to distinguish the behavior of autogenous and drying shrinkage. By calculating the water content and pore structures in the concrete under given initial environmental and curing conditions, we can predict the volume change due to shrinkage behavior from eq. (9).

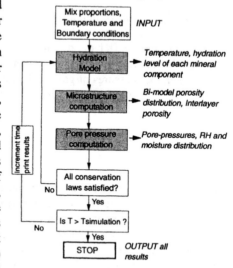

Fig.2 Coupled 3D-FEM scheme of solution for hydration moisture transport and structure formation problem in concrete

3.2 Analysis of the shrinkage behavior

For verification, prismatic specimens were used. In this case, based on the assumption of a uniform deformation in one direction, the shrinkage of the specimen can be calculated as the average of free unrestrained shrinkage strain, which would be different with each finite element.

In this computational scheme, for a given pore pressure and total water content in pores as calculated by **DuCOM**, the capillary stress σ_s can be obtained from eq. (8). As for E_s, which indicates the deformability against the capillary stress, it has been reported from past research that E_s is about 1/3-1/4 of the ordinary values of the static elastic modulus E_c of concrete under extensively applied stress under higher loading rate than the reality [5]. In this paper also, there would be some relationships between E_s and E_c, and in all cases E_s is assumed to be 1/4 of E_c. The effect of creep behavior might be included in E_s. It has also been found out and reported that a good correlation between E_c and compressive strength of concrete exists [4][10]. Therefore, to obtain elastic modulus E_c, we will try to estimate the compressive strength from the porosity of hydrated cement paste as,

$$f_c' = A \exp(-B \cdot V_{pore})$$ (10)

where, f_c' : compressive strength, V_{pore} : pore volume per unit paste volume, A,B : constant. In general, it has been reported that compressive strength is dependent on the pore volume of the cement paste, especially, comprising of whose radii lie in the 50nm - 2μm range [10]. Therefore, the total volume of pores whose radii are above 50nm is assumed to be as V_{pore} in eq. (10). Using **DuCOM**, V_{pore} can be analytically obtained from the solutions of the porosity and the pore distribution of the capillary in each element and with time.

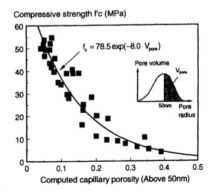

Fig.3 The relationship of measured compressive strength and computed capillary porosity

$$V_{pore} = \phi_c \cdot \exp(-B_c \cdot r)$$ (11)

where, ϕ_c is capillary porosity, B_c is the distribution parameter for capillary pore, and r is 5.0×10^{-8}[m].

Fig.3 shows the relationship of measured compressive strength and computed capillary porosity (above 50nm). The 38 specimens of mortar and concrete in Fig.2 were tested under various conditions, that is, water to cement ratio by weight: 24-51%;the aggregate volume: 40%-65%; type of powder materials: ordinary, medium heat and high-belite cement, blast furnace slag; curing conditions: sealed, submerged and exposed curing; curing temperature: 20℃, various temperature history; age of the specimen: 0.5-28days. To compute the capillary porosity of each test specimen, the same conditions were given as the experimental condition in the analysis. The computed porosity obtained by **DuCOM** involves the effects of age, curing conditions, temperature and mix proportions.

The above relationship of compressive strength and computed porosity was regressed with an exponential function, and the constants A and B in eq. (10) could be obtained. Elastic modulus of concrete E_c was calculated from the empirical formula [11] and the compressive strength relationship of eq. (10). As already mentioned, deformability against capillary stress E_s is assumed to be 1/4 of E_c in this analysis.

3.3 Verifications

3.3.1 Drying shrinkage behaviors
The powder materials, mix proportions, specimen size and experimental conditions of test specimens in this work are shown in Table 1. The target here is to verify moisture loss and drying shrinkage behaviors of mortar (No.1, 2 in Table1). In the experiment, the water to powder ratio was 28% and the size of mortar specimens were 4×4×16[cm].

Table 1. Experimental conditions of test specimens.

Specimen No.	Powder material	W/C (%)	Volume of Aggregate	Specimen size [cm]	Drying condition
1 (Mortar)	MC[*1]	28%	50%	4×4×16	66%RH after 2days
2 (Mortar)	MC	28%	50%	4×4×16	45%RH after 2days
3 (Concrete)[12]	OPC[*2]	30%	62%	10×10×120	Sealed
4 (Concrete)[12]	OPC	40%	62%	10×10×120	Sealed
5 (Concrete)[12]	OPC	30%	62%	10×10×120	50%RH after 1day
6 (Mortar)[12]	HS[*3]	30%	38%	10×10×40	50%RH after 7days
7 (Mortar)[12]	HS	30%	38%	10×10×40	Sealed
8 (Mortar)[12]	HS	50%	38%	10×10×40	50%RH after 7days
9 (Mortar)[12]	HS	50%	38%	10×10×40	Sealed
10 (Mortar)	OPC	25%	50%	4×4×16	60%RH after 7days
11 (Mortar)	OPC	25%	50%	4×4×16	60%RH after 28days
12 (Mortar)	OPC	55%	50%	4×4×16	60%RH after 28days

*1: Medium heat cement, *2: Ordinary portland cement, *3: High strength cement.

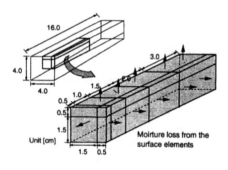

Fig.4 Mesh layout used for FE analysis

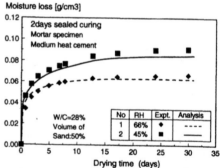

Fig.5 Moisture loss behavior for different drying cases

After 2days of sealed curing, the specimens were exposed to drying in 66%RH, 45%RH and 20℃, and moisture loss and length change with time were measured. In the FEM analysis, mix proportions and the chemical composition of the cements (C₃A, C₄AF, C₃S, C₂S, and gypsum) were given. The curing conditions and the exposed ambient conditions were also given as the boundary conditions of the target structures. All of these input values corresponded to the experimental conditions. In these tests, drying started

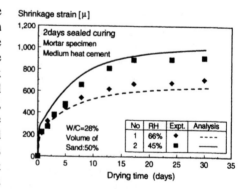

Fig.6 Drying shrinkage behavior for different drying cases

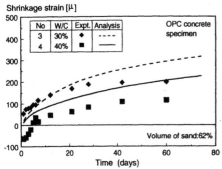

Fig.7 Autogenous shrinkage behavior of concrete for different W/C

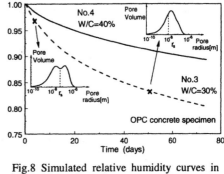

Fig.8 Simulated relative humidity curves in pore structures under sealed condition

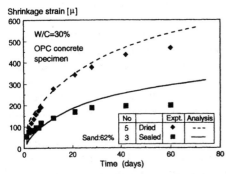

Fig.9 Prediction of autogenous and drying shrinkage behavior of concrete specimen

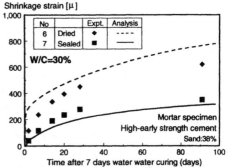

Fig.10 Autogenous and drying shrinkage behavior of mortar (W/C30%)

after 2days of sealed curing. Mesh for FE analysis is shown in Fig.4. The analysis can implicitly consider the effects of the autogenous shrinkage even under the drying conditions, while the hydration could be still significant. Of course, an effect of the delay of hydration due to a loss of free water under drying conditions can be coherently evaluated by the analysis. The computed results of moisture loss and shrinkage behavior show reasonable agreement with experimental data (Figs. 5,6).

3.3.2 Autogenous and drying shrinkage of mortar and concrete
The second case is the study of volume change due to both of the autogenous and

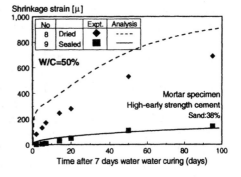

Fig.11 Autogenous and drying shrinkage behavior of mortar (W/C50%)

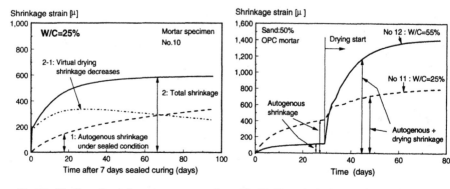

Fig.12 Nonlinearity of autogenous and drying shrinkage behavior

Fig.13 Numerical simulation of autogenous and drying shrinkage behaviors for different water to cement ratio

the drying shrinkage of mortar and concrete. As already mentioned, in the framework of analysis by **DuCOM**, the volume change can be predicted based on the micro mechanisms considering capillary tension force in a pore. No matter how the capillary tension is caused by a moisture loss due to drying or by a self-desiccation, coupled autogenous and drying shrinkage behavior can be obtained from calculating pore structure, RH in a pore, water content and deformability with each element and time domain.

First, autogenous shrinkage behaviors of concrete were studied. For verification, the experimental data by *Tazawa et al* [12] were used (No.3, 4). The size of mortar specimens was 4×4×16[cm], and the volume changes were measured under sealed condition and 20℃. In this case, the computed RH in pores and the intensity of capillary tension are dependent on the chemically combined water obtained by hydration model and the volume change of hydrated cement paste computed by pore structure formation model, since there is no moisture movement through the boundary element. Fig.7 shows the simulation of autogenous shrinkage. The calculation can reasonably follow the tendency of autogenous shrinkage strain for each W/C case. Fig.8 shows the analytically predicted average RH in pore structures of concrete, formed porosity distribution and the radius of the pore r_s in which the interface of liquid and vapor is created (eq.(6)). As shown in Fig.8, the average radius of pores in hydrated cement paste decreases as the hydration proceeds. At the same time free water for hydration is consumed. This results in the depletion of liquid water from pores. In Fig.8, this space would occupy the pores whose radii are above r_s. In **DuCOM**, considering thermodynamic equilibrium condition, RH in vacant pore space can be obtained from this radius r_s and surface tension of liquid. The calculated value of RH decreases with time. For W/C 40% and 30%, pore humidity decreases to 90% and 80%, respectively. This result seems to be rational, since the past research [13] has reported a similar quantitative trend in the decrease of pore with time. The drop of RH causes the surface tension force, and autogenous shrinkage can be obtained numerically.

Next, the authors verified the combined autogenous and drying shrinkage test for mortar and concrete specimen, which has similar mix proportion and curing condition

(Specimen No. 3-9). The one specimen was kept sealed, and the other was dried under 60% RH (Fig.9), 50%RH (Fig. 10, 11) after some curing period. The analytical results show good agreements with experimental data for different used materials, mix proportions, and curing conditions in terms of both the autogenous and the drying shrinkage.

3.3.3 Numerical simulation of coupled autogenous and drying shrinkage

Using the proposed system, several numerical sensitivity simulations about the behavior of autogenous and drying shrinkage were done.

The first case is a study of nonlinearity of autogenous and drying shrinkage behavior. Fig.12 shows computational results of shrinkage behavior of mortar, which exposed to drying after 7 days sealed curing (Specimen No.10). In Fig.12, a solid line 2 shows the total shrinkage strain, whereas a dotted line 1 shows autogenous shrinkage strain under perfect sealed condition. If total shrinkage strain can be assumed to be the linear summation of the autogenous shrinkage under sealed condition and drying shrinkage, the contribution of drying shrinkage would be the difference between total shrinkage and autogenous shrinkage, that is the line 2-1 in Fig.12. As a result, this virtual drying shrinkage strain 2-1 decreases under the assumption that the linear summation can be true. In fact, under drying condition, the autogenous shrinkage would be less than that of under sealed condition, because of the delay of hydration due to moisture loss to outside. Without considering this aspects, the contribution of autogenous shrinkage would be overestimated. Namely, for predicting the actual shrinkage behavior, it is necessary to consider the interdependance between moisture loss and hydration process.

The second case is the numerical simulations for quite different W/C ratio (Fig.13). The case of W/C55% was analyzed as ordinary concrete, whereas that of W/C25% as self-compacting concrete or high strength concrete (No.10, 11, Table1). The specimen was exposed to RH60% drying condition after 28days sealed curing. It has been reported that the shrinkage behaviors are quite different between the ordinary concrete and low W/C ratio concrete [2]. In case of the ordinary concrete, the contribution of the autogenous shrinkage to the total volume change is relatively small, compared to the drying shrinkage contribution. On the other hand, in the case of low W/C concrete, the effect of the autogenous shrinkage is not negligible compared with the drying shrinkage. The analytical results can accurately exhibit this qualitative tendency.

4 Conclusions

A micro-physical based model for autogenous and drying shrinkage of concrete based on micro mechanical was presented in this paper. The capillary stress is assumed to cause the autogenous shrinkage and the drying shrinkage. The material properties of young aged concrete were obtained by the coupled analysis considering the inter-relationship of hydration, moisture transport and pore-structure development process. This framework can be a predictive method of the volume change of concrete due to the autogenous, drying shrinkage and their combination for arbitrary mix proportions, materials, age, curing conditions and environmental conditions. The entire scheme is expected to form basis of future performance based durability design.

5 References

1. Okamura, H., Maekawa, K. and Ozawa, K. (1993) *High Performance Concrete*, Gihodo, Tokyo (in Japanese).
2. JCI, *The report of JCI Committee on Autogenous Shrinkage* (1996) (in Japanese).
3. Maekawa, K., Chaube, R.P. and Kishi T. (1995) Coupled mass transport, hydration and structure formation theory for durability design of concrete structures, *Proc. of the Int'nl workshop on rational design of concrete structures under severe conditions*, Hakodate, pp. 263-274
4. Neville, A.M. (1979) *Properties of Concrete*, ISMN, London.
5. Shimomura, T. and Maekawa, K. (1996) Analysis of the drying shrinkage behavior of concrete using a micromechanical model based on the micropore structure of concrete, *Concrete Library of JSCE No.27*, pp. 121-143.
6. Kishi, T. and Maekawa, K. (1995) Multi-component model for hydration heat of portland cement, Proc. of JSCE, No.526, V-29, pp.97-109 (in Japanese).
7. Chaube, R.P. and Maekawa, K. (1994) A study of the moisture transport process in concrete as a composite material, *Proc. of the JCI*, Vol. 16, No.1, pp.895-900.
8. Chaube, R.P. and Maekawa, K. (1996) A permeability model of concrete considering its microstructual characteristics, *Proc. of the JCI*, Vol. 18, No.1, pp. 927-932.
9. Ishida, T., Chaube, R.P. and Maekawa, K. (1996) Modeling of pore water content in concrete under generic drying wetting conditions, *Proc. of the JCI*, Vol. 18, No.1, pp. 717-722.
10. Uchikawa, H., Uchida, S. and Hanehara, S. (1988) Advances in cement manufacture and use, Engineering foundation conference, pp.271-294.
11. Okamura, H. and Maeda, S (1987) *Reinforced Concrete Engineering*, Ichigaya Syuppan, Tokyo (in Japanese).
12. Tazawa, E. and Miyazawa, S. (1994) Influence of binder and mix proportion on autogenous shrinkage of cementitious materials, *Proc. of JSCE*, No.502, V-25, pp.43-52 (in Japanese).
13. Paille, A.M., Buil, M. and Serrano, J.J. (1990) Effect of fiber addition on the autogenous shrinkage of silica fume concrete, *ACI Material Journal*, Vol.86, No.2, pp.139-144.
14. Feldman, R.F. and Sereda, P.J. (1968) A model for hydrated portland cement paste as deduced from sorption-length change and mechanical properties, *Master. Constr.*, Vol.1, 509-519.

Discussions

Modeling dimensional changes in low water/cement ratio pastes
  E.A.B. Koenders and K. van Breugel, Delft University of Technology, The Netherlands

T.Ishida   I am very interested in your research. I have one question. In the case of mass concrete or in the case of specimen which has a big cross section, the hydration level would be quite different between the surface area and inside of concrete. Volume changes owing to shrinkage would be different. How do you treat this aspect?

K.van Breugel   This was not the purpose of this study, but you are completely right. In mass concrete the shrinkage strain will differ throughout the cross section. In this study we are focusing on the mechanism behind autogenous shrinkage caused by self-desiccation. The additional effect of moisture loss to the environment, which will contribute the shrinkage strains in the exterior part of the structure, is not considered in this study.

L.Barcelo   Thank you very much Dr. Breugel for very interesting presentation. I have one question. Did you manage to correlate the relative humidity that you may calculate from your model and the relative humidity that you measure?

K.van Breugel   Yes. We have compared the results of our numerical simulations with measurements presented by Baroghel Bouny (France). In both the numerical simulations and the measurements the relative humidity varied between 70 and 80 % after about 28 days, this depending on the initial water-binder ratio.

S.Tangtermsirikul   Question is regarding your model. When you study the effect of aggregate in the cement pastes, how can you deal with the problem of the interaction between the aggregates, which means, you are considering aggregates embedded in the pastes individually with no interaction, but I believe that there are interactions between contact between the aggregates. It might be taken account in your simulation.

K.van Breugel   Yes, this is a good question. The research we did so far on this particular topic is to be considered as pilot project. It was assumed that the "beams" behave linear elastically. A next step is to consider the possibility that certain beams will collapse when a certain crack criterion is exceeded. Once this has happened to a particular beam, the calculations have to be continued while disregarding this beam. Even more important seems to me the modeling of the interfacial zone and the connection between the "beams", representing the matrix, and the aggregate. In normal density concrete this interfacial zone is generally the weakest link. So far we have not studied this point. What we did get from the present simulations, however, is at least a clear indication of stress concentrations and of points where microcracking is likely to occur and how this might affect the strength of concrete.

A study on the hydration ratio and autogenous shrinkage of cement paste
K.B. Park, T. Noguchi and F. Tomosawa, The University of Tokyo, Japan

S.Goto   You said that the autogenous shrinkage increases with decreasing the smaller pore volume. What is the practical meaning?

K.B. Park   I measured capillary pores.

E.Tazawa   Do you mean that you have some maximum volume of smaller pores at some degree of hydration?

K.B. Park   This is the measured result.

Solidification model of hardening concrete composite for predicting autogenous and drying shrinkage
R. T. Mabrouk, T. Ishida and K. Maekawa, The University of Tokyo, Japan

No question.

Micro-physical approach to coupled autogenous and drying shrinkage of concrete
T. Ishida, R.P. Chaube, T. Kishi and K. Maekawa, The University of Tokyo, Japan

B.S.M.Persson   Did you plot by any chance the relative humidity versus shrinkage?

T.Ishida	Yes we plot the relationship between relative humidity and shrinkage
B.S.M.Persson	Did you find the serious shrinkage connected with drying shrinkage?
T.Ishida	No. There is no one to one relation, because the deformability depends on w/c, micro pore structure or shrinkage stress. So if the w/c, curing condition or curing temperature is different, micropore structure or deformability against capillary stress is different, therefore, we can not find one to one relation between relative humidity and shrinkage strain.
S.Goto	You had shown that the capillary forces act on all the interface of water and capillary wall by the allows in the OHP figure. But usually the capillary force is thought to act just on the edge of water meniscus, isn't it? How do you think about it?
T.Ishida	We use the water content as a parameter of the area where the tension is applied. So strictly speaking, that parameter does not have strict physical meaning, but of course, there is some relationship and it might have some physical meaning.
S.Goto	You can calculate the pore size distribution and also the volume of water. So you are able to obtain the diameter of meniscus, and also the total length of the meniscus where the capillary tension is generating. Consequently, you can get total force for shrinkage, I think.

At first I want to point out 2 things on the mechanisms of autogenous shrinkage. One is that sometimes the autogenous shrinkage is expressed as the result of the reduction of relative humidity. It is not correct, I think. The shrinkage is the result of the force generated in the capillary meniscus whose radius is the maximum pore radius that is filled with water and the water vapor pressure (relative humidity) is equilibrated to the radius. That is to say the relative humidity is the result of the pore size distribution and the amount of the water remaining. Another one is that the capillary force act perpendicular to the capillary wall, not parallel to the wall.

Now we shall start to consider the balance of forces in the paste and the motion of materials. Three factors such as atmospheric pressure [P], capillary force [$2\pi r_t \gamma \cos\theta$] (capillary force for shrinkage is [$2\pi r_t \gamma \cos\theta$]) and bonding strength [s_t] among the particles should primarily be taken account for autogenous shrinkage. Now we assume the structure of paste as shown in Fig. 1. Here r_t is the maximum capillary pore radius filled with water at time t. In Fig.2 the force balance and materials motion are shown schematically.

When $\pi r_t^2 P$ is larger than $2 \pi r_t \gamma \cos \theta$, water in capillary will move. And when $2 \pi r_t \gamma \cos \theta$ is larger than the bonding strength, s_t, particles can be rearranged, so the chemical shrinkage can be measured completely in this case. When strength, s_t, becomes larger than $2 \pi r_t \gamma \cos \theta$, the particle displacement does not occur, and water will move alone into inside of the paste. This phenomena has already been described by Prof. Tazawa. When $2 \pi r_t \gamma \cos \theta$ becomes larger than $\pi r_t^2 P$, water can not move, and pores filled with water vapor should be created, that is, self-desiccation occurs. In this case, the radii of meniscus is determined with the pore size distribution and the amount of remaining water, and vapor pressure (relative humidity) is equilibrated to the radii through the Kelvin's equation.

Next, we will consider how the force for the autogenous shrinkage generates. Here, we have to know the pore size distribution, which changes with the degree of hydration. If we can assume that there are no entrained air void, and that the theoretical amount of water requested for hydration is 0.28, we can calculate the amount of water remains.

$$V_{remain}/1g\text{-cement unhydrated}=V_{filled\ with\ water}=W/C-0.28*a$$

From the pore size distribution and the volume of free water in the paste, we can obtain the radius of the meniscus. A force for shrinkage generates as the value of $2 \pi r_t \gamma \sin \theta$ for 1 capillary meniscus. Total force with n capillaries is obtained as $2 \pi nr_t \gamma \sin \theta$ for unit volume. So the one dimensional force per uint area will be $2 \pi nr_t \gamma \sin \theta / 3$. The $\sin \theta$ means that the capillary force does not work in the direction of parallel with the wall, but does work in the direction of perpendicular to the wall. If we can assume the pore shape as capillary but having the same length with diameter, $2r_t$, we can calculate the total edge length of meniscus, $2 \pi nr_t =L$, here, $n = \triangle v/2pr_t^3$. So total capillary force is proportional to $(\triangle v/r_t^2)* \gamma$. Here $\triangle v$ is the pore volume with radius of r_t.

The effect of relative humidity in the pore on the negative pressure in the capillary water is very small, because the differences between atmospheric pressure and the partial water vapor pressure in the pore are in a very narrow range. The partial vapor pressure of water, the maximum of which is at least 20mmHg at 20 ℃, is negligible when compared with atmospheric pressure of 760 mmHg.

As I mentioned here, the surface tension is very important to the shrinkage. In this meeting, there are some reports on the effect of temperature in which they mentioned that the curing at the higher temperature decreases the amount of autogenous shrinkage. This can be understood from the fact that the surface tension of water decreases with temperature rise. So I can expect to reduce the autogenous shrinkage by using the surface tension modifier which is already used as modifier for drying shrinkage.

K.Maekawa

Let me discuss something with Prof. Tazawa's comment in which the driving force of autogenous shrinkage and drying shrinkage are discussed and now Prof. Goto aroused some points. My understanding is like this. When the water content is larger, I understand, each micro pore has some linkage each other. When the water content is higher, it may be able to apply your models with solid skeleton micropores. In this situation, the mechanism is based upon the Pascal's principles. But I suppose around 90%, 80% or 70% RH, it will have some breakage of micropore systems.

Please consider one extreme case, when water is filled, when pore is all saturated, so I suppose, your thinking, your way, would be again something. But, when water content is so much smaller, some capillary pore structure become independent. In this situation we have to deal with three kinds of water, capillary water, gel water and interlayer water. These kinds of water have different path dependency.

Your story is for monotonic drying situation but when we consider wet and dry, small pore will be vacant and larger pore can be filled. This is the path dependency of equithermal situation. Very small pores may be kept in very strong action with gel or interlayers, so I can partially agree with you but partially I do not agree with you.

S.Goto

I had mentioned about the equilibrium state for the calculation of the force of autogenous shrinkage generated with the hydration process. When we consider the non-equilibrium state, we may not be able to calculate the force. As I mentioned above, the shrinkage force is generated by the total capillary force.

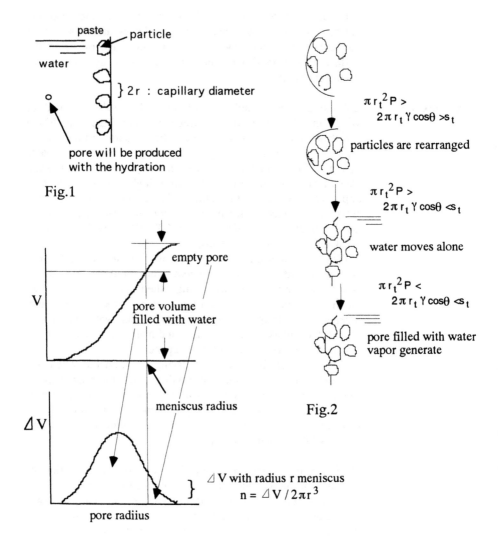

paste   particle

water

} 2r : capillary diameter

pore will be produced
with the hydration

Fig.1

$\pi r_t^2 P >$
   $2\pi r_t \gamma \cos\theta > s_t$

particles are rearranged

$\pi r_t^2 P >$
   $2\pi r_t \gamma \cos\theta < s_t$

water moves alone

$\pi r_t^2 P <$
   $2\pi r_t \gamma \cos\theta < s_t$

pore filled with water
vapor generate

Fig.2

V

empty pore

pore volume
filled with water

Δ V

meniscus radius

Δ V with radius r meniscus
$$n = \Delta V / 2\pi r^3$$

pore radiius

Fig.3

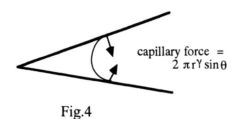

capillary force =
2 $\pi r \gamma \sin\theta$

Fig.4

23 AUTOGENOUS DEFORMATIONS AND STRESS DEVELOPMENT IN HARDENING CONCRETE

H. Hedlund and G. Westman
Luleå University of Technology, Luleå, Sweden

Abstract
Deformation measurement on concrete under temperature changes gives important information of the volume changes due to hydration. For normal concrete (high *w/c*) this behaviour is represented by two separate lines - one for the expansion phase and one for the contraction phase. Unfortunately this is not the case for high performance concrete which have a significant simultaneous shrinkage due to self-desiccation even at early ages.

This paper describes a method to separate the moisture movements - autogenous shrinkage - from the measured deformations obtained in thermal deformation tests. The result is used in stress calculations taking both thermal and moisture movement into consideration. The stress calculations are compared with results obtained in a special temperature-stress frame.
Keywords: Autogenous shrinkage, hardening concrete, maturity, stresses, thermal movements, stress measurements, isothermal temperature, stress induced deformations

1   Introduction

Stresses in concrete can, generally speaking, be caused by restrained volume changes during hardening. The volume changes are not only ruled by the degree of restraint and the thermal movements. High Performance Concrete (HPC) is usually based on high binder content mixed with small amounts of water. The low water to binder ratios leads to shrinkage at sealed conditions which has to be included in the stress analyses. For water cement ratios below about 0.40, self-desiccation results in a significant drop in the pore humidity even at early ages [1]. In contrast to the moisture flux self-

Autogenous Shrinkage of Concrete, edited by Ei-ichi Tazawa. Published in 1999 by E & FN Spon,
11 New Fetter Lane, London EC4P 4EE, UK. ISBN: 0 419 23890 5

desiccation is reflected all over the structure, and at restraint conditions this may be a significant contribution to the risk of early age cracking.

The volume changes due to temperature will during temperature rise be an expansion. At the same time the volume changes due to self-desiccation work in the opposite direction, i.e. reducing the volume with increasing temperature. During the cooling stage, the temperature and the self-desiccation shrinkage will co-operate and the rate of volume change will increase for HPC, see [2]. This is also the case when normal strength concrete is used, but the self-desiccation shrinkage is usually very small and can normally be neglected.

2   Deformations

The shrinkage of young hardening concrete competes with thermal expansion the first hours after casting. It is found necessary to record the temperature changes in the hardening concrete when shrinkage under sealed conditions is measured. For a test specimen with 100 x 100 mm^2 size in this study, the temperature rise was about 3 to 5 °C, see [2], resulting in a swelling of the test specimen.

In order to obtain the pure self-desiccation strain the thermal strain during the expansion phase, assumed to be about 10 μm/m °C (if not isothermal tests have been performed), was excluded from the shrinkage measurements. The measured strain can be expressed as

$$\varepsilon_{tot} = \varepsilon_T + \varepsilon_{SH} \qquad (1)$$

where
ε_{tot}	= measured strain, μm/m
ε_T	= thermal strain, μm/m
ε_{SH}	= true self-desiccation shrinkage, μm/m.

If the shrinkage taken from separate tests is assumed to be ruled by the equivalent time of maturity, the values of ε_{SH} can be estimated at any temperature time development. From measured or calculated temperature development the thermal strain is estimated by

$$\varepsilon_T = \Delta T \cdot \alpha_E \qquad (2)$$

where
ε_T	= thermal strain, μm/m
ΔT	= temperature difference in specimen, °C
α_E	= thermal expansion coefficient, μm/m °C

By using the maturity concept and calculated temperature development under the assumption that shrinkage under sealed conditions can be uniquely expressed as a function of equivalent time of maturity the time scale for measured values of sealed shrinkage can be recalculated, see further [2]. Similar assumptions are done in [3].

The choice of start time for separation is that both types of deformation should be valid at the same time, see Fig. 1. The evaluation of measured shrinkage starts from the actual time at the first possible individual measurement with the used test method. This means that there is some uncertainty in the shrinkage values at very early ages.

The autogenous shrinkage measured on sealed specimens have been evaluated using Eqs. (1) - (2), and fitted to the resulting final shrinkage expressed by

$$\varepsilon_{SH} = \varepsilon_{ref} \cdot \exp\left(-\left[\frac{\theta_{SH}}{t_{eq} - t_{\varepsilon0}}\right]^{\eta_{SH}}\right) \qquad (3)$$

where ε_{ref} = reference ultimate shrinkage, μm/m

θ_{SH} = empirical parameter representing the time when inclination of shrinkage changes in time, h

t_{eq} = equivalent time, h

$t_{\varepsilon0}$ = start time of measurements, h

η_{SH} = empirical constant ruling the curvature, -

The temperature equivalent time controls the shrinkage development, which is the same assumption as in [4]. In Fig. 2 an example of measured total deformation for concrete HPC 3 - $w/c = 0.30$ and $s/c = 0.10$ - is plotted together with measured shrinkage versus maturity time.

Evaluating the total measured deformation without splitting of thermal and moisture movements, i.e. to directly use the total deformation from tests, will give combined "sealed" coefficients, $\alpha_E{}^*$ and $\alpha_C{}^*$, representing the expansion phase and the contraction phase respectively. Note that these coefficients are structural related information for the specific case connected to the studied temperature-time curve.

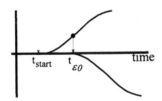

Fig. 1.   Schematically picture of chosen start time for the separation of measured deformation.

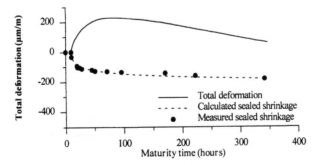

Fig. 2.   Measured total deformation and sealed shrinkage as well as calculated sealed shrinkage for HPC 3.

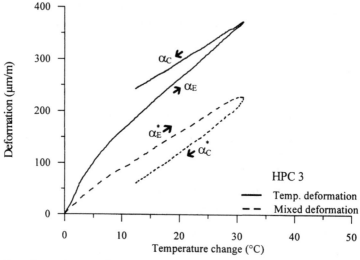

Fig. 3.   Free thermal expansion and contraction tests for HPC 3. The measurements
have started about nine hours after casting.

By subtracting the sealed shrinkage from the measured total deformation the true
thermal deformation is obtained, shown by the solid lines in Fig. 3.

The expansion coefficient, α_E, is represented by the inclination of the deformation,
when the temperature is increasing, and the thermal contraction coefficient, α_C, is
represented in the same way when the temperature is decreasing. The evaluation has
been performed from the latest start time, see Fig. 1, of either the thermal deformation
test or the start time of the shrinkage measurements.

The thermal expansion coefficients, α_E, is quite stable and is in the region of about
10 - 12 μm/m °C for both high performance concrete and normal strength concrete.
The thermal contraction coefficients, α_C, show levels about 7 - 9 μm/m °C for both
high performance concrete and normal strength concrete, see [2, 5 and 6].

Evaluated autogenous shrinkage and thermal movement coefficients are presented
in Table 1. These values will later be used in stress calculations for

• a temperature - time simulation of a 0.7 m thick wall
• a semi-isothermal temperature condition

subjected to 100% restraint conditions. Semi-isothermal conditions is here referred as a
short initial temperature wave followed by an isothermal temperature, see Fig. 4.
Comparison between measured stresses obtained in the stress test frame and computed
stresses shows good agreement, see section 4.

Table 1.   Evaluated shrinkage and thermal movements parameters for HPC 3

Concrete	ε_{ref}	θ_{SH}	$t_{\varepsilon 0}$	η_{SH}	α_E	α_C
	μm/m	hours	hours	-	μm/m °C	μm/m °C
HPC 3	-310	30	9.29	0.26	11.8	8.6

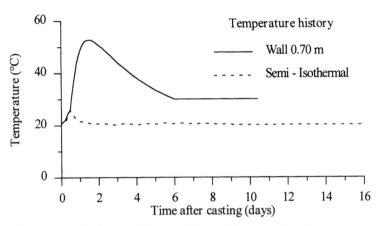

Fig. 4. Temperature history used for regulation of the stress test frame.

3 Modelling of stresses for completely restraint conditions

The presented mathematical model is a tool for extrapolating test results obtained in material tests for a specific concrete recipe. The modelling is here done in a material based way to avoid large fluctuations at applications, which might occur in modelling based on pure mathematical regression. When the concrete mix is examined and the model is applied, you have the opportunity to simulate many different types of structures cast with the evaluated concrete. It also gives you the possibility to perform calculations with other conditions in view of temperatures and environment. The main goal for this modelling is to be able to perform stress analysis for real structures cast with the analysed concrete mix. The model is not intended to be valid for concretes in general.

3.1 Constitutive equation

The uniaxial constitutive law for incremental calculations of stresses in young concrete at variable humidity and temperature can [2], [7] and [8] be expressed by

$$\Delta\sigma = E_{tot} \cdot (\Delta\varepsilon - \Delta\varepsilon_{tot}^o) \tag{4}$$

with

$$E_{tot} = E_c \cdot (1 + \gamma_d) \tag{5}$$

and

$$\Delta\varepsilon_{tot}^o = \Delta\varepsilon_{rel} + \Delta\varepsilon_\varphi^o + \Delta\varepsilon_T^o \tag{6}$$

where σ = stress in the material, Pa
ε = the total strain
Δ = denotes an increment during the actual time step
EC = fictitious elastic modulus for the time step including basic creep

effects, Pa

γ_d = incremental adjustment due to non-linear stress-strain behaviour

ε_{rel} = relaxation strain

ε_φ^o = non-elastic strain, including the stress-induced part, due to a change in humidity

ε_T^o = non-elastic strain, including the stress-induced part, due to a change in temperature

The structural condition for the test is completely restrained, i.e. a relaxation test is performed where the external deformation is identically zero ($\Delta\varepsilon \equiv 0$). For this case Eqs. (4)-(6) are reduced to

$$\Delta\sigma_{rel} = -E_{tot} \cdot \Delta\varepsilon_{tot}^o \qquad (7)$$

where σ_{rel} = relaxation stress associated with external restraint conditions, (Pa)

For loading in tension the following virgin stress-strain curve is introduced

$$\frac{\sigma}{f_{ct}} = \frac{\varepsilon_m}{\varepsilon_o} \qquad \text{for} \qquad \frac{\sigma}{f_{ct}} \le \alpha_{ct} \qquad (8)$$

and

$$\frac{\sigma}{f_{ct}} = 1 - (1 - \alpha_{ct}) \cdot \exp\left(-(\frac{\varepsilon_m}{\varepsilon_o} - \alpha_{ct})/(1 - \alpha_{ct})\right) \quad \text{for} \quad \frac{\sigma}{f_{ct}} > \alpha_{ct} \qquad (9)$$

where f_{ct} = tensile strength (Pa)

ε_m = material strain (i.e. strain related to the stress level)

α_{ct} = relative stress level (i.e. σ / f_{ct}) at which non-linear stress-strain behaviour starts

ε_o = f_{ct}/E_c = strain linearly related to the tensile strength.

The non-linear stress-strain adjustment factor is formally expressed as

$$\gamma_d = \frac{d(\frac{\sigma}{f_{ct}})}{d(\frac{\varepsilon_m}{\varepsilon_o})} - 1 \qquad (10)$$

which for monotone loading gives

$$\begin{cases} \gamma_d = \exp\left(-(\frac{\bar\varepsilon_m}{\varepsilon_o} - \alpha_{ct})/(1 - \alpha_{ct})\right) - 1 & \text{for } \frac{\sigma}{f_{ct}} > \alpha_{ct}, \text{ and } \Delta\sigma > 0 \\ \gamma = 0 & \text{for all other cases} \end{cases} \qquad (11)$$

where $\bar\varepsilon_m$ = the average material strain during the time step in question and $\Delta\sigma$ = stress increment during the time step in question (Pa). The stress-strain outlined in Eqs. (8) - (11) is illustrated in Fig. 5.

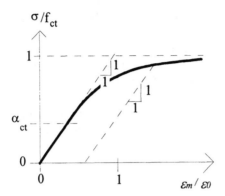

Fig. 5.   Non-linear stress-strain behaviour at tension, see Eqs. (8) - (11).

The unrestrained movements in Eqs. (4) -(6) are here expressed as stress-induced deformations according to [9]

$$\Delta\varepsilon_T^o = \Delta\varepsilon_T^{free} \cdot (1 + \rho_T \cdot \frac{\sigma}{f_{ct}} \cdot sign(\Delta\varphi))$$

(12)

$$\Delta\varepsilon_\varphi^o = \Delta\varepsilon_\varphi^{free} \cdot (1 + \rho_\varphi \cdot \frac{\sigma}{f_{ct}} \cdot sign(\Delta\varphi))$$

(13)

where $\Delta\varepsilon_T^{free}$ = unrestrained and stress-free thermal strain due to thermal changes

$\Delta\varepsilon_\varphi^{free}$ = unrestrained and stress-free moisture strain due to humidity changes

ρ_T = adjustment factor for stress-induced thermal strain

ρ_φ = adjustment factor for stress-induced moisture strain.

The change in relative humidity for simultaneous changes in evaporable water content can be expressed [7] as

$$\Delta\varphi = \bar{\mu}_T \cdot \Delta T + \frac{\partial\varphi}{\partial w_e} \cdot \Delta w_e$$

(14)

where $\bar{\mu}_T$ = average hygrothermal coefficient during the time step in question.

In general, Eq. (14) is aimed to be used to determine the term $sign(\Delta\varphi)$ in Eqs. (12) - (13). For the calculations here, i.e. for young concrete with a considerable temperature change and a change in evaporable water content only due to self-desiccation, an assumption that temperature changes dominate Eq. (15) is introduced [8] which gives

$$sign(\Delta\varphi) = sign(\Delta T)$$

(15)

Note that Eq (15) is certainly true for calculations at the temperature cooling phase, as both ΔT and Δw_e are less than zero. Another simplification [8] is that stress-induced deformations according to Eqs. (12) - (13) are not introduced until the concrete has reached an equivalent age of twelve hours. At very early ages the micro-structure of the concrete is very weak, and the essential behaviour is probably caught by high creep values at this age, see [7] where used creep functions and parameters are presented.

The calculated compressive strength is based on a piece by piece linear function in the logarithm of time, see for instance [7]. The tensile strength is then estimated in relation to the compressive strength as

$$f_{ct} = f_t^{ref} \cdot \left(\frac{f_{cc}}{f_c^{ref}} \right)^{\beta_1} \tag{16}$$

where f_{ct} = tensile strength (Pa), f_{cc} = compressive strength (Pa), f_t^{ref} = reference tensile strength (Pa), f_c^{ref} = reference compressive strength (Pa) and β_1 = exponent. In Eq. (16) f_c^{ref} can be chosen in advance (say f_c^{ref} =10 MPa). Then, f_t^{ref} is the tensile strength at the chosen reference compressive strength. The parameter β_1 expresses the curvature of the relationship, here chosen to 0.667.

The use of Eqs. (4) -(6) with the following parameters are here called *linear calculations*: $\gamma_d = \rho_T = \rho_\varphi = 0$ and the use of the following parameters are here called *non-linear calculations*: γ_d according to Eq. (11), $\rho_T \neq 0$ and $\rho_\varphi \neq 0$.

Note, to be able to use γ_d according to Eq. (11) in the non-linear calculations the parameters of Eq. (16) must be know to estimate the tensile strength.

4   Stress development at simultaneous temperature and moisture changes

To be able to calculate stresses in a satisfactory manner, it is necessary to perform stress measurements and to check the tested results with computed values. Such tests are performed for the well defined case of fully external restraint, in which the stress state is assumed to be uniform over the cross section, see Fig. 6. If this structural member is subjected to homogeneous shrinkage and a simultaneous temperature rise followed by a cooling phase, tensile stresses will develop due to the combined effect of temperature and shrinkage. If the tensile stress level approaches the tensile strength a crack may occur.

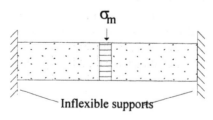

Fig. 6.   Schematically picture of a fully restrained structural member.

The assumption that the shrinkage is linear from the time, $\tau_{\varepsilon 0}$, - when the stress distributing internal structure starts to develop - until the first shrinkage measurement have been recorded at the time, $t_{\varepsilon 0}$, is introduced, see Fig. 7. Using the hypotheses of linear shrinkage at very early ages the total autogenous shrinkage can be expressed [2] as

$$\varepsilon_{SH0} = \varepsilon_{SH} + \Delta \varepsilon_{\tau} \qquad (17)$$

where ε_{SH} is expressed according to Eq. (3). The very early additional shrinkage, $\Delta\varepsilon_{\tau}$, is here chosen to be -30 µm/m for HPC 3. If the measurements of sealed shrinkage could be started four hours after casting instead of about nine hours, this early age shrinkage probably would have been reflected in the measurements. Used values for description of this additional shrinkage are presented in Table 2.

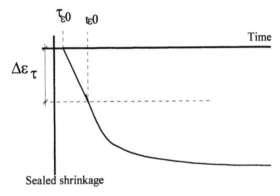

Fig. 7. Schematic picture of shrinkage behaviour at very early age.

Tensile stresses occurs in high performance concrete before any real temperature rise has started. Just a few hours after setting the concrete have developed such a rigid structure that it can distribute significantly high tensile stresses even though it still shows rather plastic behaviour. These early tensile stresses - about four hours after casting - are of the same magnitude as the developed tensile strength, see Fig. 8, and can very easily provoke a tensile failure in the structure, see also [8]. It can be seen in the figure that tensile stresses of about 0.5 MPa develops in the concrete specimen starting about four to five hours after casting.

Table 2. Used values of additional shrinkage at very early age according to Eq. (17)

Concrete	$\tau_{\varepsilon 0}$ hours	$\Delta\varepsilon_{\tau}$ µm/m
HPC 3	5.0	- 30

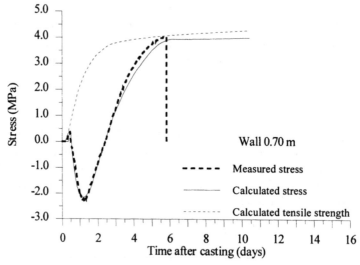

Fig. 8.   Measured and calculated stress development for a 0.7 m thick wall at total restraint conditions.

As can be seen in Fig. 8 the very early age tensile stresses can be simulated well taking moisture movements into consideration. It have been showed in [2] that this early tensile stresses have negligible effect on developed stresses in the cooling phase. Results from a semi-isothermal test at 100 % restraint condition is shown in Fig. 9. The volume changes due to temperature effects are under isothermal conditions zero and therefore, the stress state is generated solely by autogenous shrinkage. Tensile stresses are generated directly after apparent time of setting of the concrete mix.

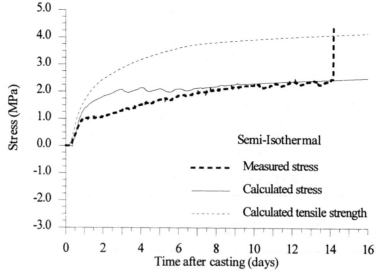

Fig. 9.   Measured and calculated stress development for semi-isothermal temperature regulation (20.5 °C) at total restraint conditions.

5 Conclusions

In order to get correct values of ultimate shrinkage, measurements have to start as soon as a internal structure has developed - about the time of apparent setting of the concrete.

Very early tensile stresses have been measured in high performance concrete before temperature rise have started. This also implies that measurements of shrinkage in high performance concrete should preferable start as early as four hours after casting.

It can be understood from the thermal expansion and contraction tests that shrinkage must be considered. If not, the evaluated thermal expansion and contraction coefficients will be representing both shrinkage and thermal movements. Which for the case of thermal and moisture differences are incorrect - i.e. case of surface cracking. In the case of through cracking the combined coefficients are not fit to be use in other cases having another temperature development then the tested.

The stress development can be modelled with stress-induced deformations and non-linear behaviour. Even if the used material model is correct, the obtained result may give totally wrong information in a stress analysis if it is implemented with incorrect data of thermal and moisture behaviour of the concrete. Therefore, great care must be used in the determination of deformation behaviour.

6 Acknowledgement

The work presented in this paper has been supported financially within the national research program for High Performance Concrete and within the research program of Crack Control. The following funds and firms have financially supported one or both of these programs: the Swedish Council for Building Research, the Swedish National Board for Technical Development, Cementa AB, Elkem A/S, Betongindustri, NCC, Skanska, Strängbetong, Swedish National Rail Administration, Foundation for Swedish Concrete Research, and the Development Fund of the Swedish Construction Industry. Special acknowledgements are directed to professor Lennart Elfgren, head of the Division of Structural Engineering at Luleå University of Technology.

7 References

1. Jonasson, J-E., Groth, P., and Hedlund, H., (1994), *Modelling of temperature and moisture field in concrete to study early age movements as s basis for stress analysis*, In Proceedings of the International Symposium on Avoidance of thermal cracking in concrete at early ages, (Edited by R. Springenschmid), International RILEM symposium on, 10 -12 October 1994, Munich, pp 45 - 52.
2. Hedlund, H. (1996), *Stresses in High Performance Concrete due to Temperature and Moisture Variations at Early Ages*, Division of Structural Engineering, Luleå University of Technology, Licentiate Thesis 1996:38L, 238 p.

3. Tazawa, E. and Miyazawa, S., (1996), *Influence of Autogenous Shrinkage on Cracking in High Performance Concrete*, 4th International Symposium on Utalization of High-strength / High-performance concrete, 29 - 31 May 1996, France, pp 321 - 330.

4. Tazawa, E., Matsuoka, Y., Miyazawa, S. and Okamoto, S., (1994*), Effect of Autogenous Shrinkage on Self Stress in Hardening Concrete*, In Proceedings of the International Symposium on Avoidance of thermal cracking in concrete at early ages, (Edited by R. Springenschmid), International RILEM symposium on, 10 -12 October 1994, Munich, pp 221 - 228.

5. Hedlund, H. and Westman, G., (1997) *Measurements and modelling of volume change and reactions in hardening concrete*, in Proceedings of an International Research Seminar in Lund, June 10, on "Self-Desiccation and its Importance in Concrete Technology", Report TVBM-3075, Lund Institute of Technology, Div. of Building Materials, pp 174-192.

6. Löfqvist, B., (1946), "Temperatureffekter i hårdnande betong", *(Temperature effects in hardening concrete)*, Dissertation, Kungliga Vattenfallsstyrelsen, Tekniskt meddelande No. 22, Stockholm 1946, 195 pp. (In Swedish)

7. Jonasson, J-E., (1994), *Modelling of Temperature, Moisture and Stresses in Young Concrete*, Division of Structural Engineering, Luleå University of Technology, Doctoral Thesis 1994:156D, 225 pp.

8. Jonasson, J-E., (1996), *ConStre - a computer package for calculations of stresses due to temperature and moisture variations*, Division of Structural Engineering, Luleå University of Technology, (In progress).

9. Bazant, Z. P. And Chern, J., (1985), *Concrete creep at variable humidity: constitutive law and mechanisms*, Material and Structures, Vol. 18, pp 1 - 20.

10. Westman, G. (1995) *Thermal Cracking in High Performance Concrete*, Division of Structural Engineering, Luleå University of Technology, Licentiate Thesis 1995:27L, 123 pp.

11. Bjøntegaard, Ø., Sellevold, E. and Hammar T. A., (1996), *High Performance Concrete (HPC) at early ages: Self Generated Stresses due to Autogeneous Shrinkage and Temperature*, In Proceedings of Nordic Concrete Research Meeting, Espoo, Finland 1996, pp 66 - 67.

24 STRESSES DUE TO AUTOGENOUS SHRINKAGE IN HIGH STRENGTH CONCRETE AND ITS PREDICTION

R. Sato, M. Xu and Y. Yang
Civil Engineering Department, University of Utsunomiya, Utsunomiya, Japan

Abstract
This paper investigates experimentally the stress due to each or simultaneous action of volume changes resulting from autogenous shrinkage, hydration heat of cement and drying shrinkage, and the contribution of each volume change to total stress. Moreover, this paper verifies the stress analysis method based on the step-by-step method and models of mechanical properties of material comparing computed results with measured data. The analysis method can predict autogenous shrinkage stress and that combined with thermal stress in non-dried concrete with satisfactory accuracy. However, this method overestimates restrained stress in concrete starting to dry at an early age when basic creep coefficient is used.
Keywords: Autogenous shrinkage, Beam theory, Creep, Drying shrinkage, FEM, High strength concrete, Hydration heat, Stress analysis.

1 Introduction

Research on high strength concrete is being actively carried out with the aim to make larger and lighter structures, and to expand its application. It is well known that autogenous shrinkage caused by the chemical reaction in high strength concrete is very high. The autogenous shrinkage research committee of JCI was established in Japan in 1994 and was active till 1997; during that time much research had been carried out. The results are included in this proceedings.

For structures made of high strength concrete the influence of cracking is more important than for usual structures, because high strength concrete is not only used for its high strength but also for its high durability and impermeability. Therefore, it is important to be able to predict the stresses and potential cracking caused by autogenous shrinkage

Autogenous Shrinkage of Concrete, edited by Ei-ichi Tazawa. Published in 1999 by E & FN Spon, 11 New Fetter Lane, London EC4P 4EE, UK. ISBN: 0 419 23890 5

accurately. In order to accomplish this, it is necessary to establish an analysis method as well as models for time-dependent strength, Young's modulus and creep etc. from very young age. This is especially important for high strength concrete because of the rapid development of shrinkage, creep and strength at an early age.

Autogenous shrinkage occurring in massive concrete structures can be superposed to the volume change caused by temperature. This should be valid for concrete structures lightly reinforced with steel bars.

However, when high strength concrete is used in reinforced concrete (RC) or prestressed reinforced concrete (PRC) structures, autogenous shrinkage stress due to reinforcement restraint could not be neglected if heavy reinforcement is present.

Except for particularly small specimens, in structures made of high strength concrete a large portion of the autogenous shrinkage develops in parallel with a volume change due to temperature changes during hydration. In cases of exposure to drying a small part of the autogenous shrinkage develops together with drying shrinkage. This indicates that in order to accurately predict the cracking of concrete it is necessary to analyze comprehensively the simultaneous action of all shrinkages. However, few experimental and analytical investigations have been carried out on the stresses caused by autogenous shrinkage, drying shrinkage and volume change due to temperature change.

Having the above mentioned situation, this paper investigates experimentally the stress due to each or simultaneous action of volume changes resulting from autogenous shrinkage, hydration heat of cement and drying shrinkage, and the contribution of each volume change to total stress. Moreover, this paper verifies the analysis method based on the step-by-step method[1] and models of mechanical properties of material comparing computed results with measured data.

2 Numerical Analysis of Stress

In this research, the two dimensional FEM method and the beam theory-based method in accordance with the principle of superposition were applied for estimating stresses due to the volume changes. They are summarized as follows:

2.1 Finite Element Analysis Method

The total strain increment of concrete $\Delta\varepsilon(t_{i+1/2},t_j)$ from the middle of j th time interval to the end of i th time interval consists of elastic and creep strain increment $\Delta\varepsilon_{e,cr}(t_{i+1/2},t_j)$ induced by restrained stress from the middle of j th time interval to the end of i th time interval, autogenous shrinkage strain increment $\Delta\varepsilon_{as}(t_{j+1/2},t_{j-1/2})$, temperature strain increment $\Delta\varepsilon_T(t_{j+1/2},t_{j-1/2})$ and drying shrinkage strain increment $\Delta\varepsilon_{ds}(t_{j+1/2},t_{j-1/2})$ developed between $t_{j-1/2}$ and $t_{j+1/2}$ as follows:

$$\Delta\varepsilon_c(t_{i+1/2},t_j) = \Delta\varepsilon_{e,cr}(t_{i+1/2},t_j) + \Delta\varepsilon_T(t_{j+1/2},t_{j-1/2})$$
$$+ \Delta\varepsilon_{as}(t_{j+1/2},t_{j-1/2}) + \Delta\varepsilon_{ds}(t_{j+1/2},t_{j-1/2}) \tag{1}$$

the strains due to the volume change of concrete were changed to equivalent nodal point loads, the stress vector and the stress related strain vector have the following relation;

$$\left\{\Delta\sigma\left(t_{i+1/2},t_j\right)\right\} = \left[D_e\right]\left\{\Delta\varepsilon_{e,cr}\left(t_{i+1/2},t_j\right)\right\} \qquad (2)$$

in which

$$\left[D_e\right] = \frac{E_e\left(t_{i+1/2},t_j\right)}{\left(1-v^2\right)}\begin{pmatrix} 1 & v & 0 \\ v & 1 & 0 \\ 0 & 0 & (1-v)/2 \end{pmatrix} \qquad (3)$$

$$E_e\left(t_{i+1/2},t_j\right) = E_c\left(t_j\right)/\left[1+\phi\left(t_{i+1/2},t_j\right)\times E_c\left(t_j\right)/E_{28}\right] \qquad (4)$$

where

$E_e\left(t_{i+1/2},t_j\right)$ = Effective modulus of the concrete including the creep effect at time $t_{i+1/2}$ after the stress develops at time t_j

$E_c\left(t_j\right)$ = Young's modulus of concrete at time t_j

E_{28} = Young's modulus of concrete cured in the water with 20 ℃ at 28 days

$\phi\left(t_{i+1/2},t_j\right)$ = creep coefficient at time $t_{i+1/2}$ when loaded at t_j

v = poisson's ratio of concrete

A truss element model is used for reinforcing bar. The stress increment $\Delta\sigma_s\left(t_{i+1/2},t_j\right)$, the total strain increment $\Delta\varepsilon_s\left(t_{i+1/2},t_j\right)$ and the strain increment due to temperature change $\Delta\varepsilon_{sT}\left(t_{j+1/2},t_{j-1/2}\right)$ are related as follows:

$$\Delta\sigma_s\left(t_{i+1/2},t_j\right) = E_s\{T\}\left\{\Delta\varepsilon_s\left(t_{i+1/2},t_j\right) - \Delta\varepsilon_{sT}\left(t_{j+1/2},t_{j-1/2}\right)\right\} \qquad (5)$$

E_s = Young's modulus of reinforcing bar;

$\{T\}$ = Transformation vector

2.2 Beam theory-Based Method

Applying the step-by-step method and considering the effect of the loading age on creep to estimate the stress history in concrete, the total strain at the end of i th time interval and at the distance z from the extreme compression fiber is given as follows[2];

$$\varepsilon_{c,i+1/2}(z) = \sigma_{c,1/2}(z)J\left(t_{i+1/2},t_{1/2}\right) + \sum_{j=1}^{i}\Delta\sigma_{c,j}(z)J\left(t_{i+1/2},t_j\right) + \varepsilon_{cs,i+1/2}(z) \qquad (6)$$

The stress at time $t_{i+1/2}$ is estimated as follows:

$$\sigma_{c,i+1/2}(z) = E_e\left(t_{i+1/2},t_i\right)\left\{\varepsilon_{c,i+1/2}(z) - \sigma_{c,1/2}(z)J\left(t_{i+1/2},t_{1/2}\right)\right.$$
$$\left. - \sum_{j=1}^{i-1}\Delta\sigma_{c,j}(z)J\left(t_{i+1/2},t_j\right) + \sigma_{c,i-1/2}(z)J\left(t_{i+1/2},t_i\right) - \varepsilon_{cs,i+1/2}(z)\right\} \qquad (7)$$

where $\varepsilon_{cs,i+1/2}(z)$ is the sum of autogenous and drying shrinkage strains at $t_{i+1/2}$, $\Delta\sigma_{c,j}(z)$ is the stress produced between $t_{j-1/2}$ and $t_{j+1/2}$, and $J\left(t_{i+1/2},t_j\right) = 1/E_e\left(t_{i+1/2},t_j\right)$.

Using the equation (7) for equilibrium, we have

For axial force:

$$0 = \int_{A_c}\sigma_{c,i+1/2}(z)dA + A_s'\sigma_{s,i+1/2}' + A_s\sigma_{s,i+1/2} + A_p\sigma_{p,i+1/2} \qquad (8)$$

For bending moment:

$$M = \int_{A_c} \sigma_{c,i+1/2}(z)(z-c)dA + A_s'\sigma_{s,i+1/2}'(d'-c)$$
$$+ A_s\sigma_{s,i+1/2}(d-c) + A_p\sigma_{p,i+1/2}(d_p-c) \qquad (9)$$

where, c is the distance from the extreme compression fiber to the centroidal axis, d' is the distance to the compressive reinforcing bar, d is the distance to tensile reinforcing bar and d_p is the distance to the prestressing bar.

The strain distribution at a section is determined by Eqs. (8) and (9), and compatibility condition that strains of reinforcing bars and prestressing bars are equal to those of the concrete at the same depths as the bars. The strain distribution is given by three variables of depth of centroid, strain at the centroid and curvature.

3   Numerical Studies

The structures or members for studying are selected so that autogenous shrinkage stress, the temperature stress due to cement hydration and the drying shrinkage stress occur independently or compositely in them.

3.1   Structure Geometry and Material properties
The geometry and material properties of the specimens used for numerical studies are shown in Table 1. The autogenous shrinkage stress was measured on sealed beams No.1 and No.4 in which the steel bars were embedded as shown in Table 1. The temperature developed differently in specimens No.1 and No.4 because of the different cross-sections of the specimens.

In the full-sized indeterminate RC specimen [4] in which the measured temperature rise due to heat of hydration was nearly 60℃, not only autogenous shrinkage stress but also temperature stress were produced by foundation restraint. Moreover, temperature stresses due to nonlinear distribution of the temperature through the section and autogenous shrinkage stress due to reinforcement restraint also develop in the specimen.

These stresses were studied by numerical analysis in this research. In the sealed specimen No.6 in which the volume change of concrete was restrained by a steel frame, temperature stresses due to temperature difference between steel frame and concrete in concrete developed in addition to the stresses due to autogenous shrinkage.

The PRC member No.3 was exposed to drying from the age of 9 days just after prestressing. Before prestressing, only autogenous shrinkage causes stresses in the PRC; but, after prestressing, the stresses are produced by the combined effect of autogenous shrinkage and drying shrinkage.

In specimen No.5 which was exposed to indoor conditions at age 1 day, stresses due to autogenous shrinkage and drying shrinkage developed. On the other hand, the stresses due to autogenous shrinkage, drying shrinkage and temperature change occurred simultaneously in No.7, which was restrained by the frame and started to dry at age 1 day.

The respective reinforcing and prestressing steel ratios ρ_s and ρ_p, based on the full cross-sectional area of the specimens are indicated in Table 1. The ratio of specimen No.5 to the pure concrete section is 10 % less than that of specimen No.7.   Restrained stress in

Table 1 Specimen Geometry and Material Properties

No. (Kind of stress)	Specimen Geometry (Unit: mm)	Concrete Properties
No.1 Sealed (σ_{as})	Reinforcing bar; 300; 2400; $\rho_s = 5.36\%$; D=32; 234; 66; 200	HESC W/B=0.25 SF/B=10% $E_{28} = 40.4 KN/mm^2$ $f_{28} = 118 N/mm^2$ $\phi(t_{i+1/2}, t_j)$ [3]
No.2 Sealed (σ_{as}, σ_T)	3400; 2900; 4400 Footing; 300; 1600; E₁=12.8N/mm²; Foundation E₂=24.5N/mm²; 4400; 6300; D=32; 850; 850; 55; 55; $\rho_s = 1.32\%$ Column; D=29; 900; 50; 50; 450; $\rho_s = 1.65\%$ Beam	HESC W/B=0.23 SF/B=10% $E_{28} = 37.4 KN/mm^2$ $f_{28} = 119 N/mm^2$ $\phi(t_{i+1/2}, t_j)$ [4]
No.3 Drying $(\sigma_{as}, \sigma_{ds})$	Reinforcing bar; PC bar; 250; 2400; $\rho_s = 1.43\%$ $\rho_p = 0.16\%$; 30; 160; 210; 200; Ds=16 Dp=7.1	HESC W/B=0.25 SF/B=10% $E_{28} = 38.7 KN/mm^2$ $f_{28} = 105 N/mm^2$ $\phi(t_{i+1/2}, t_j)$ [5]
No.4 Sealed (σ_{as}) **No.5** Drying $(\sigma_{as}, \sigma_{ds})$	Non bonded; 100; 300; 1500; $\rho_s = 6.47\%$; 50; 100; D=28.5	OPC W/B=0.25 SF/B=10% $E_{28} = 42.9 KN/mm^2$ $f_{28} = 114 N/mm^2$ $\phi(t_{i+1/2}, t_j)$ [3]
No.6 Sealed (σ_{as}, σ_T) **No.7** Drying $(\sigma_{as}, \sigma_{ds}, \sigma_T)$	Restraining frame; 150; 300; 150; 1020; $\rho_s = 7.93\%$; 100; 170; 100	OPC W/B=0.25 SF/B=10% $E_{28} = 42.9 KN/mm^2$ $f_{28} = 114 N/mm^2$ $\phi(t_{i+1/2}, t_j)$ [3]

* σ_{as}: autogenous shrinkage stress; σ_{ds}: drying shrinkage stress; σ_T: temperature stress;

** HESC: high early strength cement; OPC: ordinary portland cement.

concrete is calculated based on measured Young's modulus of the restraining steel and pure cross-section area. The history of shrinkage strains measured in concrete specimens without restraining corresponding to restraining test specimens is shown in Fig.1, which are compensated with temperature strain by assuming a concrete thermal expansion coefficient to be $10 \times 10^{-6}/°C$. The measured shrinkage strain, the temperature of the concrete and the temperature of the restraining frame were used in the numerical analysis. The restraining steel bar embedded into the concrete was assumed to be the same temperature as the concrete.

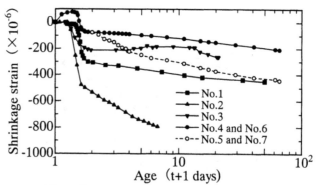

Fig.1 Time-dependent change of shrinkage strain

The empirical equation of the Young's modulus and creep coefficient used in the numerical analysis are as follows:

$$E_c(t_j) = E_{28} \times \exp\left\{ s_E \left[1 - \left((28 - a_E)/(t_j - a_E) \right)^{0.5} \right] \right\} \quad (10)$$

in which, s_E and a_E are constants, 0.077, 0.291 for concrete with water-binder ratio of 0.25 [3], and 0.098, 0.432 for concrete with water-binder ratio of 0.23 [4], respectively.

The empirical equations of creep coefficients, which were proposed in previous research [3],[4],[5] were used. These equations can be changed by adjusting the constants according to the water-binder ratio of the concrete, curing method, drying environment and other conditions. As an example of the equations, an empirical equation abstracted from the creep experiment in which loading was executed at five ages of concrete is given as follows[3]:

$$\phi(t_{i+1/2}, t_j) = \phi_0 \times \left[\frac{(t_{i+1/2} - t_j)/t_1}{\beta_H + (t_{i+1/2} - t_j)/t_1} \right]^{0.3} \quad (11)$$

$$\phi_0 = 1.10 \times \left[1 + (t_j + 0.33)^{-2.96} \right] \quad (12)$$

$$\beta_H = 0.35 t_j^2 + 8.73 t_j \quad (13)$$

in which, $\phi(t_{i+1/2}, t_j)$ = basic creep coefficient; ϕ_0 = notional creep coefficient; $t_{i+1/2}$ = age of concrete (day); t_j = temperature adjusted concrete age at stress development (day); $t_1 = 1$ day

β_H is the coefficient depending on the humidity of the environment and the size of the member as defined in CEB-FIP model code 90[6]. However, in the case of high strength concrete, it is modified to denote the effect of the density of hydrate and self-desiccation depending on the density on the rate of creep, and is expressed by temperature adjusted concrete age at loading. The relationships between $\phi_0\text{-}t_j$, $\beta_H\text{-}t_j$ and $\phi\left(t_{i+1/2},t_j\right)\text{-}t_{i+1/2}$ obtained by the experiment and regression curves are shown in Figs.2 and 3, respectively.

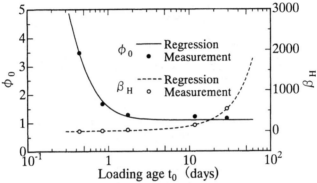

Fig.2 ϕ_0 and β_H at different loading ages

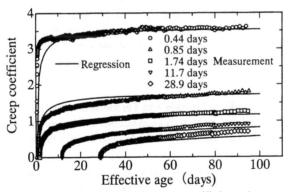

Fig.3 Comparison of creep coefficients between regression and measurement

The starting time for numerical analysis is just after casting concrete.

3.2 Results of Analysis

3.2.1 Stresses due to Autogenous Shrinkage

Measured stresses in specimens No.1 and No.4 restrained by an embedded steel bar are shown in Fig.4 together with stresses obtained by numerical analysis. Test No.1 is analyzed numerically by beam theory and FEM, and test No.4 is analyzed numerically by FEM.

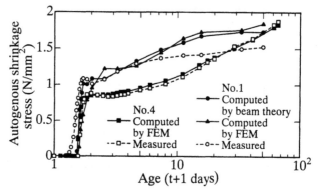

Fig.4 Time-dependent change of autogenous
shrinkage stress

Autogenous shrinkage stress in No.1 is larger than in No.4 up to 30 days and then becomes smaller. The larger stress of No.1 in the first month must be due to its autogenous shrinkage strain being larger than No.4, while the reinforcement ratio of No.1 is 1.1% smaller than No.4. The reason why the stress of No.1 develops more slowly than that of No.4 after the age of 3 days is difficult to explain by the computation, though the computation can accurately predict the stress of No.4. It might be explained by the difference between the batches of concrete for restraining test No.1 and that for shrinkage test, and by using steel strain measured only at the center section, at which Young's modulus may be lower than that for the value of concrete averaged in a test zone in the longitudinal direction.

3.2.2   Autogenous Shrinkage Stresses and Temperature Stresses

Comparison of measured and FEM-computed stresses in concrete restrained by the frame No.6 is shown in Fig.5. The mesh used in FEM consists of 119 elements and 77 nodes.

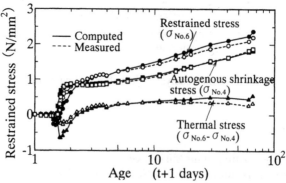

Fig.5 Components of restrained stresses (No.6)

The measured temperature with a maximum difference of 9.5℃ between the concrete and the frame was also noted. The autogenous shrinkage stresses in No.4 are also shown in Fig.5. The restrained stresses in No.6 are clearly larger than those in No.4 and this difference also exists in the analysis. This difference is conveniently defined as the temperature stress, though the reinforcement ratio of No.4 is 10% smaller than the steel frame ratio of No.6. Based on this definition, as shown in the figure, the temperature stress is 21% of the restrained stress in the experiment and 24% in the analysis at the age of 19 days. It is necessary to consider the difference of the temperature between the concrete and the restraining body to accurately estimate autogenous shrinkage stresses.

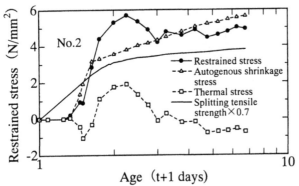

Fig.6 Components of restrained stresses
by FEM analysis

Restrained stress and its components for specimen No.2 of the reinforced concrete frame structure computed by FEM is shown in Fig.6. The Young's modulus of the footing base is 20.6 kN/mm², its foundation consists of volcanic ash with Young's modulus of 12.8 N/mm², loam with 24.5 N/mm² in the downward direction. Their thicknesses are 1.6 m and 4.4 m, respectively. The layers of volcanic ash and loam are considered as restraining bodies in the analysis. The element number and the node number are 738 and 427, respectively. The measured temperature was used in the computation. The maximum temperature rise was 66.2℃ at the center of the section located at   midheight of the column.

In the experiment, cracks were observed in the lower portion of the column and at the bottom of the beam near the beam-column joint after removing the formwork at the age of 7 days.   The restrained stresses at the bottom of the beam near the beam-column joint of the frame are shown in Fig.6. The tensile strength is defined as 70% of splitting tensile strength[7].

According to this figure, the total restrained stress exceeds the tensile strength at 0.63 days. It is also obvious that the autogenous shrinkage is the main cause for cracking. At 0.63 days, the temperature stress was compressive, which reduced the restrained tensile stress. Maximum temperature stress is about 2N/mm² accounting for 50% of autogenous shrinkage stress at the age of 1.25 days. In the stable stage of temperature, the compressive temperature stress is caused by the nonlinear distribution of temperature through the section.

3.2.3   Stress due to Autogenous Shrinkage and Drying Shrinkage

The experimental stress of No.5 is compared with the value obtained with FEM in Fig.7. At the age of 1 day, the aluminum foil sealing was removed from the specimen to allow it to dry. In this case drying shrinkage also occurred in addition to autogenous shrinkage. Its restraining stress is 2 times that of sealed specimen No.4 when crack developed at 19 days.

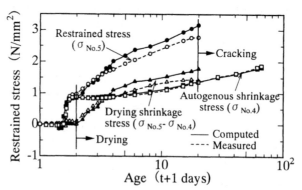

Fig.7 Components of restrained stresses (No.5)

Using shrinkage strains measured in sealed and dried specimens, the difference of stresses between specimen No.5 and No.4 was simulated by numerical analysis and the analytical results agree very well with experimental data. However, in the case of drying, the restrained stress from the analysis overestimated that from the experiment with the former being 14% larger than the latter when crack occurred. It can be explained that in this analysis the basic creep coefficient was used, while under drying condition the creep coefficient is 8 to 26% larger[3]. The drying shrinkage stress obtained by subtracting the autogenous shrinkage stress of specimen No.4 from the restrained stress of specimen No.5. is also shown in Fig.7. When cracks occur, the autogenous shrinkage stress is 49% and the drying shrinkage stress is 51% of the restrained stress from the experiment, 44% and 56% from the analysis.

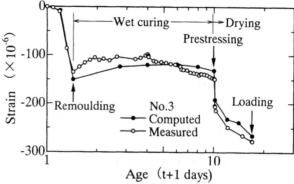

Fig.8 Time-dependent change of strain of
reinforcing bar

The strains in the tension reinforcement from the beam theory-based analysis and the experiment for specimen No.3 of the PRC beam are shown in Fig.8. In the experiment it was observed that before prestressing, the compressive strain due to autogenous shrinkage reached 150×10^{-6} at age 1 day, then was stable and thereafter decreased slightly again under moist curing . After prestressing, the strain increased rapidly due to creep, autogenous shrinkage and drying shrinkage. From the figure, it is clear that in the period from the completion of the casting the concrete to the loading, the results of the analysis agree very well with those of the experiments.

3.2.4 Stress due to Autogenous Shrinkage, Drying Shrinkage and Temperature Change

The stresses obtained by FEM analysis and by experiment for specimen No.7, which was restrained by a steel frame and dried at age 1 day, are shown in Fig.9. In this case the restrained stress consists of temperature stress due to the difference between the frame and the concrete temperatures, autogenous shrinkage stress, and drying shrinkage stress.

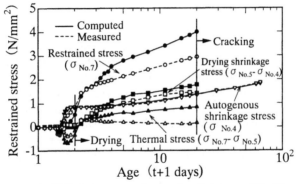

Fig.9 Components of restrained stresses (No.7)

The autogenous shrinkage stress is measured in No.4 and the stress induced by the autogenous shrinkage combined with the drying shrinkage is measured in No.5. Based on superposition, the stress increase due to the effect of drying is $\sigma_{No.5} - \sigma_{No.4}$, and the stress due to the effects of hydration heat under drying is $\sigma_{No.7} - \sigma_{No.5}$. These stresses are also shown in Fig.9.

The computed stress for No.7 is 36% larger than the observed stress at cracking at the age of 19 days. The difference between both is significant, compared with test No.5. One reason for this is due to applying basic creep coefficient to the specimen dried during restraining, which results in an underestimation of the creep effect for stress relaxation. The underestimation for No.7 should be more remarkable than No.5, because three stress components are included in No.7 ,while only two are included in No.5.

Another reason could be the smaller V/S (V: volume, S: surface area exposed to air) of No.5 due to the embedment of a steel bar and coating for strain gauges, than those of specimen for measuring shrinkage strain and No.7. Smaller V/S should increase shrinkage strain and shrinkage stress in the specimen even in case of high strength concrete, getting close to computed value.

4 Conclusion

(1) The analysis method based on the superposition principle for concrete creep can predict autogenous shrinkage stress and that combined with thermal stress in non-dried concrete with satisfactory accuracy. This shows that the step-by-step method using the appropriate mechanical model is valid for the analysis of stress due to autogenous shrinkage and/or hydration heat of cement in non-dried concrete.

(2) Computation showed that full-sized indeterminate reinforced concrete structure made by high strength concrete would crack within one day after placing concrete mainly due to autogenous shrinkage, while cracks were observed at 7 days of age when the formwork was removed.

(3) Restrained stress in concrete starting to dry at an early age was overestimated when basic creep coefficient was used, which showed that modeling of creep especially if exposed to drying at early ages should be reflected in an exact analysis for high strength concrete.

5 References

1. Sato, R., Dilger, W. H. and Ujike, I. (1994) Deformations and Thermal Stresses of Concrete Beams Constructed in Two Stages, Proceedings of the International Symposium on Thermal Cracking in Concrete at Early Ages, RILEM, pp.313-320.
2. Yamamoto, H., Sato, R., Wakui, H. and Ochiai, M. (1994) On Long-Term Deformation of Prestressed   Reinforced Concrete Flexural Members, Proceedings of the Japan Concrete Institute Vol.16, No.2, pp.973-978 (in Japanese).
3. Yang, Y., Sato, R., Xu, M. and Tezuka, M. (1997) Creep and Shrinkage of High Strength Concrete, Proceeding of The 7th Symposium on Development in Prestressed Concrete, pp.817-822 (in Japanese).
4. Japan Concrete Institute (1996) Report of Autogenous Shrinkage Research Committee.
5. Tezuka, M., Sato, R., Xu, M. and Ochiai, M. (submitting) Research on Time-Dependent Flexure Deflection of High Strength Concrete Member, Journal of Structural Mechanics and Earthquake Engineering (in Japanese)
6. Euro-International Committee for Concrete (1991) CEB-FIP Model Code 1990.
7. Makizumi, T. and Tokumitsu, Y. (1983) Study on Shrinkage Cracking of Concrete, Proceedings of the Japan Concrete Institute Vol.5, pp.185-188 (in Japanese).

25 THE INFLUENCES OF QUALITY OF COARSE AGGREGATE ON THE AUTOGENOUS SHRINKAGE STRESS IN HIGH-FLUIDITY CONCRETE

H. Matsushita and H. Tsuruta
Department of Civil and Structural Engineering,
Kyushu University, Fukuoka, Japan

Abstracts
This study examines how the autogenous shrinkage stress in high-fluidity concrete is affected by the aggregate crushing value of coarse aggregate and its volume per unit volume of concrete. The following findings were obtained. The autogenous shrinkage stress is not affected by aggregate crushing value, but is reduced when the coarse aggregate's volume per unit volume of concrete increases due to the restraining effect of coarse aggregates. Autogenous shrinkage is reduced by using light-weight aggregate, which has high absorbtion of water.
Keywords: Aggregate crushing value, autogenous shrinkage stress, high-fluidity concrete, light-weight aggregate, quality of coarse aggregate.

1 Introduction

Concrete has remarkably been sophisticated in recent years, as a result of the development of superplasticizer and the adoption of ground granulated admixture that led to enhanced flow properties of concrete and lower W/C. On the other hand, it has been reported that the enhancement in concrete performance entails the rapid development of autogenous shrinkage caused by the hydration reaction of binder [1]. It has also been reported that the compressive strength of high-strength concrete varies depending on the type of coarse aggregate[2]. Addressing the issue of the influence of coarse aggregate's quality on the autogenous shrinkage strain of concrete, the present authors reported in their past research that the use of high-quality coarse aggregates would reduce the strain induced by autogenous shrinkage[3]. However, one of the most problematic issues in actual structures is the shrinkage stress that causes cracks, and hence it would be more important to

Autogenous Shrinkage of Concrete, edited by Ei-ichi Tazawa. Published in 1999 by E & FN Spon, 11 New Fetter Lane, London EC4P 4EE, UK. ISBN: 0 419 23890 5

consider the stress induced by autogenous shrinkage.

In view of the above, this study examines how the autogenous shrinkage stress in high-strength, high-fluidity concrete is affected by the quality of coarse aggregate and its volume per unit volume of concrete(Vg). Particularly authors paid attention to aggregate crushing value(ACV) as quality of coarse aggregate. We consider that ACV shows crushing strength and interlocking properties of coarse aggregate in a situation that aggregates are packed in the mold, it may be effective to estimate shrinkage of concrete. Generally, Young's modulus of rock of coarse aggregate should be effect on shrinkage of concrete. But it takes time to measure and understand it. On the other hand, ACV is measured by simple test, so we paid attention to it.

2   Outline of tests

2.1   Materials
The following materials were used in the tests carried out in this study: normal portland cement (density: 3.15 g/cm^3, specific surface area: 3300 cm^2/g); sea sand (specific gravity: 2.58, fineness modulus: 3.08) as fine aggregates; three types of crushed stone and the artificial light-weight aggregate as coarse aggregates (see Table 1 for the physical properties of coarse aggregates). The crushed stones came from the northern region of Kyushu Island of Japan, and the artificial light-weight aggregate was crushed-and-coated type made of expanded shale. These aggregates were selected so as to have different aggregate crushing values(ACV), while the maximum size was set at 20 mm. Note that their crushing values had been measured by applying 400 kN loading complying with BS 812. Specific gravity, absorption and percentage of abrasion in coarse aggregate were measured according to JIS A 1110 and JIS A 1121. Ground granulated blast-furnace slag (GGBFS) (specific gravity: 2.90, specific surface area: 6000 cm^2/g) was used as admixture to control the rise in the heat of hydration caused by increase in the amount of cement. Also, polycarboxylic-acid based superplasticizer (specific gravity: 1.04) was used as chemical admixture.

Table 1.   The physical properties of coarse aggregates

Types of coarse aggregates	Specific gravity	Absorption (%)	ACV (%)	Percentage of abrasion(%)
Andesite	2.73	0.72	9	10.0
Crystalline Schist	2.81	0.52	13	15.7
Amphibolite	2.73	1.00	17	21.6
Light-weight aggregate	1.46	12.6	36	36.5

2.2   Test parameters and variables
The parameters in the autogenous shrinkage stress test were the quality of coarse aggregate and the volume per unit volume of concrete(Vg). The variables set for

these parameters are shown in Table 2.

Table 2. Test parameters and variables

Test parameters	Variables
Types of coarse aggregates	4 types of coarse aggregates in Table 1.
Coarse aggregate's volume per unit volume of concrete(Vg)($1/m^3$)	0, 130, 230, 330

When varying the volume per unit volume of concrete, amphibolite was designated as the constant, i.e. the standard type of coarse aggregate. On the other hand, when varying the quality of coarse aggregate, the volume per unit volume of concrete(Vg) was fixed at 330 ($1/m^3$).

2.3 Mixture proportions

The mixture proportions of concrete used in the test are shown in Table 3. For all the concrete mixtures, the replacement ratio of slag, the proportion of the superplasticizer(SP) and goal of air content in concrete were set at constant values, that is 50%, 1% of the weight of binder and 2%, respectively. To study how the quality of coarse aggregate would affect the autogenous shrinkage stress, four types of coarse aggregate were tested with No. I of mixture proportion, fixing the aggregate's volume per unit volume of concrete at 330 ($1/m^3$). On the other hand, amphibolite only was tested with No. I to IV of mixture proportion varying the volume of amphibolite per unit volume of concrete, to study how the coarse aggregate's volume per unit volume of concrete would affect the autogenous shrinkage stress. Mixing was carried out with a twin-screwed force mixer, into which the coarse aggregate, binder and fine aggregate were fed in this order. After 30 seconds of dry mixing, water mixed with the superplasticizer was fed into the mixer, which was followed by a further mixing of 3 minutes.

Table 3. Mixture proportions(The case of Amphibolite)

No.	W/B (%)	s/a (%)	Vg ($1/m^3$)	Quantity of material per unit volume of concrete (kg/m^3)					
				W	C	GGBFS	S	G	SP
I	28	45.7	330	170	304	304	717	901	60.4
II	28	58.3	230	196	350	350	831	628	70.0
III	28	73.8	130	222	396	396	944	355	79.2
IV	28	100	0	256	457	457	1086	0	91.4

2.4 Specimens used for the measurement of shrinkage stresses

Specimens[4] were square columns of 10 × 10 × 150 cm, one each of which was prepared for each variable in the measurement of shrinkage stresses that was two fold, i.e. the measurement of autogenous shrinkage stress and that of drying shrinkage stress. Fig.1 shows the conditions of the specimens and mold. The mold was composed of steel side panels and wooden bottom and end panels. After assembling the mold, Teflon sheets (1 mm thick) were attached inside the bottom, side and end panels, so that the free deformation of specimen would not be restricted. Furthermore, on top of the Teflon sheets, polyester films (0.1 mm thick) were attached to prevent the direct contact between concrete and the mold.

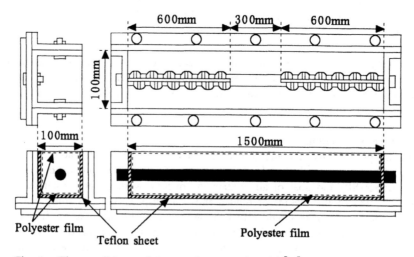

Fig. 1. The conditions of the specimens and mold.[4]

To provide restraining reinforcement, a deformed bar(D32, nominal diameter: 31.8 mm) was inserted. In the central 30 cm portion of the deformed bar, ribs and sections were removed to make a circular cross section of about 30 mm in diameter. A strain gauge (gauge length: 5 mm) was attached to the central portion for measuring the strain of steel bar. This was followed by the application of a Teflon sheet (1 mm thick) which was wound trebly around the central circular section of steel bar so as to break the bond between the concrete and steel bar in that portion.

The specimens were prepared in a room with steady temperature and humidity of 20 ± 2 °C and 60 ± 5%, respectively. Wet cloths and vinyl sheets were used to wrap the specimens to prevent the moisture loss. The mold was removed in 24 hours after placement. In the case of the specimens to be used for measuring autogenous shrinkage stress, the removal of mold was immediately followed by the sealing of the specimens all over with aluminum foil tapes, so as to prevent the moisture loss. They were then left in this way for sealed curing in a room with steady temperature and humidity of 20 ± 2 °C and 60 ± 5%, respectively. Meanwhile, after the mold was removed, the specimens to be used for measuring drying shrinkage stress were left as they were for air-dry curing in the same steady temperature and humidity room.

2.5 Shrinkage stress measurement procedures
The measurement of shrinkage stresses started soon after concrete placement in a room with steady temperature and humidity of 20 ± 2 °C and 60 ± 5%,

respectively. With the strain gauge attached to the central portion of steel bar, the strain produced by shrinkage stress was measured at the following intervals: every 1 hour during the first 24 hours from the removal of mold; every 4 hour up to three days; and once a day afterwards. At the same time, the temperature of concrete was also measured with a thermoelectric thermometer installed at the central portion of the specimens. Using the value of the strains thus measured, the stress induced by autogenous shrinkage was obtained from the following formula:

$$\sigma_c = \frac{E_s \times \varepsilon_s \times A_s}{A_c}$$

where, σ_c: Autogenous shrinkage stress induced in concrete (N/mm^2)
$\quad$ Es: Elastic modulus of steel bar (N/mm^2)
$\quad$ ε_s: Strain of steel bar
$\quad$ As: Cross sectional area of steel bar (mm^2)
$\quad$ Ac: Net sectional area of concrete (mm^2)

3 Test results and discussions

3.1 Physical characteristics of concrete at the time of placement
The physical characteristics of each concrete mixture at the time of placement, i.e. the slump flow value, flow time, air content and temperature, are shown in Table 4. Larger slump flow values were observed when the volume per unit volume of concrete was varied, but there found no segregation between the coarse aggregate and mortar.

Table 4. The data of used concrete

Case	Slump flow (mm)	Time of flow (sec)	Air content (%)	Temperature of concrete(°C)
Andesite	730x740	69.8	0.5	16.5
Crystalline Schist	660x705	56.7	0.4	16.5
Amphibolite	775x770	45.0	0.1	16.0
Light-weight aggregate	975x910	71.1	0.4	16.0
Vg=330(l/m³)	775x770	45.0	0.1	16.0
Vg=230(l/m³)	830x780	63.6	0.4	15.5
Vg=130(l/m³)	970x900	62.7	0.2	15.0
Vg= 0 (l/m³)	1050x980	95.3	0.3	15.0

3.2 Influence of the quality of coarse aggregates
Fig. 2 shows the changes with time in the autogenous shrinkage stress induced in concrete, with reference to each type of coarse aggregate. The origin of X axis of

this figure is time that concrete placement has started. As seen from the figure, none of the three types of crushed stone indicated clear influence on the autogenous shrinkage stress.

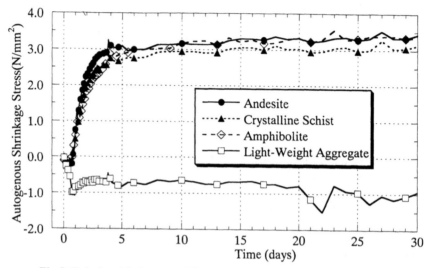

Fig.2 Relationship between Time and Autogenous Shrinkage stress

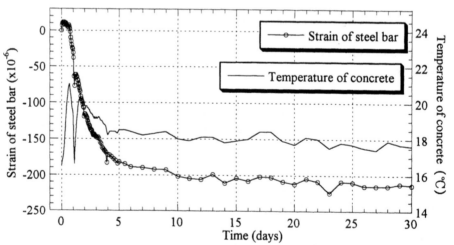

Fig.3 Relationships between strain of steel bar, temperature of concrete and time (The case of Amphibolite)

On the other hand, it is shown that no autogenous shrinkage stress was induced in the case of the light-weight aggregate. The light-weight aggregate had a absorption ten times larger than that of the crushed stones, and it is believed that, the fact that it was used after prewetting for 24 hours had a considerable impact on the autogenous shrinkage stress here. If the water infiltration speed cannot keep pace with the development of pores that are newly developed within concrete due to its hydration reaction, water cannot refill the pores and the humidity inside the pores will be reduced eventually. This would occur regardless of the availability of an external or peripheral source of water, and indeed, is the mechanism of self desiccation of concrete. In other words, when the water infiltration speed is accelerated by wet-curing so as to reach the speed of the pores being developed due to hydration reaction, the self-desiccated state of concrete will be relieved, which will in turn reduce the speed of shrinkage. Given this, it must have been the case that, having been subjected to prewetting for 24 hours, the light-weight aggregate provided an internal source of water as soon as concrete was placed, and thereby relieved the self desiccation state of concrete. Now, in terms of changes with time in the temperature of concrete and in the strain of the steel bar, all the three types of crushed stone again showed similar tendencies. For this reason, Fig. 3 shows the results of one type of crushed stone, i.e. amphibolite only as a typical example.

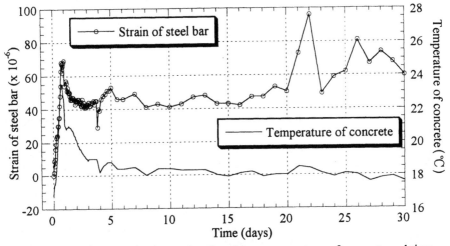

Fig.4 Relationships between strain of steel bar, temperature of concrete and time
(The case of Light-weight aggregate)

Fig. 4, on the other hand, shows the results of the specimen with the light-weight aggregate. It is demonstrated clearly that the specimen with the light-weight aggregate expanded greatly in accordance with the initial rise in the concrete temperature, compared with the specimens with crushed stones. The reason for this is that, again, having fully been absorbed in water, the light-weight aggregate must have provided necessary water and accelerated the hydration reaction of cement. Shrinkage forces started working after the maximum expansion strain was induced in the specimen. However, because the strain converged in a short period of time, the specimen remained in the expanded state, and therefore the shrinkage force did not grow to be influential.

Little influence of the quality of coarse aggregates was observed in terms of the autogenous shrinkage stress. The reasons for this can be discussed by comparing the results of the present study with those of the authors' past research that identified the

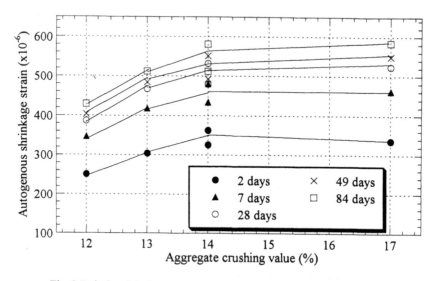

Fig.5 Relationship between aggregate crushing value and
autogenous shrinkage strain [3]

influence of the crushing value of coarse aggregates on the autogenous shrinkage strain[3]. In this past research, specimens ($10 \times 10 \times 24$ cm) were designed to have the following properties: W/B= 28%, s/a = 45%, W = 160 kg/m^3, and B = 572 kg/m^3, setting the coarse aggregate's volume per unit volume of concrete at 345 (1/m^3). The same superplasticizer was used, and the autogenous shrinkage strain was measured with an embedded strain gauge (gauge length: 100 mm). The proportions of concrete mixes were similar to those adopted in this study, but the range of aggregate crushing value was rather short, i.e. between 12 and 17. The relationship between the autogenous shrinkage strain and crushing value obtained in this past test is shown in Fig. 5. The figure indicates that the smaller the crushing value, the smaller the autogenous shrinkage strain became. The differences between the autogenous shrinkage strain of a type of aggregate with the largest crushing value and that of another type with the smallest crushing value were about 110 to 160 $\times$ 10^{-6} for all ages. Despite this, the present study found no difference between different types of coarse aggregate in terms of the autogenous shrinkage stress.

This discrepancy seem to be explained by the following three factors: (i) difference between the autogenous shrinkage strain in an unrestrained specimen and that in a restrained specimen; (ii) the elastic modulus of the matrix at an early age when the increase in the autogenous shrinkage strain is rapid; and (iii) the influence of creep.

The first factor is concerned with the fact that, when using concrete mixtures prone to a great extent of autogenous shrinkage, the autogenous shrinkage strain becomes much smaller in a restrained specimen than that in an unrestrained specimen. This means that the differences in the measurement values of strain do not directly represent the differences in the stress. The second factor is concerned with the fact that the elastic modulus of matrix is still small at early ages. Therefore, the differences in strain cannot be reflected clearly on stress, even if such strain difference is created. The third factor is concerned with the fact that the influence of creep was small because high-strength concrete with a low W/B was used here.

3.3 Relationship between the stress induced by autogenous shrinkage and the strength of concrete

Table 5 gives the types and aggregate crushing values of coarse aggregates, and compressive strength, tensile strength, elastic modulus and autogenous shrinkage stress of concrete at the age of 28 days. With reference to the compressive strength and elastic modulus, it is shown that they have good correlations with the aggregate crushing value. This is explained by the fact that the crushing value is the index of coarse aggregate's strength. On the other hand, the autogenous shrinkage stress has no correlation with the aggregate crushing value.

Table 5 The strength of concrete, elastic modulus and autogenous shrinkage stress at the age of 28 days

Types	Compressive strength (N/mm^2)	Tensile strength (N/mm^2)	Elastic modulus (GPa)	Autogenous shrinkage stress (N/mm^2)
Andesite	85.2	5.0	43.3	3.33
Crystalline Schist	82.3	4.4	43.7	3.03
Amphibolite	75.6	5.1	36.2	3.38
Light-weight aggregate	67.3	4.1	33.8	-1.16

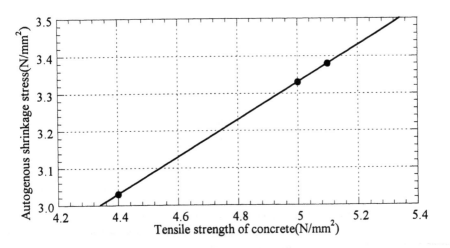

Fig.6 Relationships between tensile strength of concrete and autogenous shrinkage stress

Fig. 6. shows the relationship between the tensile strength and autogenous shrinkage stress in the specimens with three types of crushed stones (Table 4). Autogenous shrinkage stress is smaller than   tensile strength of concrete at the age of 28 days. Therefore we can see from this that cracks are not caused by autogenous shrinkage stress under this condition.

3. 4 Relationship between autogenous shrinkage stress and drying shrinkage stress

Fig. 7. shows changes with time in the stress induced by the drying shrinkage of specimens with coarse aggregates, which were measured to be compared with the result of the autogenous shrinkage test. In contrast with the autogenous shrinkage test, development of cracks is observed here. However, each type of coarse aggregates again showed no difference in terms of stress, as was the case in the autogenous shrinkage stress test.

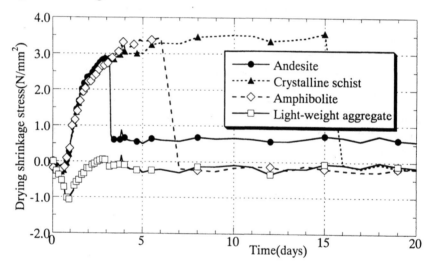

Fig.7 Relationship between time and drying shrinkage stress

Fig.8. shows the proportion of the autogenous shrinkage stress to that of drying shrinkage stress. In actual structures, both autogenous and drying shrinkage develop at the same time. Strictly speaking, therefore, it is impossible to isolate autogenous shrinkage from drying shrinkage in a single specimen. Nevertheless, because the strain induced by drying shrinkage comprises the strain induced by autogenous shrinkage, it would be fruitful to examine the proportion of the latter to the former. For this purpose, the stress values measured in the drying shrinkage test were compared with those measured in the autogenous shrinkage test. As seen from the figure, the proportion of the autogenous shrinkage stress to the drying shrinkage stress was not consistent at early ages, but it gradually converged to about 0.8 to 1 after a day had passed, i.e. after the removal of mold. This result indicates that, under the test conditions assumed in this study, about 80% of the drying shrinkage stress is induced by autogenous shrinkage.

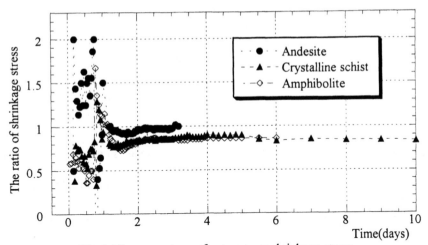

Fig.8 The percentage of autogenous shrinkage stress
to that of shrinkage stress

3.5 Influence of coarse aggregate's volume per unit volume of concrete on autogenous shrinkage stress

Fig. 9. shows changes with time in the autogenous shrinkage stress for various volumes of coarse aggregate per unit volume of concrete. It is shown that the larger the volume per unit volume of concrete was, the smaller the autogenous shrinkage stress became. This is caused by the fact that the volume of paste changed according to the changes in the coarse aggregate's volume per unit volume of concrete. Here,

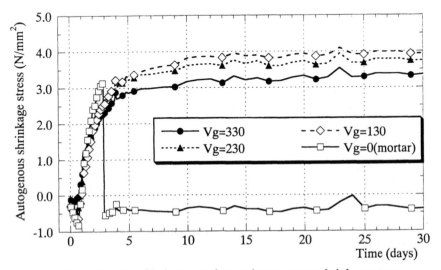

Fig.9 Relationship between time and autogenous shrinkage stress

the coarse aggregate's restraining effect on shrinkage has well been demonstrated. When the volume per unit volume of concrete was 0 (i.e. mortar), the autogenous shrinkage stress became the greatest due to a larger amount of cement paste. Furthermore, cracks were developed as there was no restraining aggregate, which was then followed by the release of stress after about 3 days.

4 Conclusions

This study examined how the autogenous shrinkage stress in high-strength, high-fluidity concrete is affected by the quality of coarse aggregate and its volume per unit volume of concrete(Vg). As a result, the following insights have been obtained thorough the present study:

1. The aggregate crushing value of coarse aggregates has an effect on autogenous shrinkage strain but not on stress;
2. Use of the light-weight aggregate that has been subjected to through water absorption relieves the self-desiccating process of concrete, and thus reduces the stress induced by autogenous shrinkage;
3. Autogenous shrinkage stress was smaller than tensile strength of concrete at the age of 28 days. Therefore we could see from this that cracks were not caused by only autogenous shrinkage stress under this condition;
4. Under the test conditions assumed in the present study, the proportion of the stress induced by autogenous shrinkage to drying shrinkage stress is about 0.8 to 1, i.e. the stress induced by autogenous shrinkage accounts for about 80% of the drying shrinkage stress; and
5. The autogenous shrinkage stress is reduced when the coarse aggregate's volume per unit volume of concrete increases due to the restraining effect of coarse aggregates.

5 References

1. Tazawa,E., Miyazawa,S. and Sato,T., (1992) Autogenous Shrinkage of Concrete, Proceedings of the Japan Concrete Institute, Vol. 14, No.1, pp. 561-566
2. Noguchi,T., Onoyama,K. and Tomosawa,F., (1993) Influence of Coarse Aggregate on Compressive Strength of High-strength Concrete, Lectures in the 47th cement technology forum, pp.720-725
3. Nakae,K., Matsushita,H., Makizumi,T. and Tsuruta,H., (1997) Experimental Study on Shrinkage of High-Fluidity, High-Strength Concrete, Proceedings of the Japan Concrete Institute, Vol.19, No.1, pp.721-726
4. Japan Concrete Institute, (1996) JCI research report of autogenous shrinkage, pp.199-201

26 RESEARCH OF TEST METHOD FOR AUTOGENOUS SHRINKAGE STRESS IN CONCRETE

Y. Ohno and T. Nakagawa
Department of Architecture Engineering, Osaka University, Suita City, Osaka, Japan

Abstract
In this paper, the effect of length and configuration of specimens and restraint steel areas on autogenous shrinkage stress were investigated.   Based on measured distributions of steel strain, anchorage length and configuration of specimen were discussed.   Shrinkage stress of concrete was 2.24 N/mm^2 at 28 days, which was 0.44 of split tensile strength, when using a deformed bar 32 mm in nominal diameter as the restraint steel bars. Anchorage length was 30 cm, so specimen length of 1m in this test was considered long enough to be used.   Shrinkage stress in specimens with enlarged section in anchorage zone was greater than in specimens with uniform section.   Theoretical values calculated by an iterative approach method showed a good agreement with experimental data.
Keywords: Autogenous shrinkage, shrinkage stress, restraint test, configuration of specimens, creep analysis, degree of restraint.

1   Introduction

Autogenous shrinkage will be a big problem in high strength concrete of which the water-cement ratio is usually very low and unit cement content is always very high.   If shrinkage strains of concrete members are restrained by steel bars, molds etc., tensile stress occurs.   When the tensile stress caused by autogenous shrinkage is greater than tensile strength of concrete, crack occurs in the early age.   There are complicated relationships between tensile stress and autogenous shrinkage, restraint, elastic modulus, and creep properties.   To measure these complicated performance, the autogenous shrinkage committee of JCI has presented a testing method.   In this method, 10 cm $\times$ 10 cm $\times$ 1.5 m prisms with deformed bars 32 mm in nominal diameter embedded in the middle of section are used as specimens.   The shrinkage stress is supposed to be

Autogenous Shrinkage of Concrete, edited by Ei-ichi Tazawa. Published in 1999 by E & FN Spon, 11 New Fetter Lane, London EC4P 4EE, UK. ISBN: 0 419 23890 5

calculated from the steel strain which is measured within the middle 30cm of specimens where bond affect has been removed.

The length of specimens are usually decided by the anchorage length of concrete and steels. In the method of JCI, specimens' length was considered too long. And the section areas of specimens in the middle 30 cm remain same as section area in the anchorage zone. Consequently, in the procedure of testing, there is no insurance that cracks occur in the range of middle 30 cm.

In this paper, with considering configuration of specimens and area of restraint steel bars, the JCI testing method had been used to research the following factors which include configuration and length of specimens, area of restraint steel bars, tensile stress and occurrence time of tensile stress.

2   Testing outline

2.1 Materials and types of specimen
To make these specimens, Portland cement, fine aggregate mixed by sea sand and mountain sand, coarse aggregate mixed by crushed stones and high effective dispersing agent were used. Water cement ratio is 0.24 and unit cement content is 729 kg/m^3. The results of strength test are shown in Table 1.

Fig.1 shows the configuration of specimens, and types of specimens are presented in Table 2. Deformed bars 32, 25 and 19 mm in nominal diameter (D32, D25 and D19)

Table 1 Strength of concrete

Age (days)	Compressive strength (N/mm^2)	Splitting tension strength (N/mm^2)	Young's modulus (kN/mm^2)
1	29.9	2.9	25.4
3	63.5	-	33.5
7	64.2	3.5	36.6
14	74.1	4.5	37.4
28	83.9	5.1	40.8

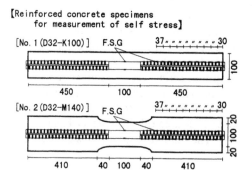
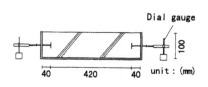

Fig. 1. Shape of specimens

Table 2 Test program

Number	Specimens	Size of steel	Shape of specimens	Width of anchorage range
1	D32-K100	D32	Uniform	100mm
2	D32-M140	D32	Tapered	140
3	D25-K100	D25	Uniform	100
4	D25-M140	D25	Tapered	140
5	D25-M120	D25	Tapered	120
6	D19-M140	D19	Tapered	140

D32, D25 and D19 are deformed bars 32, 25 and 19 mm in nominal diameter, respectively.

used as restraint steel. Within the middle 10 cm of steels, ribs and knots should be removed to get rid of bond. The steel ratio in this range were 6.78% (D32), 4.41% (D25), 2.34% (D19) respectively.

In the JCI testing method, the length of steel without bond in the middle 30 mm is decided according to the length of external restrained specimens of which strain are measured by mechanical strain gauges of which effective measuring length are 20-30 cm. However, in the internal restrained specimen for autogenous shrinkage stress, steel strain is measured by wire strain gauge set on steel bars. In this experiment, the measuring range was shorten to be 10 cm, and the length of the specimens was chosen to be 1 m. The configuration of specimens were three types: 1) the section area remains same (10 cm×10 cm) along the full length. 2) the section area in the measuring range (the middle 10 cm) is 10 cm×10 cm, the section area in the anchorage range is 10 cm×12 cm. 3) the section area in the measuring range (the middle 10 cm) is 10 cm×10 cm, the section area in the anchorage range is 10 cm×14 cm. Three specimens of each kinds had been prepared. Section width of anchorage range in the proposed method had been enlarged because if the section area remains same, it is very possible that cracks occur outside of the measuring range.

Two prisms of 10 cm×10 cm×50 cm and one prism of 10 cm×14 cm×50 cm were made as the specimens to test free autogenous shrinkage. Four prisms of 10 cm×10 cm ×50 cm were made for the creep test.

2.2 Testing procedure
The experiment was conducted in a temperature (20±2℃) and humidity (65±5%) constant room. The autogenous shrinkage experiment and specimens' making were conducted according to the requirement of autogenous shrinkage committee of JCI: 0.1 mm thin polystyrene sheet had been spread on the inner bottom of the constructed mold, and then, all the inner area of the mold (including polystyrene sheet) should be spread by polyester film to make sure that mold and concrete can not contact directly. When the cast of concrete finished, finishing should be given immediately. Then, the surface of concrete should be covered by a layer of 0.1 mm thin polyester film and a layer of wet fabric. After 24 hours, specimens were removed from molds and sealed with 0.05 mm thin aluminum adhesive tape.

Autogenous shrinkage stress in restraint autogenous shrinkage experiment was calculated using average strains of steel bars which had been measured by foil strain gauges (F.S.G) set on both sides of steel bars. Steel strains in the anchorage range were measured by 10 pairs of F.S.G set between ribs on both sides of steel bars. Free autogenous shrinkage strain was calculated from the displacement change which measured by electrical strain gauge set between gauge plugs embedded in both ends.   In creep experiment, sustained stress of 6 N/mm² were applied on the specimens at 1 day and 3 days of concrete age.   Displacement was measured by contact strain gauges (C.S.G) set on every 30 cm of specimens' four side surfaces.

Temperature of concrete was measured by a thermocouple embedded in the center of concrete section. In those experiments concerning strain and temperature, setting time were also measured to determine the starting time which would be the basic time when to start measuring.

3   Results and Discussion

3.1 Temperature in concrete, free autogenous shrinkage and creep coefficient
The time-dependent change of concrete's temperature is shown in Fig.2.   The temperature was 18℃ at casting, it reached its peak (29~31℃) at 0.8 day.   From this figure, we can see that the smaller the steel area, the higher the temperature.

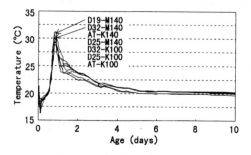

Fig. 2. Temperature in Concrete

The time-dependent change of free autogenous shrinkage strain from the basic time is shown in Fig.3.   The curves in the figure is approximate curves approximated by the

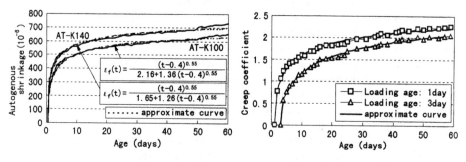

Fig. 3. Free autogenous shrinkage   Fig. 4. Creep coefficient

method of least squares. The shrinkage strain of specimen 10 cm×14 cm was greater than that of specimen 10 cm×10 cm by 25% at 1 day and 10% at 7 days.

Fig.4 shows the time-dependent change of creep coefficient. The curves in Fig.4 were approximated by the method of least squares.

3.2 Anchorage length

Fig.5 shows distributions of steel strain in anchorage zone. Fig.5(a) and (b) show the time-dependent change of steel strain distribution in and after the first day respectively. From Fig.5(a), we can see that compression strain of steel bars began to develop during

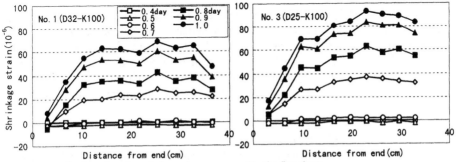

Fig. 5(a). Distribution of steel strain in anchorage range (in the first day)

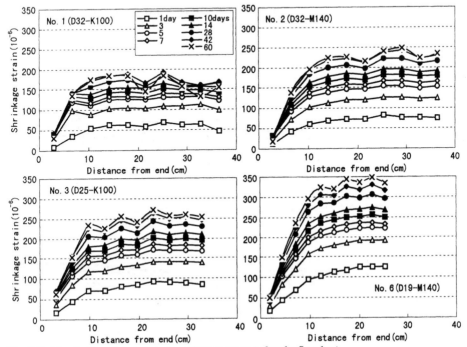

Fig. 5(b). Distribution of steel strain in anchorage range (after the first day)

0.6~0.7 day although concrete had already shrunk before steel bars were compressed, that is, strength and elastic modulus of concrete are too small in the early age to cause restraint stress.

For specimen No.1 (D32-K100), we can see the concrete was fixed completely at the location of the fourth gauge from the end of specimen at one day, and it kept unchangeable when age increases. The bond strength of concrete became larger with time.

Location of the fourth gauge of specimen No.1 was 14.9 cm from the end. In this location, strain of steel bars was 62.9×10^{-6} and 133×10^{-6} at 1 day and 7 days respectively. So the average bond stress in the anchorage range (14.9 cm) was 0.69 N/mm^2 and 1.47 N/mm^2 respectively, that is 1/43 and 1/44 of compressive strength at 1 day and 7 days respectively.

The smaller the steel ratio, the greater the steel stress, and the longer the anchorage length. The previous experimental results showed that the anchorage length of 30 cm in specimens with D32 was not long enough [1]. The length should depend on the properties of concrete and the surface condition of steel bar. In this experiment, the anchorage length is 15 cm in specimen No.1 and 25 cm in specimen No.6. From these results, it may be stated that 30 cm is long enough for the anchorage length in the range of this experiment.

3.3 Strain at middle of steel bars

The time-dependent strain change at the middle of steel bars of which ribs and knots had been removed was shown in Fig.6. As diameters of steel bars become small, stresses of steel bars become great because restraints cause by steel bars become small. Shrinkage strain of the uniform section specimens (10 cm×10 cm) was smaller than that of those specimens which have enlarged section in anchorage range (10 cm×14 cm) because the

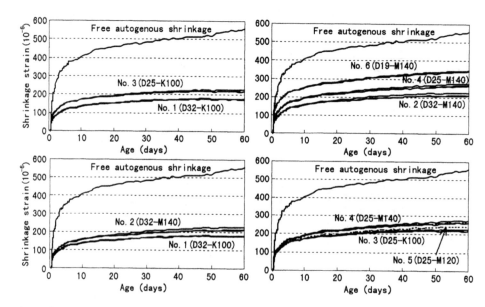

Fig. 6. Strain in the middle of steel bars

anchorage ends is not in the ends of the middle 10 cm, but in the anchorage range. Based on the condition that concrete strain change between two anchorage ends equal to the strain change of steel bars, the steel ratio of the enlarged section is smaller so that steel strain turns to be large.   Consequently, the effect of specimen configuration on the steel strain and concrete stress should be noted.

3.4   Shrinkage stress of concrete
The time-dependent change of shrinkage stress in concrete calculated by using strain of middle steel bars is shown in Fig.7.   Shrinkage stress increased rapidly in the first 6 days corresponding to autogenous shrinkage.   Shrinkage stress of specimen No.1 which have uniform section with D32 embedded was 2.24 N/mm^2 at 28 days, and shrinkage stress of specimen No.3 with D25 embedded was 1.84 N/mm^2 .   Shrinkage stress of specimens which have enlarged section in anchorage range presented a larger tendency than that of specimens which have uniform section.   The maximum shrinkage stress was reached in specimen No.2(D32-M140).   The stress was 0.52 of split tensile stress at 28 days. Analysis curve approximated by iterative approach method [1], [4] is shown in Fig.7.

Equation (1) which was approximated by using experimental autogenous shrinkage data of 10 cm × 10 cm specimens until 60 days in Fig.3 was used for creep analysis. Creep coefficient equation (2) was obtained from the 1day-age creep coefficient curve shown in Fig.4.   We also used the age coefficient formulation (3) by using reference [1], [4]. Elastic modulus equation (4) in CEB-FIP Model Code 1990 had been used for analysis considering the setting time [1].

$$\varepsilon_f(t) = (t - 0.65)^{.055} / (2.99 + 1.56(t - 0.65)^{0.55}) \qquad (1)$$

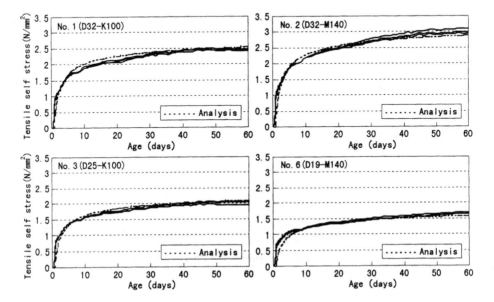

Fig. 7. Tensile self stress

382 *Ohno and Nakagawa*

$$\phi_t(t) = \kappa(t) \cdot (t-\tau)^{0.6} / (0.977 + 0.367(t-\tau)^{0.6}) \tag{2}$$

$$\kappa(t) = 6/(5+\sqrt{\tau}), \kappa(1) = 1 \tag{3}$$

$$E_c(t) = E_{28} \cdot \exp(0.0602(1-((28-0.65)/(t-0.65))^{0.5})) \tag{4}$$

where t : age

τ : loading age

$\kappa(\tau)$: age coefficient

E_{28}: elastic modulus of concrete at 28days

Analysis values had a good agreement with experimental data. However, the anchorage length of those specimens which have enlarged section in anchorage zone had been considered starting from 9 cm of the ends of the middle unanchorage range.

4 Conclusion

With using the concrete which has a an autogenous shrinkage strain 257×10^{-6} and 569×10^{-6} at 1 day and 28 days, considering the affect of specimens' configurations and area of restraint steel bars to shrinkage stress test, general conclusions and observations based on the test results are as follows:

1) Shrinkage stress of concrete in specimen with uniform section was 0.72 N/mm^2 and 2.24 N/mm^2 at 1 day and 28 days respectively, which was 0.14 and 0.44 of split strength, when using D32 as the restraint steel bars.
2) Anchorage length in the two ends of specimens was about 30 cm, specimen length of 1 m in this test was considered long enough to be used.
3) Shrinkage stress of those specimens which have enlarged section in anchorage range was greater than that of specimens which have uniform section.
4) Analysis values which calculated by an iterative approach method had a good agreement with experimental data.

5 Acknowledgments

The authors wish to thank T.Hayashida for the experiment.

6 References

1. Report (1996.11) Autogenous Shrinkage Research Committee of Japan Concrete Institute.
2. Takeuti,Y. (1997) Autogenous shrinkage stress of high strength concrete, *Proceedings of Japan Concrete Institute.* Vol.19, No.1, pp.751-756.
3. Ohno,Y., Nakagawa, T. and Yamamoto,S. (1997) Investigation of test method of self stress due to autogenous shrinkage in concrete, *Proceeding of Cement Concrete.* No.51, pp.630-635.
4. Neville, A.M. (1972) Properties of Concrete, *Pitman Publishing.*

Discussions

Autogenous deformations and stress development in hardening concrete
H. Hedlund and G. Westman, Lulea University of Technology, Sweden

R. Sato | Stress-strain relation shows non-linearity at high stress region, and creep behavior closely depends upon the stress level. Do you consider the creep model which depends upon the stress level?

H.Hedlund | Yes we consider. The creep model we used depends upon the stress level.

Stresses due to autogenous shrinkage in high strength concrete and its prediction
R. Sato, M. Xu and Y. Yang, Utsunomiya University, Japan

K.van Breugel | Did you consider the effect of microcracks in your analysis?

R.Sato | No, not considered. The stress at failure is about 70% of splitting tensile strength of concrete. This should be accumulation of micro cracks under sustained tensile stress. In this case large diameter reinforcing steel bars was used. Therefore micro cracks should develop more significantly. I think that it is necessary to consider the development of micro cracks.

The influences of quality of coarse aggregate on the autogenous shrinkage stress in high-fluidity concrete
H. Matsushita and H. Tsuruta, Kyushu University, Japan

Ø.Bjøntegaard | It seems that you obtain the same stresses even with very different autogenous shrinkage. To my opinion, this face must be counteracted by "equally" differences in stiffness and creep properties. Have you measured the elastic modulus or creep properties of the tested concretes?

H.Tsuruta | I haven't measured the elastic modulus or creep properties of concrete. I think that the influence of elastic modulus of concrete at early ages is very strong.

E.Tazawa | I don't understand the conclusion (5).
Why the autogenous shrinkage stress is reduced when the coarse aggregate volume is increased? This is probably because you measure the stress only from steel strain by restraining the autogenous shrinkage. You have local stress field around the

aggregate particles that also act as self stress. This is caused by different kind of restraint. So the actual failure is under the effect of the combination of local stresses around aggregate and overall stress restrained by reinforcement. How do you think? I think that the chairman has some comment.

K.van Breugel   The point here is that on the macrolevel you will measure less shrinkage of your concrete if you have the high aggregate content. On the microlevel, however, the paste has experienced the autogenous deformation, but these deformations have been partly restrained by the aggregate and even microcracking may have occurred already.

Research of test method for autogenous shrinkage stress in concrete
Y. Ohno and T. Nakagawa, Osaka University, Japan

R.Sato   According to JCI investigations, the anchorage length 30 cm is not enough.
the anchorage length 60 cm can give the reliable data, which means that the anchorage length depends upon the type of concrete. As a testing method , we have responsibility for giving the reliable results to users. How do you think?

Y.Ohno   In this experiment, the distribution of steel strain in the anchorage zone was measured. Based on the measured steel strain, I made a decision. Of course, the anchorage length depends on the concrete property. The larger autogenous shrinkage is, the longer the anchorage length is. I think the anchorage length of 30 or 40 cm is adequate for this test.

R.Sato   In order to solve this problem, much more data using many kind of concrete should be needed. But at present, longer anchorage can give reliable results. That is my opinion.

Y.Ohno   I agree with your opinion. But a 1.5 m length specimen is too long to handle. A shorter specimen is better to handle. I would like to determine the necessary and sufficient length of the specimen.

D.Van Gemert   Shrinkage stress of tapered specimen is larger than that of uniform specimen. Why did the JCI committee propose the uniform specimen?

Y.Ohno   The reason is that it is easy to calculate the stress and make the uniform specimen. The uniform specimen is suited for measuring the stress. The tapered specimen is better if cracking tendency is studied.

PART EIGHT

DISCUSSIONS AND CONCLUSIONS

8.1   Concluding session

Dr. Tazawa

First, I would like to ask you to have some sheets for enquete and fill in the enquete. When you go back, please place it at the entrance.   If you don't mind, please identify yourself.   If you mind, it's all right.   In this concluding session, I would like to have four subjects.   Number one, I would like to introduce one paper by Mr. Kohno.   His paper is on the front desk.   You can take one if you want.   His subject is related to the effect of lightweight aggregate on autogenous shrinkage.   On this subject, we have two contributors, one from Dr. Breugel, Delft University, and one from Dr. Tsuruta, Kyushu University.   They all notice that lightweight aggregate reduces autogenous shrinkage.   And we also have the same paper presented by Mr. Kohno.

Mr. Kohno reported results of his experimental work on autogenous shrinkage of lightweight aggregate concrete.   Using three types of lightweight aggregate (LA, HLA1, HLA2) and one normal aggregate (Table 1), it has been shown that the autogenous shrinkage of lightweight aggregate concrete decreases with increasing moisture content.   When higher degree of water absorption is attained by boiled immersion, less shrinkage is observed.   In some cases autogenous expansion is observed for long curing age more than two months. (Fig. 1, Fig. 2 and Fig. 3)

Table 1 Mixture properties

No.	W/C (%)	W (kg/m³)	C (kg/m³)	s/a (%)	G ** Type *	G ** Moisture cont. (vol. %)	G ** Unit cont. (L/m³)	Slump (cm)	Air cont. (%)
1					LA	22.3 [Immersed]		17	5.5
2				43.3	LA	0.00 [Oven dried]	350	15.5	4.6
3						35.4 [Boiled]		18	4.9
4				48.2			320	12	5.6
5	32	166	518	38.5		4.42 [Immersed]	380	18	4.4
6					HLA1			12.5	5.2
7				43.3		0.00 [Oven dried]	350	18	4.6
8						9.67 [Boiled]		12	4.8
9				43.3	HLA2	4.79 [Immersed]	350	14	5.7
10				43.3	CS	1.94 [Immersed]	350	12.5	4.1

       *   Specific gravity, LA: 1.27, HLA1: 1.17, HLA2: 0.94, CS: 2.62
     ** Maximum size 15 mm

Dr. Tazawa
Do you have any question for this presentation?

Dr. Barcelo
Thank you very much for your presentation.   How can you explain that even the concrete with totally dried lightweight aggregate is swelling during the first day?
Mr. Kohno
The specimen with absolutely dried aggregate swelled during the early stage.   This fact means the aggregate absorbed the mixing water until the end of the initial setting.

Dr. T. A. Hammer
We have also tried to measure exactly the same with initially dried aggregate. As you said, aggregate absorbed mix water drying mixing and curing. This water is enough to actually fill the paste during the first day. Then we also have seen expansion like you have.

Dr. Tazawa
But you see the difference in the initial forced absorption has some effect. How do you think? They have three different lines. The highest expansion is observed for the highest moisture content. And the lowest one for the lowest.

We will move to the next stage. Now we would like to talk about the definition of technical words, discussion of testing methods, discussion of mechanism of autogenous shrinkage. To begin with, I will introduce the name of contributors for the definition of terminology, there are many comments.

I could comment personally on each question, but the formal answer is needed for these questions. So, I would like to have some time to discuss your questions in the committee and later we would like to have some communication with you on each discussing points. So at present I would like to present my personal opinion.

We didn't have any question on the definition of terminology in advance. But

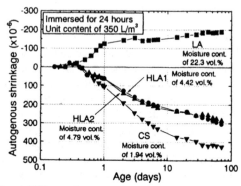

Fig. 1 Effect of type of immersed coarse aggregate on autogenous shrinkage

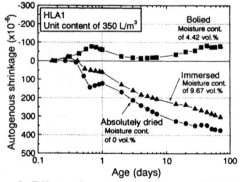

Fig. 2 Effect of moisture content of HLA1 on autogenous shrinkage

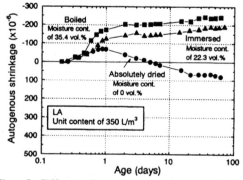

Fig. 3 Effect of moisture content of LA on autogenous shrinkage

we have some questions from different organizations in advance for the questionnaire which were sent a month ago. For chemical shrinkage, we have reply from French people, LAFARGE Company. They discussed together, they had some opinions as a Institute. Their question is related to the depth of sample, and it should be thin enough for the water to penetrate completely. The second point is some effect of entrapped water under the rubber plug. These points can be taken care of. In our proposal, we fix the water-cement ratio to 0.50. We made different measurements on different thickness of specimen and we are quite sure that no self-desiccation can occur for the depth proposed if we use water-cement ratio 0.5. If you would like to make experiment for the water-cement ratio less than 0.5, the thickness of sample becomes very important. In another question, which we had on chemical shrinkage measurement, is the final expression of the chemical shrinkage. We use the percent expression in stead of the amount of chemical shrinkage in terms of per gram of cement. Question is that this definition is quite different from the definition in nature. We think this value can be converted to percentage expression using the following equation.

$$S\,(\%) = \frac{\text{Amount of chemical shrinkage (ml)}}{\text{Volume of cement paste (ml)}} \times 100\,(\%)$$

$$S'\,(\text{mm}^3/\text{g}) = \frac{\text{Amount of chemical shrinkage (mm}^3)}{\text{Gram of cement (g)}} \times 100\,(\text{mm}^3/\text{g of cement})$$

For conversion of S (%) into S' (mm^3/g)

$$S'\,(\text{mm}^3/\text{g}) = S\,(\%) \times (W/C + 1/G_c) \times 10 \quad (\text{mm}^3/\text{g of cement})$$

So the exactly same thing we are measuring.

Dr. Aïtcin
I would like to make a comment about the polyester film. Recently I visit a food processing research institute and they are much concerned by absorption and movement of water and vapor water through the film. And I will get report where they have tested all kind of a film. And there are presently films which are absolutely unpermeate to vapor and to water. And I will be able to send you this report. We can solve the problem of the polyester film.

Dr. Tazawa
Thank you very much for nice suggestion. For these amendments, we are basically at this stance.
1. The methods should provide correct data for everyone, that is, for not very skilled man who wants to use them.
2. Controlling test conditions in narrow range would cost much. From engineering point of view, the testing conditions should not be too loose, nor too strict. The required accuracy is dependent upon the object for which measured value is utilized. On these bases, we will discuss the amendments of testing conditions. Is it all right for you? Any objection? If you agree we will do these process.

Is it all right?   OK.   Then we will proceed to the next.

There are many opinions that we should include the measurement at plastic stage. As we explained, the basic idea is the difference between horizontal and vertical directions.   For two directions, deformation should be discussed separately, because deformation to the vertical direction is under the effect of gravitational force.   So we might not have deformation to the horizontal direction.

Question No.1: Is self-desiccation possible for fresh concrete?   When concrete is fresh, water layer always exists between solid particles or flocculates.

No.2: Even if it is possible, only upper surface portion of concrete might be self-desiccated.   In this case, self-desiccation can be prevented by supplying water to the top surface, as in the case of plastic shrinkage caused by rapid drying from the top surface.   So, for the engineering purposes, we don't have to define deformation during plastic stage to the horizontal direction.   For the vertical direction, we have definition of subsidence that we often use to control cracking at early stage.   May be it is enough.   But somebody does not agree with this.   We will find the way anyway. Please write down in the enquete.

Next is a comment for Technical Committee Report, and test methods.   Size effect of volume of cement paste, container, and paraffin rap.   For the third point, it is a very good idea to have liquid paraffin on the top to a pipette to a certain depth.   But in this case, the pipette should be long enough.   Because, if we want to add some water afterward, we have trouble.   Some mixing can occur.   If we could use a longer pipette, it is possible.   But the longer pipette we will break when we use it.   How to handle is very important.   It depends on situation, and we have to check that.

Dr. Aïtcin

If you add water, you can use a syringe.

Dr. Tazawa

That's good.   We can take care of this.

As to the evaluation of chemical shrinkage, we think like this.

1. Chemical shrinkage provides information how much void is created in hardened cement paste.
2. Chemical shrinkage can be used to evaluate relative hydration speed under a given curing condition and also can evaluate mixtures with respect to hydration speed. And also it provides rough and macroscopic information on degree of hydration.
3. But in this case, chemical shrinkage provides no information on each cement minerals but can be used for engineering purposes.
4. For the same material or mixture proportion, chemical shrinkage can evaluate overall effectiveness of curing conditions.   But also in this case, information on microstructure can not be provided.

These are the evaluation of chemical shrinkage.   Is there any objection for this?

In relation to this, we had some presentation for Dr. Boivin, Dr. Acker and Dr. Justnes, the rubber bug method measurement.   In Dr. Boivin's paper, the importance of free penetration of water into the sample was recognized for the measurement of chemical shrinkage by weighing method.   Weighing method itself is not bad.   But preventing water from penetrating to the inner sample is the problem.   Dr. Justnes uses a rotating sealed sample.   In this case, penetration of water into microvoids was prevented.   So he named the shrinkage as external shrinkage.   If we prevent water penetration into the microvoids, we measure different physical value. Until the time of re-absorption of bleeding water t_{ra}, this value indicates chemical shrinkage. And after t_{ra}, it measures macroscopic shrinkage that is equal to autogenous shrinkage of cement structure. Therefore the measured value is the addition of chemical shrinkage up to t_{ra} plus autogenous shrinkage after t_{ra}. And t_{ra}, namely, the time when bleeding water disappears, could be before, during or sometime after the setting time. It is not exactly equal to the setting time but close to the setting time. In the rubber bag method, in which water penetration is prevented, the measurements are indescribable values. It depends on situation, when we started measuring. So up to this time, t_{ra}, this value is absolutely equal to the chemical shrinkage, but after that it means nothing. Because the total amount can not be used for the crack prevention. That is our stance.

Dr. Justnes
We would like to see the definition of autogenous shrinkage from the beginning not from the knee point. So our gross rubber bag method, it is to measure, according to our definition, autogenous shrinkage. We don't like the JCI definition. We think it should start from time zero not from initial set.

Dr. Tazawa
Autogenous shrinkage as we define it, deformation during plastic is quite different from deformation after setting. So we have to divide if we want to use the measured value for some engineering purpose.

Dr. Justnes
It's not only settling in plastic stage, we can also have crack formation.

Dr. Tazawa
When you rotate, you measure deformation to the horizontal and vertical at the same time, and mix them.

Dr. Justnes
I have been measured it fundamental purpose, not according to engineering purpose.

Dr. Tazawa
We don't have such case in actual job site.
Relative humidity measurement, we have many presentations for this. In both papers by Dr. Persson and Dr. Hanehara, crashed samples were stored in air-tight containers and ambient humidity equilibrated with the samples was measured. Some time lag was pointed out by chairman between the time of negative pressure generation determined by deformation measurement and the time of self-desiccation obtained by this method. Dr. Holt and Dr. Leivo reported similar situation caused by vertical movement of interstitial water as in the case of Dr. Hammer, both measurements are based on pore pressure. My personal opinion is very close to Mr. Ishida's opinion. He pointed out relative humidity is influenced by microstructure of cement paste, relative humidity of atmosphere that is in equilibrium with vapor pressure of water existing in microstructure in cement paste.
So it is effective in evaluating relative intensity of self-desiccation or degree of hydration.

If there is any distribution of water across the section, sampling from different points might be needed on each measurement. For different pore size or pore structure of cement paste, does it always provide the same degree of hydration for the same degree of water saturation? It depends on the average size of microstructures.

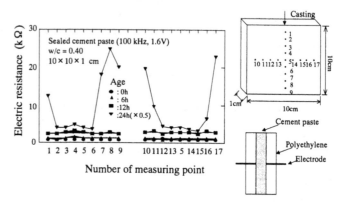

Fig. 4 Electric resistance observed for sealed cement paste

This aspect was discussed by Dr. Ishida in the last presentation.

As for this, I had a few additional experiments. One example is shown in Fig. 4. We put some cement paste between the plastic plates. We have electrodes from here to here at exactly the same position, and we measured electric resistance between these two electrodes. The electrode was distributed like this, one to nine, ten to seventeen. We measured electric resistance at the beginning, after 6 hours, 12 hours and 24 hours. For the measurements for 24 hours, we have very big variation so we divided it into half, then plotted in this figure. At first we had a very constant resistance. Then a little bit raising upward. Afterwards we had very rapid increase for the No. 1 that is here, No.7, 8, 9 that is around here. And also No.10, 11 and 16, 17 that is around here. This means periphery portion of the specimen is dried within one day. Water moves from the peripheral sites to the central portion of the specimen. But if we increase specimen size, this time we used 22cm by 22 cm and thickness is 3 cm. The previous data were taken by 1 cm. And in this case, for example, we measured 1 to 11 like this, but after certain time we have disconnection, not uniform. This kind of phenomenon was pointed out by Dr. Maekawa in the question in which he argued the continuity of capillary water. In this case we have disconnection around here, and also sometime we have non-uniform distribution over the section at this point.

This experiment was done by the suggestion of Dr. Jahren from

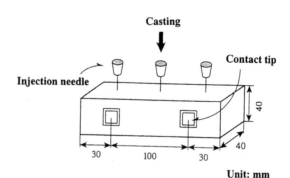

Fig. 5 Specimen for autogenous shrinkage measurement with injection needles

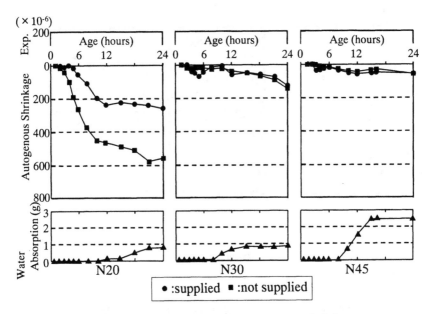

Fig. 6 **Effect of water supply on autogenous shrinkage**

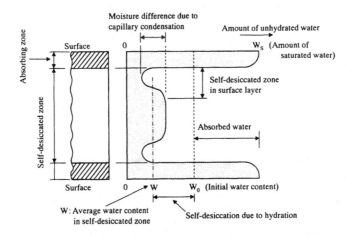

Fig. 7 **Distribution of water content in cement paste**

Norway. He suggested to do this experiment (Fig. 5). We used a injection needle to supply water into the middle part of the specimen. In Fig. 6, these figures here, denote water-cement ratio, 0.20, 0.30 and 0.45. For these two cases, we had no big differences at initial stage. But, in here, how much water was absorbed to the central portion is plotted. Also these two cases, we have some water supply. But only for this case, we had big difference in autogenous shrinkage. For these cases, strain difference was observed later and for longer age.

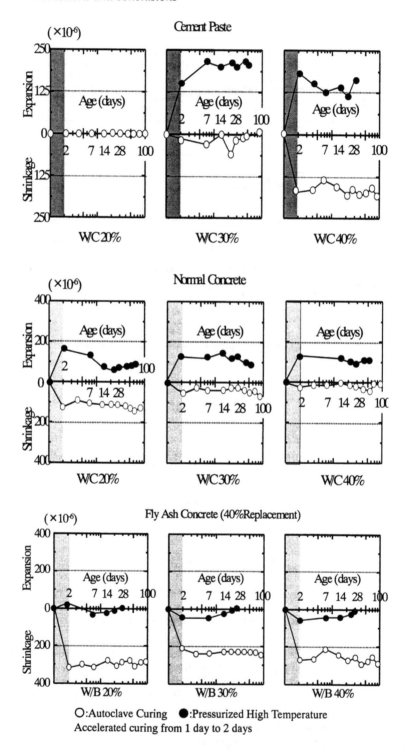

Fig. 8 Effect of accelerated curing on autogenous shrinkage

So if we store the specimen in some case we could have this kind of water distribution (Fig. 7). In the surface we have higher water content and in the middle we have higher. On the surface portion, some self-desiccation at periphery can occur. But it depends on situation. How much water can penetrate from surface to the central portion is dependent on how is the water content distribution around here. Sometime may be we could have this kind of simple water distribution and in some case the lowest water content comes around here. It is possible. But this kind of distribution we have not checked yet.

Next, the effects of microstructure are discussed by Dr. Hanehara and Dr. Tangtermsirikul.

Finer microstructure in larger mount increases autogenous shrinkage. This can be attained by selecting materials, for example, by choosing finer particles for the same constant chemical component. As for this point, I would like to introduce some experiment which we have done recently. We measured specimen in two different ways. We used normal autoclave curing and pressurized high temperature curing. Temperature versus time curve is quite the same and pressure is quite the same. Only difference is that the specimen is stored completely in water or in vapor. That is the only difference. In that case, we have great difference in autogenous shrinkage, as shown in Fig. 8. For cement paste, normal concrete, and fly ash concrete, as you can see, autogenous shrinkage is less and autogenous expansion is larger for high temperature water cured specimen. And Fig. 9 shows the case in which we use expansive concrete. Expansion of expansive concrete cured under pressurized high temperature curing is about twice as that obtained by autoclave curing. If we cure specimens in this condition from here to this point, after this point we cured sealed curing, underwater curing or dried curing. In this case, for all specimens we didn't have any volume change. But for this case we increase expansion at under water curing, and for sealed curing and dried curing, no volume change. You can see for the same expansive agent content, the expansion is quite different. This is about the half the expansion obtained for this. So, we need less expansive agent for this type. This reason is considered to be the effect of the microstructure. When we cure the specimen under the condition, every time chemical shrinkage occurs in this case water penetrates from outside so we always have high temperature water in the interstitious space of cement paste. So the hydration products have probability to distribute in everywhere, uniform for this case. But in autoclave curing we generate microvoids during hydration according to the chemical shrinkage which proceeds gradually. So in this case, in vacant space we can have no probability to have hydration product for the space filling. This might be the basic reason for this difference. The data symbolizes the effect of space filling or microstructure effect of volume change.

Does anybody have any question?

Dr. Barcelo
According to the definition of autogenous shrinkage, how do you define the setting time?

Dr. Tazawa
We use the normal Japanese Standard, penetration method to determine the initial setting time.

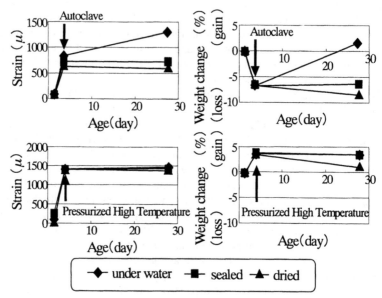

Fig. 9 Effect of curing method on expansion of expansive concrete

Dr. Barcelo
Do you think there is good correlation between this method and the moment when autogenous plastic shrinkage start to diverge from chemical shrinkage?

Dr. Tazawa
We are not sure. But normally, a little bit later than the initial point we observed the initial start of autogenous shrinkage. So the definition might be correct. Unless we have solid material, the deformation is only shear deformation not related to the internal stress. That's why we define the autogenous shrinkage from the beginning of setting time.

8.2 Suggestions on the terminology and the test methods proposed by JCI

T.A. Hammer, H. Justnes, SINTEF, Norway
Ø. Bjøntegaard and E.J. Sellevold, The Norwegian University of Science and Technology, Norway

We suggest defining **autogenous deformation** as all the external volume change taking place without any mass change. The chemical shrinkage is the main driving force, but any re-absorption of bleed water and/or absorbed water in the aggregates as well as the aggregate restraint will influence the relation between autogenous deformation and chemical shrinkage. The autogenous deformation may be measured volumetric or linear. In linear measurements the autogenous deformation may be both contraction and expansion, see below.

The proposed definition of autogenous shrinkage, given in your Fig.1.2 seems a bit artificial from a fundamental point of view: The autogenous deformation starts practically from time zero and it may contribute to cracking both before and after the time of setting (before setting: cracking of **horizontal** surfaces as plastic shrinkage does). In fluid pastes-mortars-concretes the autogenous deformation development coincides with the chemical shrinkage development some time from mixing and on as shown in the attached figure. However, recent results show that the diversion of the two curves may start much earlier than initial setting, particularly when using low w/c (see the paper by TA Hammer), all the way from "zero" to several hours depending on the initial fluidity, material composition and height of the sample. We propose therefore, from a fundamental point of view, to use the **time when the two curves divert** as shown in the attached figure rather than **initial setting time**.

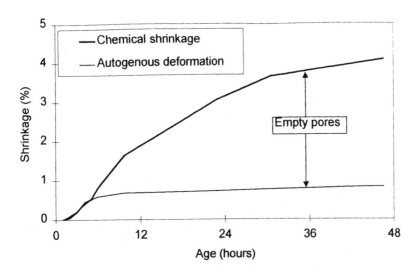

Fig.10 Chemical shrinkage and autogenous shrinkage

From a design point of view we propose to distinguish between the **"plastic phase"**(or **"initial phase"**) starting from time zero (plastic shrinkage type of cracking and cracking during slipforming), and the **"hardening phase"** (or **"thermal phase"**) starting around the time of initial setting.

The most important thing from a design point of view is the autogenous shrinkage development after setting (hardening phase), i.e. the driving force for "stress generating shrinkage".   Thus, the diversion point in the attached figure (around initial setting) could be considered as the "zero point". However, the "zero point" may be hard to define because re-absorption of any bleed water has shown to play an important role, causing reduced shrinkage rate or even expansion (see the papers by TA Hammer and Ø.Bjøntegaard, EJ Sellevold).   The test method should, therefore, include measurement under **external** absorption of any bleed water.

In your proposed test method for chemical shrinkage it is important that the thickness of cement paste is limited to about 10mm to avoid segregation of capillary pores and secure water access to all contraction pores independently of w/c and harening time.   In order to maximise mass and minimise thickness of paste, a flask of Erlenmeyer shape is recommended.   Instead of the paraffin wrap, we recommend a drop of liquid paraffin on top of the water column in the pipette to avoid vacuum effects and secure free movement of the water column.

P.Acker, L.Barcelo, S.Boivin, LAFARGE Laboratoire Central de Recherche, France
V.Baroghel Bouny, LCPC, France

A.Terminology
1.Autogenous shrinkage
For us, autogenous shrinkage is taken from the time of casting of materials. It is composed of <<plastic autogenous shrinkage>> before setting and <<selfdesiccation shrinkage>> after setting.   To take initial setting time as the beginning of autogenous shrinkage appears not possible to us because:
1) it is difficult to define precisely the moment when the paste is rigid enough to oppose chemical shrinkage and this depends on different parameters(type of cement, W/C---).
2) <<plastic autogenous shrinkage>>may be one of the key parameters influencing cracking in the plastic phase.
We generally define Autogenous Volume Change without specification of shrinkage or expansion.

Two types of units can be used to express autogenous volume changes. On one hand, if the behaviors of different types of cement have to be compared, then we generally express autogenous volume changes in mm3 per gram of cement(since those volume changes are related to cement hydration).   On the other hand, when autogenous volume changes (measured by a volumetric method) have to be compared with autogenous shrinkage strains (measured by a linear method), then the autogenous volume changes can be expressed as the percentage of volume reduction.

2.Chemical shrinkage

We generally express chemical shrinkage in mm^3 per gram of cement, since chemical shrinkage is related to cement hydration.

B. Testing methods

During the plastic stage, autogenous strains are not isotropic (orthotropic behavior). Consequently, linear measurement is not sufficient to characterize the material. We use volumetric method during this initial period (Test method used by Pr. Sellevold seems to be a good way to avoid bleeding artifacts). When the paste hardens, autogenous stains become progressively isotropic. It is then possible to use linear measurement method (as the one proposed by technical committee of JCI for example). The beginning of setting could be good criterion to begin linear measurements. In this approach, volumetric and linear measurements are complementary.

1.Test method for chemical shrinkage of cement paste

Concerning the procedure:

Sample thickness: We recommend a limitation of cement paste thickness in the vessel. A thickness of about 6cm seems too important for several reasons:(1)this thickness may induce a size efect at some point in the measure (caused by a reduced permeability of the paste during hydration),(2) it is more difficult to remove air bubbles from to fresh mix(and the presence of air bubbles in the mix can greatly affect the measure), (3) with big samples, the thermal effects may induce errors in the measure.

Concerning the problem of size effect, it is important to evaluate it with high reactivity cements (they give more rapidly a denser microstructure).

Insertion of plug: we generally insert the silicon plug under water so as to prevent the entrapment of air bubbles below the plug.

Storage of samples: we generally store the samples in a thermo-regulated bath.

Concerning the calculation:

We generally express chemical shrinkage in mm^3 per gram of cement, first of all because chemical shrinkage is directly related to cement hydration. Moreover, when chemical shrinkage is expressed as a percentage of cement paste volume, the bleeding water is included in the volume of paste that is considered. This error may be important for high W/C pastes which bleed a lot.

2.Test method for autogenous shrinkage stress of concrete:

In the proposed test method, no limitation on the change in mass of the specimens can be found (therefore, limitation of water evaporation from the concrete specimen is not specified).

B.S.M. Persson, Lund Institute of Technology, Sweden

• Test method on chemical shrinkage
1) Parallel vessel with only water required to detect temperature changes. Please see my comments.
2) Parallel test on hydration should be carried out to obtain

$$K = 1 - \frac{\delta w_n}{w_n}$$

δw_n : chemical shrinkage
w_n : hydrated water

on the same sample!
3) Silica fume has an effect on K in blended cement.
4) The mixing, storage and reading of the volume must take place at the same temperature in a climate chamber 0.1
5) Maximum thickness of sample: 7mm
6) Volume unimportant.
7) The pipette must be calibrated.
8) Scale of 0.001g required.

• Test method for autogenous shrinkage
1) Polyester film absorbs and transports moisture so it cannot be used.   There should be aluminum sheet also in the mould.
2) Cover by aluminum sheet also.
3) No wet cloth.   Will give additional moisture effect and temperature decrease due to evaporation heat.
4) Storage at temperature (climate chamber) ±0.5℃
5) Two thermocouples to be cast in the mould.
6) Thermocouples must be calibrated.
7) Measurement dials to be calibrated.
8) Coefficient of contraction is smaller than expansion.
9) Weight losses: maximum 0.2%
10) Scale of ±0.1g required.
11) Measurement for at least to 90days.
12) Storage room ±0.1℃

8.3   Result of questionnaire

At the end of the workshop, the attendants were asked to express their opinions on the definitions of chemical shrinkage and autogenous shrinkage etc., the test methods for autogenous shrinkage and autogenous shrinkage stress etc., and the mechanisms of self-desiccation etc., which have been proposed by the JCI Committee.   This vote was not done for decision by majority but for knowing opinion distribution of the attendants.   The result is show in Table 2.

Table 2   The result of questionnaire

	yes	Not voting	no	other	tatal
1. Definition					
Chemical shrinkage	11	1	0	1	13
(%)	*84.6*	*7.7*	*0*	*7.7*	*100*
Autogenous shrinkage	7	0	5	1	13
(%)	*53.8*	*0*	*38.5*	*7.7*	*100*
Self desiccation	11	0	1	1	13
(%)	*84.6*	*0*	*7.7*	*7.7*	*100*
Subsidence	7	4	2	0	13
(%)	*53.8*	*30.8*	*15.4*	*0*	*100*
2. Test methods					
Chemical shrinkage	5	2	4	2	13
(%)	*38.5*	*15.4*	*30.8*	*15.4*	*100*
Rubber bag method	4	3	2	4	13
(%)	*30.8*	*23.1*	*15.4*	*30.8*	*100*
Autogenous shrinkage	8	1	1	3	13
(%)	*61.5*	*7.7*	*7.7*	*23.1*	*100*
Autogenous shrinkage stress	5	3	3	2	13
(%)	*38.5*	*23.1*	*23.1*	*15.4*	*100*
3. Mechanism					
Self desiccation	12	0	0	1	13
(%)	*92.3*	*0*	*0*	*7.7*	*100*
Movement of capillary water	10	2	0	1	13
(%)	*76.9*	*15.4*	*0*	*7.7*	*100*
Negative pressure	7	3	1	2	13
(%)	*53.8*	*23.1*	*7.7*	*15.4*	*100*

Many comments on the JCI proposal were also given on the questionnaire sheet as follows.

1.Definition
(a)Chemical Shrinkage
B.S.M.Persson   Chemical shrinkage should be mm^3/g.

anonym   Chemical shrinkage is fundamental, i.e. not dependent on bleeding.

(b)Autogenous Shrinkage

anonym   • Autogenous deformation in "fresh concrete" (before setting) :
                An pore water under pressure has been reported (Holt, Leivo, Hammer)   suggesting that self-desiccation may take place far before the time of setting. This coincides with the measured shrinkage development. The magnitude of the shrinkage may vary by concrete depth (the weight of the concrete may exceed the strain capacity at a certain point). The practical consequence is uncertain, but should be further investigated.

anonym   Autogenous deformation is driven by the chemical shrinkage but dependent on bleeding, specimen size   (very early stage) , absorbed water in aggregate.

anonym   Should start before setting.

E.E.Holt   Include immediate reactions, as early as possible   (can not say Δ Shy~0)

H.Justnes   Define from adding water, not from initial set.

L.Barcelo   Why for answer "no" ; because we think we have to consider it from the beginning of hydration.

(c)Self Desiccation

E. Holt   In post-presentation discussion, it is unsure about words on self-desiccation influences.   Of course, it is possible at very early ages, especially at low W/C when the cement can not get enough water. The internal pores show self-desiccation.   So how can your overhead slides question the existence of self-desiccation?   Even with vertical movements (subsidence) the self-desiccation can play a role (around setting stage)

(d)Subsidence

Ø.Bjøntegaard   • Subsidence = All vertical deformation except autogenous shrinkage
                Subsidence = bleeding (+ escape of air)
                No bleeding means no subsidence, but a vertical movement due to autogenous shrinkage only.

B.S.M.Persson   Subsidence does not occur when the concrete has an ideal grading.

anonym   OK with definition. Not sure about English wording

2.Measuring Method
(a)Chemical Shrinkage
anonym   Adopted, see pg68

E.E.Holt   Use test method which is valid for various W/C ratio. Therefore, need thin paste(10mm?)

L.Barcelo   Why for answers "no"; because we think the thickness of the sample is too high.

(b)Rubber bag method
anonym   If use, clearly define if rotate or not.

(c)Autogenous Shrinkage

Ø.Bjøntegaard   1-a) In Norway we focus an autogenous shrinkage from time " zero" due to cracking problems " before and during" setting.
1-b) For calculation purposes, the autogenous shrinkage from the time of setting must be emphasized, and the measurement should therefore start approximately at setting (this point is not always easy to determine and it may occur very early for low W/C-concrete). When demoulding and measuring as late as 24h. (as proposal) a major part of the deformations may already have occurred.
1-c) Lastly, the end plate of the "Rig" should be removable in order to allow expansion.

E.E.Holt   Method is good, but···
     - include early age before setting also
     -    possible to measure with capillary pressure
     -    At this time, we may not have enough support, but in the future I think we can provide a clear point of "transition" which is based on pressure, earlier than setting time.

L.Barcelo   Why for answers "no"; because it is only a linear measurement. It is OK for the test method. But as cement paste behavior is not isotropic at the beginning of hydration, we also need a volume test method.

(d)Shrinkage Stress

Ø.Bjøntegaard   It seems that the "frame" gives higher stresses than the "reinforcing bar" method. This is probably due to a higher degree of restraint of the "frame". For standardized method, the degree of restraint must be known.   100% restraint   (which gives a direct relation between (free) deformations and (restraint) stresses is probably obtained only by use of "feedback"-systems keeping $\Delta L=0$ within the specimen. In other method or set-ups we will probably obtain a high but, nevertheless, unknown degree of restraint. The degree of restraint in stress-rigs without a $\Delta L \equiv 0$ facility will change continuously during the hydration process due to the change in the ratio of concrete stiffness to the stiffness of the steel (degree of restraint = $1- E_{con} \cdot A_{con.} / E_{steel} \cdot A_{steel}$)

B.S.M.Persson   · The upper edge of the specimen must be sealed in a better way.
· T_o short anchorage length; perhaps 30d

H.Hedlund   Use the stress frame, young concrete does not have the enough bonding to tension the re-bar loosing early stress information.

anonym   Autogenous deformation stress

E.E.Holt   But why have such high stress?

3.Mechanism
(a)Self Desiccation
B.S.M.Persson   · Means decline for chemical shrinkage of the internal relative humidity in concrete especially pronounced at low w/c owing to the fine pore size distribution.

(b)Negative Pressure

B.S.M.Persson   Negative pressure is impossible from the physical point of view.

H. Justness   Under Pressure

4.Other

Ø.Bjøntegaard   We must agree upon some primary questions:
A. Does bleeding water or water in the aggregate influence our test?
 For instance, if autogenous shrinkage is dependent on porosity of the aggregate (working as water supply for the binder phase) differences in autogenous shrinkage can not be attributed directly to the binder phase composition. A change of aggregate type may, in this case, totally alter conclusions.
And availability of water to the binder phase will also alter the relationship between measurements in paste and concrete.
B. For calculation purposes, there is a need for material models based on measurement of deformations (autogenous shrinkage + thermal dilation)at realistic temperatures. From my point of view, generalized models of autogenous shrinkage and thermal dilation based on isothermal tests are not trustworthy today.

8.4 Answer by the JCI Committee

JCI Committee has been suggested by LAFARGE and SINTEF that autogenous shrinkage should be taken from the time of casting. Plastic shrinkage, of course, may be one of the key parameter influencing cracking in plastic phase. We think that deformations before setting are very important for prevention of cracking. It is necessary that the segregation and subsidence are considered in definition of deformations before setting. In our opinion, it is not easy to express the early shrinkage before setting briefly in one or two sentences. And crack formation before setting can be prevented by considering the curing method or the execution method such as water supply on the top surface and hand trowelling, as is not the case of autogenous shrinkage after setting. Autogenous shrinkage is generally used for prediction of stress related cracking in hardened concrete, the strain generated in a period when cementitious material is fresh is excluded. Therefore, the time of initial setting of cement is specified as the start point of autogenous shrinkage measurement.

As is discussed in the Concluding Session, we have measured chemical shrinkage for cement paste with different thickness in order to determine test conditions. Some examples of the test results are shown in Fig. 11. For cement paste with 0.5 and 0.6 of water-cement ratio, the same result is obtained for different thicknesses of samples from 50 to 200 mm, whereas the size effect is observed with cement pastes with low water-cement ratio (w/c=0.26 and 0.40). The purpose of this test is presumed to investigate the degree of hydration for different type of cements with and without chemical and/or mineral admixtures. So, in the JCI test method, water-cement ratio is fixed to 0.5 in order to supply sufficient water for the tested materials to obtain their potential hydration.

406   *Discussions and conclusions*

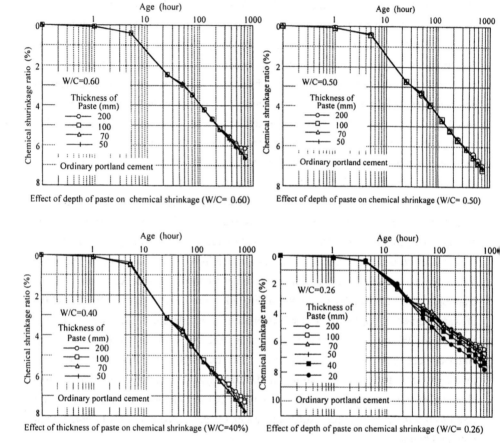

Fig. 11 Test results for chemical shrinkage of cement paste with different thickness

AUTHOR INDEX

SUBJECT INDEX

This index is compiled from the keywords provided by the authors of these papers, edited and extended as appropriate. The numbers refer to the first page of the relevant papers.

Milton Keynes UK
Ingram Content Group UK Ltd.
UKHW021830071024
449327UK00021B/1475